高等职业教育机械类专业系列教材
TEXTBOOKS FOR MACHINERY SPECIALITY OF HIGHER VOCATIONAL EDUCATION

三维数字化设计与制造
3D Digital Design and Manufacturing

梁宇栋 编
Edited by Yudong Liang

本书以典型产品为载体，介绍三维数字化设计与制造的技术流程，包括三维逆向扫描、逆向方式选择及逆向建模、数控加工编程与仿真加工等。

本书包含 4 个项目，共 15 个任务，项目案例由简单到复杂，难度逐步提高，主要内容涉及利用扫描设备及配套软件进行数据采集、基于 Geomagic Design X 或 Geomagic Wrap 进行数据处理、基于 Geomagic Design X 或 Siemens NX 进行逆向建模、基于 Geomagic Control X 进行产品分析比对，以及基于 Siemens NX 进行数控加工编程。

针对教学的需要，本书配套有由杭州浙大旭日科技开发有限公司提供的教学资源，由杭州学呗科技有限公司提供的信息化教学工具（学呗课堂），使教学内容更丰富、形式更多样，教学更简单，可以更好地提高教学的效率、强化教学效果。

本书可作为高等职业院校"三维数字化设计与制造"课程的教材，还可作为各类技能培训的教材，也可供相关工程技术人员参考。

本书配有电子课件，凡选用本书作为教材的教师，可登录机械工业出版社教育服务网（www.cmpedu.com）免费下载。咨询电话：010-88379375。

图书在版编目（CIP）数据

三维数字化设计与制造：英汉对照/梁宇栋编．—北京：机械工业出版社，2022.5（2024.2 重印）
高等职业教育机械类专业系列教材
ISBN 978-7-111-70453-9

Ⅰ.①三⋯ Ⅱ.①梁⋯ Ⅲ.①计算机辅助设计—应用—软件—高等职业教育—教材—英、汉 Ⅳ.①TP391.72

中国版本图书馆 CIP 数据核字（2022）第 050863 号

机械工业出版社（北京市百万庄大街 22 号 邮政编码 100037）
策划编辑：于奇慧　　　　　责任编辑：于奇慧
责任校对：史静怡　刘雅娜　封面设计：陈　沛
责任印制：刘　媛
涿州市般润文化传播有限公司印刷
2024 年 2 月第 1 版第 3 次印刷
184mm×260mm・23.75 印张・582 千字
标准书号：ISBN 978-7-111-70453-9
定价：69.00 元

电话服务　　　　　　　　　网络服务
客服电话：010-88361066　　机　工　官　网：www.cmpbook.com
　　　　　010-88379833　　机　工　官　博：weibo.com/cmp1952
　　　　　010-68326294　　金　书　网：www.golden-book.com
封底无防伪标均为盗版　　　机工教育服务网：www.cmpedu.com

前　　言

　　三维数字化设计与制造是一项综合技术，涉及基于数字化设计软件进行数字化建模、逆向造型、数控编程与加工等，在航空、航天、汽车、通用机械和电子等多个工业领域得到广泛应用。目前，企业对掌握三维数字化设计与制造技术的人才的需求量越来越大。

　　本书采用"项目驱动，任务引领"的编写模式，选取较为典型的产品作为案例，并结合国内高等职业院校相关赛项实例及要求，详细介绍了三维数字化设计与制造的技术流程，内容包括：

1）使用通用扫描设备进行产品扫描，以获取扫描数据。

2）使用 Geomagic Design X 或 Geomagic Wrap 软件对扫描数据进行处理，以满足逆向建模的需要。

3）使用 Geomagic Design X 或 Siemens NX，利用处理过的扫描数据进行逆向建模。

4）使用 Geomagic Control X 软件进行产品分析比对，通过扫描数据与逆向建模的模型的分析比对，更直观地观察逆向建模结果。

5）使用 Siemens NX 软件完成零件的数控编程与加工。

　　本书作者从事 CAD/CAM/CAE 教学和研究多年，具有丰富的三维数字化设计与制造的工作经验和教学经验。在编写本书的过程中，借鉴了杭州浙大旭日科技开发有限公司等多家公司多位资深造型工程师的经验，对本书的结构及内容进行了改进和完善。本书包含 4 个项目，共 15 个任务，将相关的知识、技能与方法进行有机结合，让学生在学习的同时，能够进行同步练习，以提高其实际操作能力，做到学以致用。

　　本书可作为高等职业院校"三维数字化设计与制造"课程的教学与实训教材，也可用作各类职业院校参加相关赛项的实训教材，还可供模具、机械、家电等行业制造企业，以及工业设计公司的工程师参考。

　　本书由天津机电职业技术学院梁宇栋编写，在编写过程中得到了杭州浙大旭日科技开发有限公司单岩教授的帮助。

　　限于编者的水平，书中必然存在需要进一步改进和提高的地方。期望读者及专业人士提出宝贵意见与建议，以便今后不断加以完善。

编　者

目　　录

前言

项目1　FIFA世界杯奖杯工艺品 ··· 1
任务1　数据采集 ·· 1
1.1.1　扫描仪标定 ·· 3
1.1.2　数据扫描 ··· 5
任务2　数据处理 ·· 9
任务3　加工编程 ·· 14
1.3.1　初始设置 ··· 15
1.3.2　粗加工 ··· 20
1.3.3　二次粗加工 ··· 23
1.3.4　半精加工 ··· 26
1.3.5　精加工 ··· 30
1.3.6　仿真加工 ··· 32

项目2　涡轮叶轮 ··· 35
任务1　数据采集 ·· 35
2.1.1　扫描仪标定 ··· 35
2.1.2　贴标记点 ··· 37
2.1.3　扫描标记点 ··· 38
2.1.4　扫描激光点 ··· 39
任务2　数据处理 ·· 42
任务3　逆向建模 ·· 47
2.3.1　主体建模 ··· 47
2.3.2　叶片建模 ··· 53
任务4　加工编程 ·· 66
2.4.1　加工分析 ··· 66
2.4.2　编程准备 ··· 66
2.4.3　多叶片粗加工 ··· 71
2.4.4　主叶片精加工 ··· 74
2.4.5　分流叶片精加工 ··· 75

2.4.6	轮毂精加工	76
2.4.7	圆角加工	77
2.4.8	复制程序	78
2.4.9	仿真加工	80

项目3 挤牙膏器 — 81

任务1 数据采集 — 81
- 3.1.1 贴标记点 — 81
- 3.1.2 扫描标记点 — 82
- 3.1.3 扫描激光点 — 82

任务2 数据处理 — 85
- 3.2.1 导入模型 — 85
- 3.2.2 上盖扫描数据处理 — 86
- 3.2.3 下盖扫描数据处理 — 87

任务3 逆向建模 — 89
- 3.3.1 导入模型 — 89
- 3.3.2 确定坐标系 — 90
- 3.3.3 下盖逆向建模 — 92
- 3.3.4 上盖逆向建模 — 102
- 3.3.5 细节处理 — 104

任务4 产品分析比对 — 105
- 3.4.1 数据的导入 — 105
- 3.4.2 初始模型对齐 — 105
- 3.4.3 最佳拟合对齐 — 105
- 3.4.4 3D比较 — 106
- 3.4.5 2D比较 — 107
- 3.4.6 比较点 — 107
- 3.4.7 横截面 — 108
- 3.4.8 标注功能 — 109
- 3.4.9 生成报告 — 113
- 3.4.10 输出报告 — 114

项目4 游戏手柄 — 115

任务1 数据采集 — 115
- 4.1.1 扫描仪标定 — 116
- 4.1.2 贴标记点 — 120
- 4.1.3 数据扫描 — 122

任务2 数据处理 — 124
- 4.2.1 手动拼接 — 124

4.2.2　数据封装 …………………………………………………………… 128
任务3　逆向建模 ……………………………………………………………… 130
　　4.3.1　确定坐标系 ………………………………………………………… 131
　　4.3.2　铺面 ………………………………………………………………… 133
　　4.3.3　绘制大致形状 ……………………………………………………… 138
　　4.3.4　绘制细节特征 ……………………………………………………… 142
任务4　加工编程 ……………………………………………………………… 144
　　4.4.1　工艺分析 …………………………………………………………… 144
　　4.4.2　编程准备 …………………………………………………………… 145
　　4.4.3　反面粗加工 ………………………………………………………… 150
　　4.4.4　反面半精加工 ……………………………………………………… 155
　　4.4.5　反面精加工 ………………………………………………………… 160
　　4.4.6　正面几何体设置 …………………………………………………… 163
　　4.4.7　正面粗加工 ………………………………………………………… 164
　　4.4.8　正面半精加工 ……………………………………………………… 167
　　4.4.9　正面精加工 ………………………………………………………… 167
　　4.4.10　仿真加工 ………………………………………………………… 174

参考文献 ……………………………………………………………………… 175

项目1　FIFA世界杯奖杯工艺品

FIFA世界杯奖杯（又称大力神杯，后同）是足球世界杯的奖杯，是足球界的最高荣誉的象征。奖杯由意大利艺术家西尔维奥·加扎尼加（Silvio Gazzaniga）设计。奖杯的外形是两个大力士托举着地球。雕刻线条从底座开始向上延展，以螺旋式的曲线上升，两名运动员以大力士的形象也随之浮现，他们向上伸展身体，托举起整个地球。在这个充满动态的、紧凑的杯体上，雕刻出两个胜利后激动的运动员的形象。

本项目以大力神杯模型（工艺品）（图1-1）为载体，使用桌面式扫描仪EinScan-S进行三维扫描，使用Geomagic Design X 2016进行数据处理，最后在Siemens NX 10软件中完成数控加工仿真。项目实施流程如图1-2所示。

图1-1　大力神杯实物模型

图1-2　项目实施流程

任务1　数据采集

采集奖杯模型的数据，使用的是一款高精度白光桌面3D扫描仪EinScan-S，其扫描精度在0.1mm以内，有两种扫描模式：转台扫描和固定扫描。在转台扫描模式下，360°转台扫描一圈只需2min，最大扫描范围是200mm×200mm×200mm。在固定扫描模式下，最大扫描范围是700mm×700mm×700mm。EinScan-S总重3.5kg，轻巧便携。

使用桌面式扫描仪EinScan-S进行数据采集的操作流程如图1-3所示。

透明或反光的物体不能直接扫描，需要喷涂反差增强剂。奖杯模型外表面有一层电镀膜，金光闪闪，有反光现象。如果不喷涂反差增强剂，扫描一遍后仅能获得420个点和267个面片，扫描效果不好，如图1-4a所示。喷涂反差增强剂之后，扫描一遍能够获得334207个点和322738个面片，如图1-4b所示，扫描效果好很多。

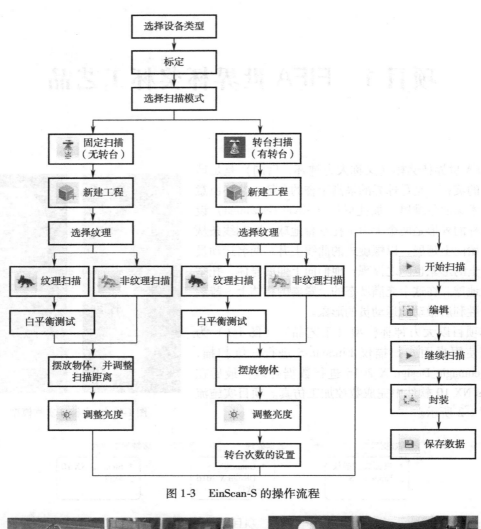

图 1-3　EinScan-S 的操作流程

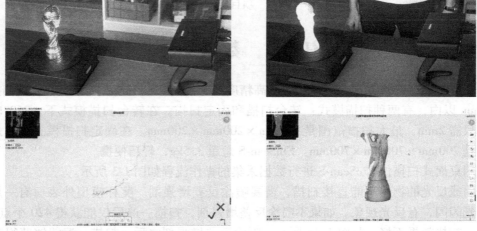

a）喷涂前　　　　　　　　　　　　　　　b）喷涂后

图 1-4　反差增强剂喷涂前后的扫描效果对比

1.1.1 扫描仪标定

安装扫描仪软件后,第一次扫描前需先进行标定,不标定则无法进入扫描模式。若没有标定,软件会提示"没有标定数据,请先进行标定"。进行标定前,需注意标定板的边需与图 1-5 中手指处的线对齐,同时也需与光线的边对齐。

图 1-5 标定板的放置

首次使用 EinScan-S 软件时,软件会提示选择设备类型。选择【EinScan-S】后单击【下一步】,自动进入标定界面。标定界面如图 1-6 所示。

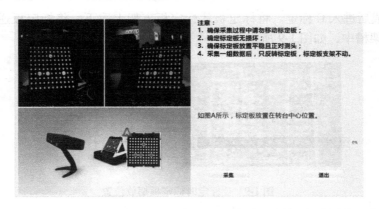

图 1-6 标定界面

EinScan-S 需要标定 3 个方位,所以标定板需要摆放 3 次,摆放位置根据软件向导提示进行。

首先根据软件向导提示,调整扫描仪与标定板之间的距离,将扫描仪十字对准标定板并调整至十字清晰。将标定板支架放置在转台中心位置,标定板的第一个摆放的方位如图 1-6 所示。确保标定板放置平稳且正对测头后,单击【采集】,转台自动旋转一周并采集数据,采集过程中请勿移动标定板。

采集完毕后,转台停止不动,软件界面提示进行 B 标定,如图 1-7 所示。

如图 1-8 所示,将标定板从标定板支架上取下,将标定板逆时针旋转 90°,嵌入标定板支架槽中。注意:只反转标定板,标定板支架不动。

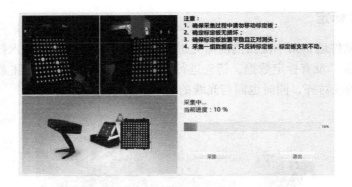

图 1-7　软件界面提示进行 B 标定

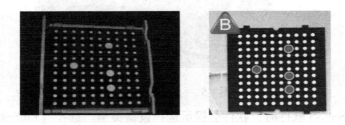

图 1-8　B 标定时标定板摆放位置

B 标定完成后进入 C 标定。将标定板从标定板支架上取下，将标定板逆时针旋转 90°，嵌入标定板支架槽中，如图 1-9 所示。

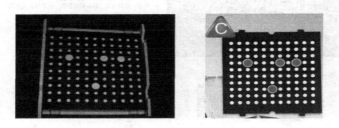

图 1-9　C 标定时标定板摆放位置

采集完成后，进行标定计算，如图 1-10 所示。

图 1-10　标定计算

标定成功后,会自动返回扫描界面。如果提示标定失败,请按照上述步骤重新进行标定。

初次标定时,必须按照流程标定。以下情况需要再次进行标定:

1)扫描仪初次使用,或长时间放置后使用。

2)扫描仪在运输过程中发生严重振动。

3)扫描过程中频繁出现拼接错误、拼接失败等现象。

4)扫描过程中,扫描数据不完整,数据质量严重下降。

1.1.2 数据扫描

标定完成后即可进行扫描,这里选择转台扫描模式,具体步骤如下。

1)将奖杯模型(工艺品)放置在转台上,如图 1-11 所示。调整设备与物体间的距离至合适的工作距离(290~480mm),以投影出的十字在扫描物体上清晰为最佳扫描距离。

图 1-11 将奖杯模型放置在转台上

2)打开桌面式扫描仪 EinScan-S 的配套扫描软件,选择设备为【EinScan-S】后单击【下一步】,如图 1-12a 所示。

3)选择扫描模式为【转台扫描】后单击【下一步】,如图 1-12b 所示。

a)选择扫描设备　　　　b)选择扫描模式

图 1-12 选择扫描设备和扫描模式

4)单击【新建工程】后输入工程名称,如图 1-13 所示。

5)选择【非纹理扫描】后单击【应用】,如图 1-14 所示。

6)根据物体的明暗,选择合适的亮度设置。可以在左侧预览窗口中实时查看当前亮度,以呈现白色或者稍许泛红为宜。设置为最亮,如图 1-15 所示,再单击【应用】。

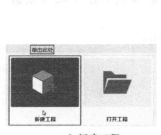

a）新建工程　　　　　　　　b）输入名称

图 1-13　新建工程并输入名称　　　　　　　图 1-14　选择非纹理扫描

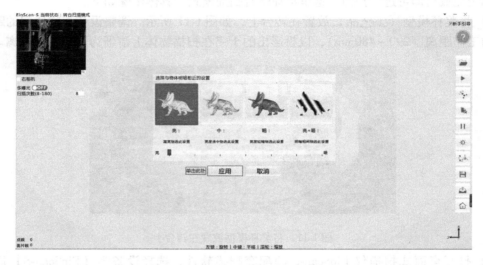

图 1-15　调整亮度

7）扫描次数为默认值，即 8 次。单击"扫描"按钮 ▶ 开始扫描，如图 1-16 所示。扫描过程中请勿移动物体和设备。

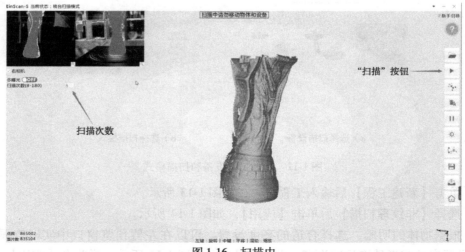

图 1-16　扫描中

8）扫描结束后界面如图 1-17 所示，屏幕下方出现一组编辑工具，这 4 个工具从左到右分别表示：①撤销选择；②反选；③删除选中；④撤销删除。每扫描一组数据，可对当前扫描的单组数据进行编辑，可删除数据多余部分或杂点，数据和标记点均可进行编辑。单击右下方的按钮 ✔ 保存扫描数据。

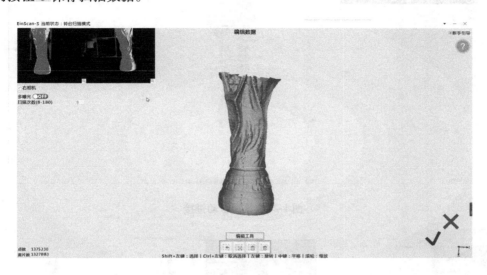

图 1-17　保存单组扫描数据

9）通过观察发现由于奖杯模型的顶部已经超出相机范围，没有扫描获得该部分的数据。需调整奖杯模型的摆放位置，将其平放在转台上，如图 1-18 所示，然后单击"扫描"按钮 ▶ 继续扫描，如图 1-19 所示。

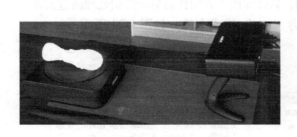

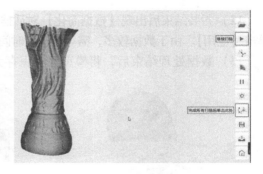

图 1-18　调整奖杯模型在转台上的摆放位置　　　　图 1-19　继续扫描

10）扫描结束后，单击界面右下方的按钮 ✔，两组扫描数据会自动拼接在一起，如图 1-20 所示。

11）数据扫描完成后，单击"生成网格"按钮，对数据进行封装处理。封装处理包括封闭和非封闭两种模式。封闭封装可直接用于 3D 打印模型。这里选择【封闭模型】，如图 1-21 所示，然后选择【高细节】，如图 1-22 所示，软件界面下方会显示数据封装的进度条。

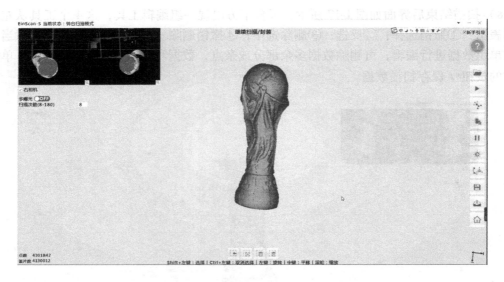

图 1-20 数据自动拼接

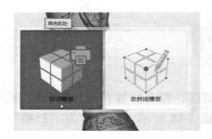

图 1-21 选择封闭模型

图 1-22 选择高细节

12）封装结束后出现【数据简化】对话框，如图 1-23 所示。这里不对数据进行简化，直接单击【应用】。由于数据较多，需等待一定时间，软件界面下方同样会显示数据处理的进度条。

13）数据处理结束后，将模型数据保存为".stl"格式文件，如图 1-24 所示。

图 1-23 【数据简化】对话框

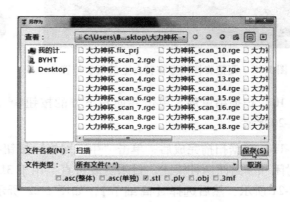

图 1-24 保存为".stl"格式文件

14）随后出现数据尺寸缩放窗口，如图 1-25 所示，这里缩放比例保持默认值 100，即不缩放数据。数据尺寸缩放只对扫描数据的体积尺寸进行缩放，不会减少三角面片的数量及数据的大小。

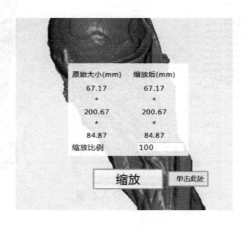

图 1-25　数据尺寸缩放窗口

至此，大力神杯模型（工艺品）的数据采集完成。

任务 2　数据处理

数据处理是指对扫描数据进行优化，例如去除扫描过程中模型表面上的杂点及外界环境带来的噪声，填补标记点带来的破洞及扫描仪获取不到的数据等。这里选择 Geomagic Design X 作为数据处理软件，数据处理流程如图 1-26 所示，具体步骤如下。

图 1-26　大力神杯模型的数据处理流程

1）导入 STL 模型。打开 Geomagic Design X 2016 软件，单击界面左上方的"导入"按钮，选择之前扫描生成的大力神杯模型的 STL 文件，单击【仅导入】，如图 1-27 所示。

2）创建平面。选择【模型】选项卡中的【平面】命令，在弹出的对话框中选择【方法】为【选择多个点】，然后选择大力神杯模型底面上的 3 个点创建一个平面。单击"OK"按钮，完成平面的创建，如图 1-28 所示。

3）创建直线。选择【模型】选项卡中的【线】命令，在弹出的对话框中选择【方法】为【检索圆柱轴】，然后在大力神杯模型底部圆柱面上选择一些区域，使生成的线在圆柱面中心位置，如图 1-29 所示。

图 1-27　导入数据

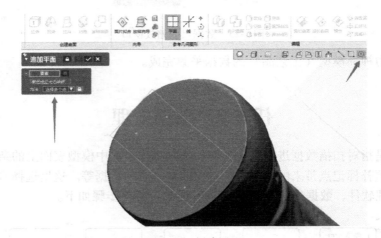

图 1-28　创建平面

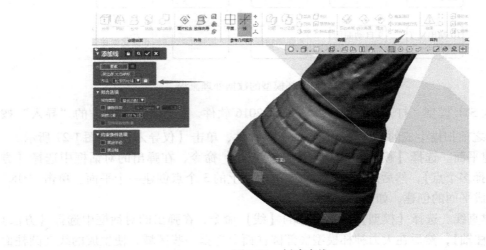

图 1-29　创建直线

4)手动对齐。通过定义扫描模型中的基准特征与世界坐标系中的坐标轴或坐标平面相匹配,使模型与世界坐标系对齐。选择【对齐】选项卡中的【手动对齐】命令,如图1-30所示,在弹出的对话框中单击"下一步"按钮,如图1-31所示。

图1-30 手动对齐　　　　　　　　　　　　图1-31 单击"下一阶段"按钮

选择【X-Y-X】对齐方式,该方式所使用的要素为3条直线或2条直线及1个原点。【位置】选择之前创建的平面及直线(注意:选择一个特征后,需按住<Ctrl>键再选择另外一个特征),表示以平面和直线的交点作为坐标原点。【Z轴】选择之前创建的平面,表示以平面的法向作为Z轴方向,然后根据实际情况确定是否单击Z轴箭头使Z轴反向,如图1-32所示。最后单击"OK"按钮,完成坐标系对齐,使大力神杯模型的底面和轴线与绝对坐标系对齐,如图1-33所示。

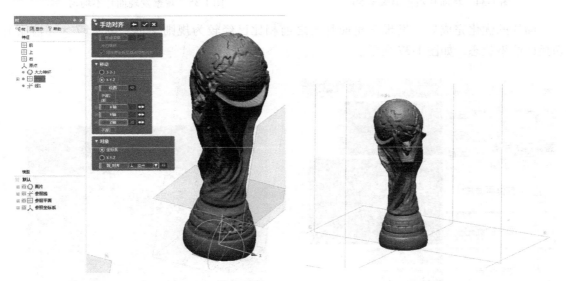

图1-32 指定位置与Z轴进行手动对齐　　　　图1-33 完成手动对齐

5)消减。消减指在保证几何特征形状的同时,通过合并单元顶点的方式减少面片或选定区域的单元面的数量。通过扫描得到的大力神杯模型的单元面数量为4699088个,数据量较大,使用【多边形】选项卡中的【消减】命令可使其消减50%,如图1-34所示。(消减所需的时间因计算机性能差异会有所不同,普通性能的计算机一般需要几分钟的时间。)

6)优化面片。根据面片的特征形状,通过设置单元边线的长度和平滑度来优化面片。在默认快捷键状态下按下<F8>,进入边线渲染模式,放大后观察面片,可以发现三角面片并不均匀,如图1-35所示。选择【多边形】选项卡中的【面片的优化】命令,弹出的对

话框中参数保持默认值，单击"OK"按钮☑，进行面片优化，如图1-36所示。（优化过程同样需耗费一定时间，需耐心等待。）

图1-34 将面片数量消减至50%　　　　图1-35 观察发现面片不均匀

面片的优化完成后，发现三角面片与之前相比已经较为规则匀称。按＜F7＞（渲染）切换回渲染状态，如图1-37所示。

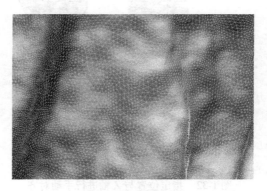

图1-36 进行面片优化　　　　图1-37 观察优化后的面片

7）加强形状。加强形状用于锐化面片上的尖锐区域（棱角），同时平滑平面或圆柱面区域，以提高面片的质量。选择【多边形】选项卡中的【加强形状】命令，弹出的对话框中参数保持默认值，单击"OK"按钮☑，如图1-38所示。

8）删除特征。删除特征用来删除面片上的特征形状或不规则的突起，重建单元面。观察发现奖杯工艺品上有小凸点，选择【多边形】选项卡中的【删除特征】命令，在套索选择模式下选取凸点特征。单击"OK"按钮☑，删除凸点特征，如图1-39、图1-40所示。用同样的方法处理其他位置处的凸点特征。

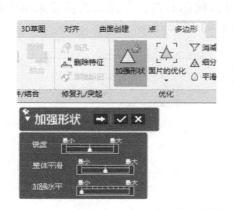

图 1-38 通过【加强形状】进行锐化、整体平滑

图 1-39 【删除特征】命令

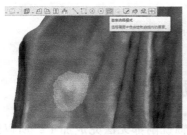

a) 凸点删除前

b) 凸点删除后

图 1-40 删除特征的前后对比

9) 自动曲面创建。自动曲面创建用于自动生成 NURBS 曲面。选择【曲面创建】选项卡中的【自动曲面创建】命令，弹出的对话框中参数保持默认值，单击"下一阶段"按钮，如图 1-41a 所示。观察如图 1-41b 所示的面片是否符合要求，确认无误后单击"OK"按钮完成曲面创建。

自动曲面创建完成后，可以看到左侧的模型树中会生成一个实体，如图 1-42 所示。

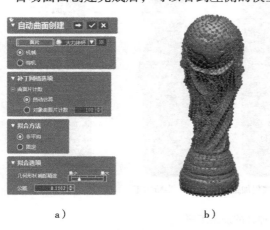

a) b)

图 1-41 自动曲面创建

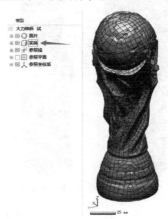

图 1-42 模型树中生成对应的实体

10）输出 x_t 模型。选择模型树中的实体后单击鼠标右键，在弹出的快捷菜单中选择【输出】，再选择【保存类型】为"*.x_t"，如图1-43所示。

图1-43　输出 x_t 模型

任务3　加工编程

大力神杯形状比较复杂，采用车削或者三轴铣削都无法完成加工，所以采用多轴加工。大力神杯的多轴加工有4道工序，分别是粗加工、二次粗加工、半精加工和精加工，如图1-44所示。

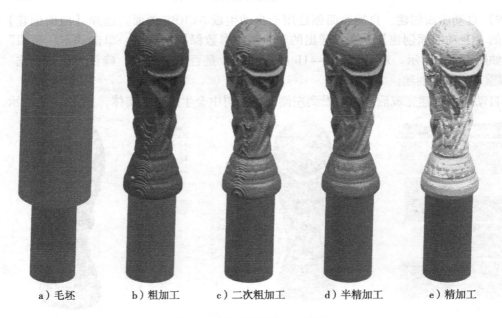

a）毛坯　　b）粗加工　　c）二次粗加工　　d）半精加工　　e）精加工

图1-44　大力神杯的加工工序

1.3.1 初始设置

导入模型。使用【文件】/【导入】/【Parasolid】命令，导入".x_t"格式的大力神杯数据模型。

在创建多轴加工程序之前需要进行初始设置，如图1-45所示，具体步骤如下。

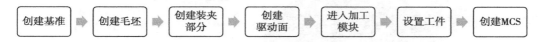

图1-45 初始设置的主要步骤

1）创建基准。进入【建模】应用模块，选择【插入】/【基准/点】/【基准CSYS】命令，弹出【基准CSYS】对话框。单击【确定】，直接在原点上创建基准坐标系，如图1-46所示。

图1-46 创建基准

2）创建毛坯。使用【拉伸】命令，在【拉伸】对话框中单击【截面】选项组中的"绘制截面"按钮，进入草图模式。在XC-YC平面绘制一个圆，此圆需包覆住整个工件，直径取整数80mm，然后退出草图模式。拉伸的高度需包住整个工件，输入200mm，单击【确定】后，得到一个80mm×200mm的圆柱毛坯，如图1-47所示。

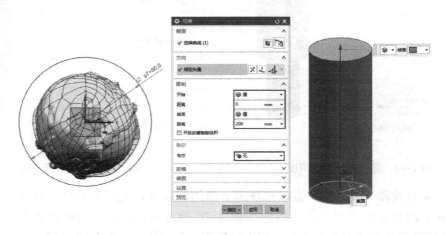

图1-47 创建毛坯

3）创建装夹部分。工件需要装到机床上，所以需要创建装夹部分。同样采用【拉伸】的方法进行创建，在 XC-YC 平面绘制一个直径为 50mm 的圆，向下拉伸 100mm，【布尔】运算选择【无】，如图 1-48 所示。

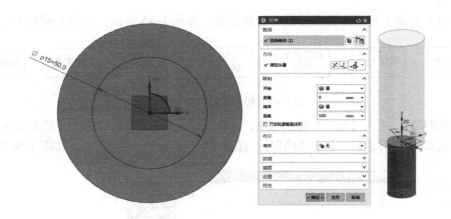

图 1-48　创建装夹部分

使用【编辑对象显示】命令或者按下快捷键 < Ctrl + J >，更改毛坯和装夹部分的颜色和透明度，如图 1-49 所示。

4）创建驱动面。选择【旋转】命令，单击【旋转】对话框中【截面】选项组中的"绘制截面"按钮，弹出【创建草图】对话框。选择 XC-ZC 平面，如图 1-50 所示，单击【确定】后进入草图模式。

图 1-49　更改颜色和透明度　　　　　　图 1-50　选择草图平面

绘制一条直线和一段圆弧，需注意草图上部的圆弧终点及圆心需约束在坐标轴上，如图 1-51 所示，然后退出草图模式。指定矢量为 ZC 轴方向，指定点为坐标原点，单击【确定】，旋转所得的几何体将会作为多轴加工时的驱动几何体，如图 1-52 所示。

项目1 FIFA世界杯奖杯工艺品

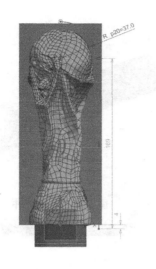

图 1-51 创建草图

图 1-52 旋转草图生成驱动几何体

5)进入加工模块,在弹出的【加工环境】对话框中选择【CAM 会话配置】为【cam_general】,选择【要创建的 CAM 设置】为【mill_contour】(轮廓铣),单击【确定】,如图 1-53 所示。

6)设置工件。单击【工序导航器】中的"几何视图"图标,然后单击【MCS_MILL】前面的"+"号,展开后如图 1-54 所示。双击【WORKPIECE】,弹出图 1-55 所示的【工件】对话框,依次指定"部件""毛坯"和"检查",如图 1-55 所示。

图 1-53 【加工环境】对话框

图 1-54 工序导航器-几何视图

7)创建 MCS(机床坐标系)。该工件在粗加工时,先在其中一面粗加工,旋转 180°后,再次进行粗加工,如图 1-56 所示,所以需要创建两个机床坐标系。

17

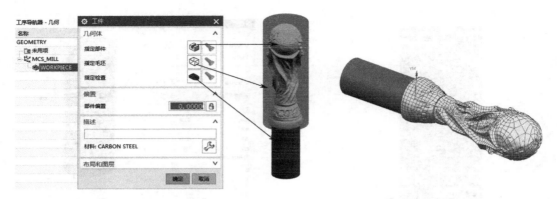

图 1-55 设置工件　　　　　　　图 1-56 分析工件

单击【创建几何体】命令,在弹出的对话框的【几何子类型】中选择"MCS",【几何体】选择【WORKPIECE】,输入【名称】为"MCS-1",如图 1-57 所示。单击【确定】后弹出【MCS】对话框,如图 1-58 所示。单击【机床坐标系】组中的"CSYS 对话框"按钮 ,弹出【CSYS】对话框。旋转坐标系使 ZM 轴向上,XM 轴朝向大力神杯顶端,如图 1-59 所示。【特殊输出】选择【使用主 MCS】,输入【安全距离】为"100",如图 1-60 所示。至此,完成第 1 个机床坐标系的创建,如图 1-61 所示。

图 1-57 【创建几何体】对话框

图 1-58 【MCS】对话框

图 1-59 MCS 设置

图 1-60　MCS 参数设置

图 1-61　创建第 1 个机床坐标系

第 2 个机床坐标系的创建方法与第 1 个相同，不同的是需要将坐标系方位调整至 ZM 轴向下，XM 轴朝向大力神杯顶端，确保第 2 个机床坐标系的 ZM 轴方向与第 1 个机床坐标系的 ZM 轴方向相反，如图 1-62 所示。

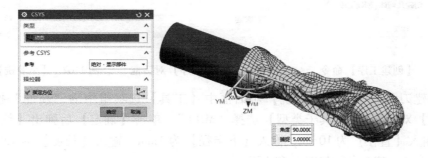

图 1-62　旋转坐标系

如图 1-63 所示，双击【工序导航器-几何】中的【MCS_MILL】，在弹出的【MCS 铣削】对话框中，【用途】选择【主要】，【安全距离】输入"200"，如图 1-64 所示。

图 1-63　机床坐标系设置完成

图 1-64　MCS 设置

1.3.2 粗加工

粗加工采用型腔铣,即固定轴等高加工,这种加工方法的材料去除效率最高,编程也较为简单。基本思路是先在 MCS-1 中编制其中一侧的粗加工程序,然后将其复制到 MCS-2 中,得到加工另一侧的加工程序,从而完成整个工件的粗加工。具体步骤如下。

1)创建工序。单击图 1-65 所示的【创建工序】命令,弹出如图 1-66 所示的对话框。【类型】选择【mill_contour】,【工序子类型】选择"型腔铣",【几何体】选择【MCS-1】,单击【确定】后弹出图 1-67 所示的【型腔铣】对话框。

图 1-65　【创建工序】命令　　图 1-66　【创建工序】对话框　　图 1-67　【型腔铣】对话框

2)创建刀具。单击【型腔铣】对话框中【工具】选项组中的"新建"按钮,弹出【新建刀具】对话框。【刀具子类型】选择"MILL",单击【确定】后弹出【铣刀-5 参数】对话框。输入【直径】为 10mm,输入【下半径】为 1mm,输入【长度】为 150mm,输入【刀具号】为 1,其余参数如图 1-68 所示。

a)　　　　　　　　　　　　b)

图 1-68　创建刀具

3）创建刀具夹持器。单击【铣刀-5 参数】对话框中的【夹持器】选项卡，单击【库】选项组中的"从库中调用夹持器"按钮，创建刀具夹持器的步骤如图 1-69 所示。

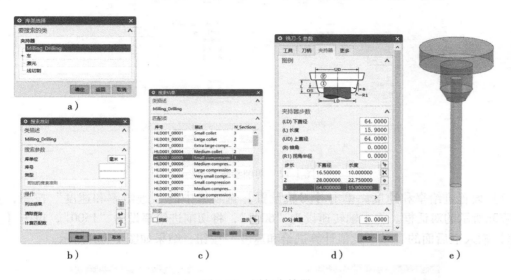

图 1-69　添加夹持器

4）刀轨设置。在【型腔铣】对话框中的【刀轨设置】选项组中，将【切削模式】设置为【跟随周边】，每刀切削深度的【最大距离】设为"1mm"，如图 1-70 所示。

5）设置切削层。单击【刀轨设置】选项组中的"切削层"按钮，弹出图 1-71 所示的对话框。毛坯直径是 80mm，所以初始深度为 80mm，由于分两侧进行两次粗加工，为了保证中间不留残余，范围深度适当大于 40mm，将【范围深度】设为 42mm，如图 1-71 所示。

图 1-70　刀轨设置

图 1-71　切削层设置

6）设置切削参数。单击【刀轨设置】选项组中的"切削参数"按钮，弹出图 1-72 所示的对话框。勾选【策略】选项卡中的【岛清根】，在【余量】选项卡中将【部件侧面余量】设为"0.5"。

a) b)

图 1-72　切削参数设置

7）设置进给率和速度。单击【刀轨设置】选项组中的"进给率和速度"按钮，弹出图 1-73a 所示的对话框。将主轴转速设为"6000"，将切削进给率设为"1500"，单击【主轴转速】输入框后面的"基于此值计算进给和速度"按钮，结果如图 1-73b 所示。

a) b)

图 1-73　非切削移动设置

8）生成刀轨。单击【操作】选项组中的"生成"按钮，生成图 1-74 所示的刀轨。该刀轨将会切除工件的上半部分，完成其中一侧的粗加工。

图 1-74　生成刀轨

9）复制刀轨。复制在【MCS-1】中创建的粗加工程序【CAVITY_MILL】，将其【内部粘贴】到【MCS-2】中，如图1-75所示。双击【CAVITY_MILL_COPY】，将切削深度改为42mm，其余参数保持不变，如图1-76所示。单击【操作】选项组中的"生成"按钮，生成另一侧的刀轨。

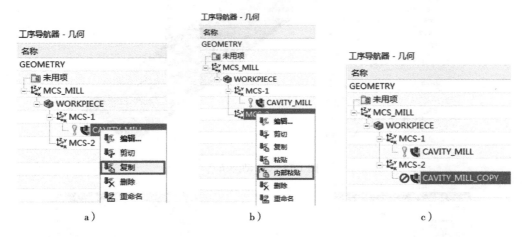

图1-75 复制刀轨

10）确认刀轨。选择工序导航器几何视图中的【MCS_MILL】，然后单击【确认刀轨】命令，弹出【刀轨可视化】对话框。单击【3D动态】选项卡中的"播放"按钮，可以查看切削效果，如图1-77和图1-78所示。

图1-76 切削层设置

图1-77 确认刀轨

1.3.3 二次粗加工

粗加工后在工件的一些位置处仍然有较大的余量，接下来使用【剩余铣】工序进行二次粗加工。具体步骤如下。

1）创建工序。使用【创建工序】命令打开相应对话框。【工序子类型】选择"剩余铣"，【几何体】选择【MCS-1】，其余参数设置如图1-79所示。单击【确定】后，弹出图1-80所示的【剩余铣】对话框。

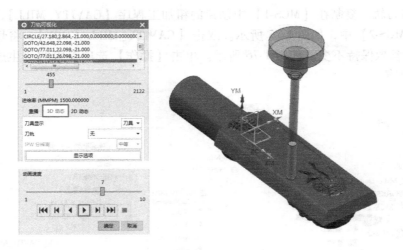

图 1-78 3D 动态仿真

图 1-79 【创建工序】对话框　　　　图 1-80 【剩余铣】对话框

2）创建刀具。单击【剩余铣】对话框中【工具】选项组中的"新建"按钮,弹出【新建刀具】对话框。【刀具子类型】选择"MILL",单击【确定】后弹出【铣刀-5 参数】对话框。输入【直径】为"6",输入【下半径】为"3",输入【长度】为"125",输入【刀具号】为"2",其余参数如图 1-81 所示。

3）创建刀具夹持器。在【夹持器】选项卡中从【库】中选择刀具夹持器,如图 1-82 所示。

4）设置刀轨。【切削模式】选择【跟随周边】,输入【最大距离】为"1",如图 1-83 所示。

5）设置切削层。单击【刀轨设置】选项组中的"切削层"按钮,输入【范围深度】为"44",此值略大于粗加工时的深度,如图 1-84 所示。

6）设置切削参数。单击【刀轨设置】选项组中的"切削参数"按钮,在【余量】选项卡中输入【部件侧面余量】为"0.3",如图 1-85 所示。

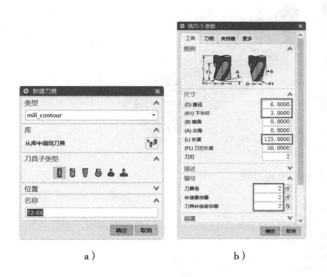

a）　　　　　　　　b）

图1-81　创建直径6mm刀具　　　　　图1-82　添加夹持器

图1-83　设置刀轨　　　　图1-84　设置切削层　　　　图1-85　设置切削参数

7）设置进给率和速度。单击【刀轨设置】选项组中的"进给率和速度"按钮,将主轴转速设为"6000",将切削进给率设为"1000",单击【主轴转速】输入框后面的"基于此值计算进给和速度"按钮。

8）生成刀轨。单击【操作】选项组中的"生成"按钮,生成图1-86所示的刀轨。二次粗加工后工件上各处的加工余量会较为均匀。

9）复制刀轨。复制在【MCS-1】中创建的剩余铣程序【REST_MILLING】,将其【内部粘贴】到【MCS-2】中,如图1-87所示。双击【REST_MILLING_COPY】,将切削深度改为"44",其余参数保持不变。单击【操作】组中的"生成"按钮,生成另一侧的刀轨。

10）确认刀轨。选择工序导航器几何视图中的【WORKPIECE】,然后单击【确认刀轨】

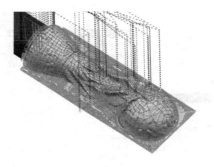

图 1-86　生成刀轨

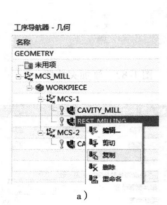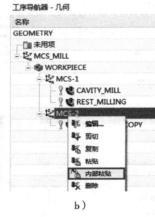

a)　　　　　　　　　　　b)　　　　　　　　　　　c)

图 1-87　复制刀轨

命令，弹出【刀轨可视化】对话框。单击【3D 动态】选项卡中的"播放"按钮，可以查看切削效果，如图 1-88 所示。

图 1-88　确认刀轨

1.3.4　半精加工

由于采用固定轴方式进行粗加工，部分区域会留下不均匀且较多的余量，需使用多轴加

工的方式进行半精加工，具体步骤如下。

1）创建工序。使用【创建工序】命令打开【创建工序】对话框。【类型】选择【mill_multi-axis】，【工序子类型】选择"可变轮廓铣"，【刀具】选择【T2-B6】，【几何体】选择【WORKPIECE】，其余参数设置如图 1-89 所示，单击【确定】，打开【可变轮廓铣】对话框。

2）设置驱动方法。在【可变轮廓铣】对话框中，【驱动方法】选项组中的【方法】选择【曲面】，如图 1-90 所示，然后指定图 1-91 所示的驱动几何体。

图 1-89　创建可变轮廓铣工序　　　图 1-90　驱动方法选择曲面　　　图 1-91　指定驱动几何体

为便于观察切削方向，选择【切削模式】为【往复】，选择【步距】为【数量】，输入【步距数】为"10"，单击"显示"按钮，观察到现在的切削方向为竖向，如图 1-92 所示，需将其改为一圈一圈往上绕的横向方式。

图 1-92　观察切削方向

单击"切削方向"按钮，选择图1-93b所示的驱动方向。再次单击"显示"按钮进行观察，如图1-93c所示。采用【数量】方式，两个面的宽度相差较多时，会出现一边较密集一边行数较少的情况。

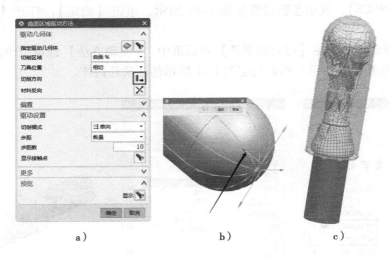

图1-93　切削方向修改

将【切削模式】改为【单向】（保证按同一方向进行旋转，不会出现一圈正转一圈反转的情况），将【步距】改为【残余高度】，将【最大残余高度】设为"0.1"。再次单击预览中的"显示"按钮，如图1-94所示（为方便观察可将旋转体隐藏）。单击【确定】，完成驱动方法设置。

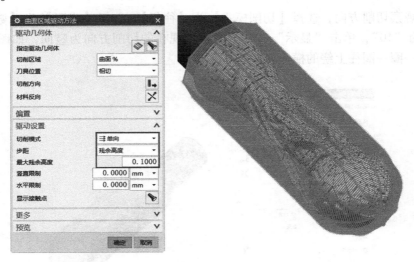

图1-94　驱动方法设置

3) 设置刀轴。在【刀轴】选项组中选择【轴】为【垂直于驱动体】，表示刀轴将垂直于驱动面，如图1-95所示。

4) 设置切削参数。单击【刀轨设置】选项组中的"切削参数"按钮。在【切削参数】对话框中的【余量】选项卡中输入【部件余量】为"0.1"，如图1-96和图1-97所示。

项目1　FIFA世界杯奖杯工艺品

图1-95　设置刀轴　　　图1-96　设置切削参数　　　图1-97　设置部件余量

5) 设置非切削移动。单击【刀轨设置】选项组中的"非切削移动"按钮。将【进刀类型】设为【无】，将【退刀类型】设为【与进刀相同】，将区域内的【移刀类型】设为【直接】，如图1-98所示。

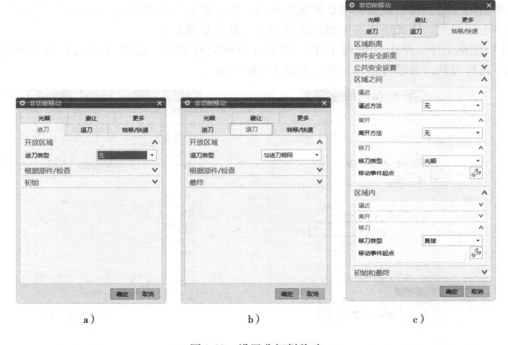

图1-98　设置非切削移动

6) 设置进给率和速度。单击【刀轨设置】选项组中的"进给率和速度"按钮。将主轴转速设为"8000"，将切削进给率设为"1200"，单击【主轴转速】输入框后面的"基于此值计算进给和速度"按钮。

7)生成刀轨。单击【操作】选项组中的"生成"按钮,如图1-99a所示,生成图1-99b所示的刀轨。

8)确认刀轨。单击【操作】选项组中的"确认"按钮,弹出【刀轨可视化】对话框。单击【3D动态】选项卡中的"播放"按钮,可以查看切削效果,如图1-99c所示。

 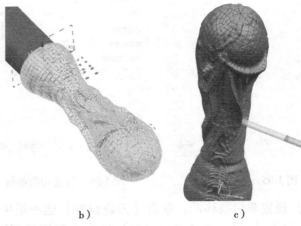

a) b) c)

图1-99 生成并确认刀轨

1.3.5 精加工

半精加工后,可以观察到模型的基本形状,但还留有一定的余量,特别是一些凹坑中的余量可能还较大,因此还需要更精细的精加工,具体步骤如下。

1)创建工序。单击【创建工序】命令打开相应对话框。选择【刀具】为【NONE】,表示不选择刀具,后续再创建刀具,如图1-100所示。

图1-100 【创建工序】对话框   图1-101 【可变轮廓铣】对话框

2)创建刀具。单击【可变轮廓铣】对话框中【工具】选项组中的"新建"按钮(图1-101),弹出【新建刀具】对话框。【刀具子类型】选择"BALL MILL",如图1-102a所示。单击【确定】后弹出【铣刀-球头铣】对话框。输入【球直径】为"2",输入【长

度】为"25",输入【刀具号】为"3",其余参数如图 1-102b 所示。单击【刀柄】选项卡,刀柄参数如图 1-102c 所示。

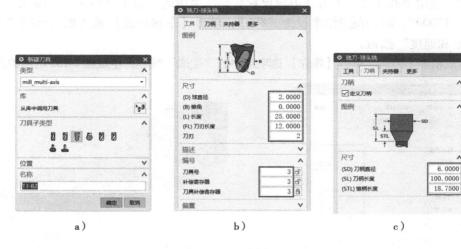

图 1-102 创建刀具

3)创建刀具夹持器。从【库】中选择刀具夹持器,如图 1-103 所示。

4)设置驱动方法。选择【曲面】驱动方法,如图 1-104a 所示。驱动体与切削方向的指定与半精加工中相同。将【切削模式】改为【螺旋】(为螺旋式下降,没有层间的切换),将【步距】设为【残余高度】,【最大残余高度】设为"0.01",其余参数设置如图 1-104b 所示。

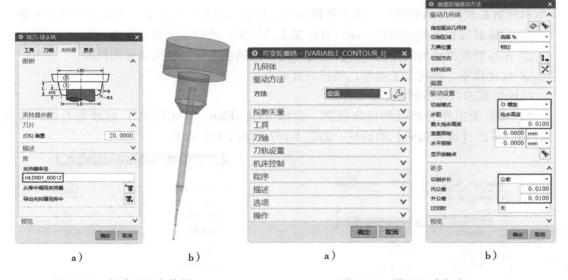

图 1-103 创建刀具夹持器　　图 1-104 设置驱动方法

5)设置刀轴。在【刀轴】选项组中选择【轴】为【垂直于驱动体】,表示刀轴将垂直于驱动面。

6)设置切削参数。单击【刀轨设置】选项组中的"切削参数"按钮,在【切削参数】对话框中的【余量】选项卡中输入【部件余量】为"0",如图 1-105 所示。

7)设置非切削移动。由于采用螺旋方式,只有一次进刀及退刀,其余均为连续。因此可以采用默认值。

8)设置进给率和速度。单击【刀轨设置】选项组中的"进给率和速度"按钮。将主轴转速设为"12000",将切削进给率设为"800",单击【主轴转速】输入框后面的"基于此值计算进给和速度"按钮。

9)生成精加工刀轨。单击【操作】选项组中的"生成"按钮,生成图1-106所示的刀轨。

图1-105 设置切削参数

图1-106 生成精加工刀轨

1.3.6 仿真加工

仿真加工是利用 Siemens NX 10 建立机床的仿真模型,对刀具、工件和夹具进行装配,读取机床数控代码并解释执行。对工件的加工过程进行三维运动仿真,主要解决以下问题:

① 验证数控程序的正确性,减少首件加工的风险,增加程序的可信度。

② 模拟数控机床的实际运动,检查潜在的碰撞错误,降低机床碰撞的风险。

③ 优化程序,提高加工效率,延长刀具寿命。

机床仿真的具体步骤如下。

1)切换至工序导航器的机床视图。选择【GENERIC_MACHINE】后单击右键,选择【编辑】,弹出【通用机床】对话框,如图1-107和图1-108所示。

图1-107 工序导航器机床视图

图1-108 【通用机床】对话框

2）单击"从库中调用机床"按钮，弹出【库类选择】对话框。选择【MILL】，单击【确定】，如图1-109所示。

3）弹出【搜索结果】对话框，选择【sim06_mill_5ax_sinumerik_mm】，单击【确定】，如图1-110所示。

图1-109 【库类选择】对话框　　　　　　　图1-110 【搜索结果】对话框

4）弹出【部件安装】对话框，【定位】选择【使用装配定位】，再选择部件、毛坯、夹持部分和驱动面，单击【确定】，如图1-111所示。

5）弹出【添加加工部件】对话框，【定位】选择【通过约束】，如图1-112所示。

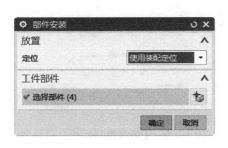 

图1-111 【部件安装】对话框　　　　　　　图1-112 【添加加工部件】对话框

6）弹出【装配约束】对话框，【类型】选择【同心】，然后选择毛坯夹持部分的底部圆边和工作台的圆边，通过两者的圆心进行定位，单击【确定】，如图1-113和图1-114所示。

7）此时的工序导航器机床视图如图1-115所示。单击【机床仿真】命令，弹出【仿真控制面板】对话框。勾选【显示3D除料】，单击"播放"按钮，如图1-116所示。

8）仿真加工过程如图1-117所示。

图 1-113 【装配约束】对话框

图 1-114 同心定位

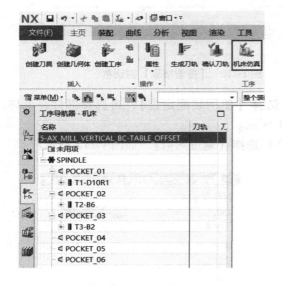
图 1-115 工序导航器机床视图

图 1-116 【仿真控制面板】对话框

图 1-117 仿真加工过程

项目2　涡轮叶轮

本项目以涡轮叶轮实物（图2-1）为基础，使用手持激光扫描仪 BYSCAN 510 进行数据采集，使用 Geomagic Design X 2016 软件进行数据处理并完成逆向建模，最后在 Siemens NX 10 软件中完成数控加工仿真。项目实施流程如图2-2 所示。

图2-1　涡轮叶轮实物

图2-2　项目实施流程

任务1　数据采集

本项目的数据采集使用手持激光扫描仪 BYSCAN 510，配套扫描软件 ScanViewer，如图2-3 所示。

2.1.1　扫描仪标定

当设备长期不用或者经过长途运输发生过抖动时，需快速标定。一般情况下，一次标定后可以长期使用。标定步骤如下。

第1步，单击【快速标定】命令，如图2-4 所示。单击后出现快速标定界面，如图2-5 所示。

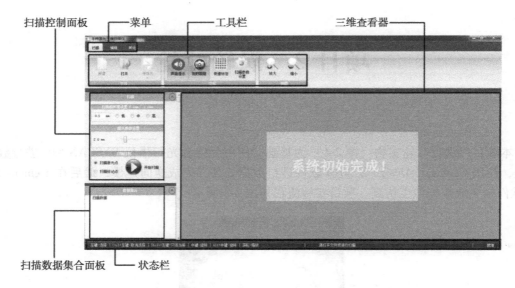

图 2-3 ScanViewer 软件界面介绍

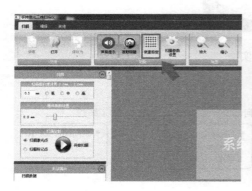

图 2-4 单击【快速标定】

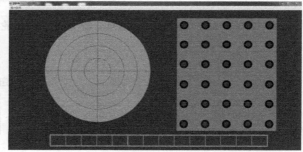

图 2-5 快速标定界面

第 2 步，将标定板放置在稳定的平面上，扫描仪正对标定板，距离为 30cm，按下扫描仪由上往下的第 3 个按键（以下称扫描键），发出激光束（以 3 条平行激光为例），如图 2-6 所示。

第 3 步，控制扫描仪角度。调整扫描仪与标定板的距离，使左侧的圆形阴影重合在保证左侧圆形阴影基本重合的状态下，在扫描仪所处的水平面上不改变角度，水平移动扫描仪，使右侧的梯形阴影重合，如图 2-7 所示，进入下一步标定。

第 4 步，右侧 45°标定。将扫描仪向右倾斜 45°，激光束保持处在第 3 行与第 4 行标记点之间，使阴影重合，进入下一步标定，如图 2-8 所示。

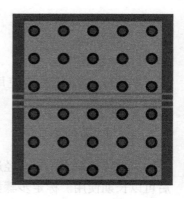

图 2-6 按下扫描键

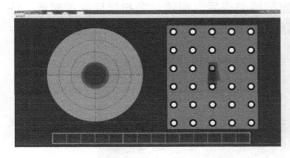 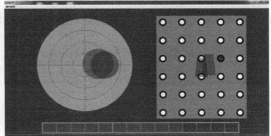

　　图 2-7　标定第 3 步　　　　　　　　　图 2-8　标定第 4 步

　　第 5 步，左侧 45°标定。将扫描仪向左倾斜 45°，激光束保持处在第 3 行与第 4 行标记点之间，使阴影重合，进入下一步标定，如图 2-9 所示。

　　第 6 步，上侧 45°标定。将扫描仪向上倾斜 45°，激光束保持处在第 3 行与第 4 行标记点之间，使阴影重合，进入下一步标定，如图 2-10 所示。

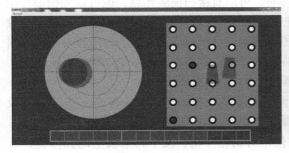 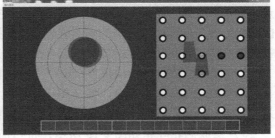

　　图 2-9　标定第 5 步　　　　　　　　　图 2-10　标定第 6 步

　　第 7 步，下侧 45°标定。将扫描仪向下倾斜 45°，激光束保持处在第 3 行与第 4 行标记点之间，使阴影重合，如图 2-11 所示。

　　完成第 7 步之后，界面提示如图 2-12 所示。单击右上角，关闭标定窗口。标定到此完成。

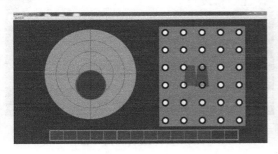 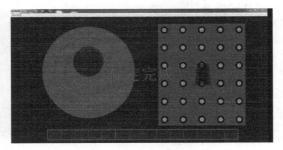

　　图 2-11　标定第 7 步　　　　　　　　　图 2-12　标定完成

2.1.2　贴标记点

　　开始扫描前，先在涡轮叶轮的正、反面上都贴上标记点，如图 2-13 所示，并且在正面

和反面的过渡区域贴上 5~6 个标记点。当正面扫描完成后，翻转过来扫描反面的时候，注意不要碰落或移动过渡区域的标记点，否则可能会影响拼接的准确性。

a）正面标记点　　　　　　　　　　b）反面标记点

图 2-13　贴标记点

2.1.3　扫描标记点

使用手持激光扫描仪 BYSCAN 510 扫描物体时，可以直接扫描激光点，也可以先扫描标记点再扫描激光点，后者的扫描精度更高，而且扫描过程中过渡方便。涡轮叶轮的正、反面过渡区域是垂直于底面的比较窄的圆周区域，实际操作表明，直接扫描激光点的方式很容易造成拼接失败，所以这里采用先扫描标记点再扫描激光点的方式，并且在扫描标记点时使用辅助板。

1) 打开手持激光扫描仪 BYSCAN 510 的配套扫描软件 ScanViewer。

2) 在扫描控制面板中，将【扫描解析度设置】设为"1mm"，将【曝光参数设置】设为"1ms"，选中【标记点】选项后单击【开始】按钮，如图 2-14 所示。

3) 摆放涡轮叶轮如图 2-15 所示，将扫描仪正对涡轮叶轮，按下扫描仪上的扫描键，开始扫描。扫描过程中的软件界面如图 2-16 所示。

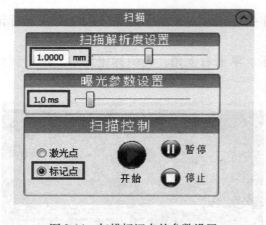 

图 2-14　扫描标记点的参数设置　　　　　图 2-15　摆放涡轮叶轮

4) 涡轮叶轮正面的标记点扫描完成之后，在工作台上放置辅助板并调整涡轮叶轮的摆放位置，使其正、反面的标记点都能扫描到，如图 2-17 所示。由于涡轮叶轮的正、反面呈 90°过渡，如果直接扫描正面和反面，则正、反面标记点的相对位置关系不便于确定，所以这里使用了辅助板。首先扫描涡轮叶轮的正面和辅助板，由此确定涡轮叶轮正面的标记点和辅助板上标记点的相对位置关系。然后扫描涡轮叶轮的反面和辅助板，由此确定涡轮叶轮反

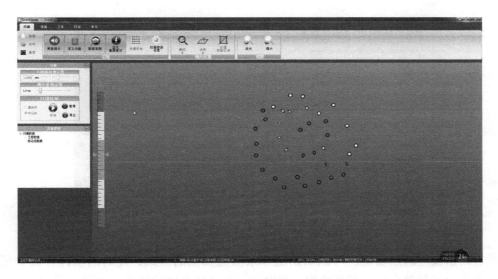

图 2-16　扫描标记点时的软件界面

面的标记点和辅助板上标记点的相对位置关系。这样，以辅助板上的标记点为"桥梁"，确定了涡轮叶轮正面和反面标记点的相对位置关系。

5）最后，删除扫描的辅助板上的标记点，如图 2-18 所示，标记点扫描完成。

图 2-17　使用辅助板

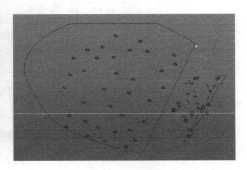

图 2-18　选择辅助板上的标记点并删除

2.1.4　扫描激光点

标记点扫描完成之后，进行激光点扫描。涡轮叶片之间的区域比较难扫描到，为便于扫描，可以采用单束光扫描的方式。

1）在扫描控制面板中，将【扫描解析度设置】设为"0.5mm"，将【曝光参数设置】设为"1ms"，选中【激光点】选项后单击"开始"按钮，在弹出的下拉列表中选择【红光】，如图 2-19 所示。

2）摆放好涡轮叶轮，使其正面朝上，将扫描仪正对涡轮叶轮，按下扫描仪上的扫描键，开始扫描。扫描激光点时的软件界面如图 2-20 所示。

图 2-19　扫描激光点的参数设置

图 2-20 扫描激光点时的软件界面

双击扫描仪上的扫描键,可以切换到单束激光扫描的模式,此时的软件界面如图 2-21 所示。多束激光和单束激光交替使用,有助于扫描叶片之间的区域,如图 2-22 所示。

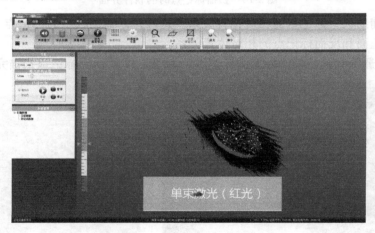

图 2-21 单束激光扫描

a)多束激光扫描　　　　　　　　b)单束激光扫描

图 2-22 多束激光与单束激光扫描

按下扫描仪上的视窗放大键，可以放大视图，便于观察扫描区域，如图 2-23 所示。同样地，按下扫描仪上的视窗缩小键，可以缩小视图。

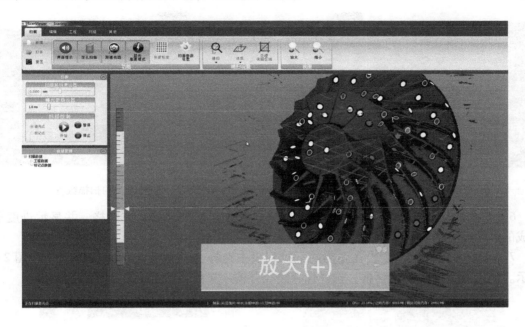

图 2-23　放大视图

3）正面扫描完成之后，按下扫描仪上的扫描键停止扫描，再单击扫描软件上的"暂停"按钮。使用套索工具选中扫描得到的无关数据，按下键盘上的 <Delete> 键将其删除，如图 2-24 所示。

图 2-24　选中并删除无关数据

4）调整涡轮叶轮的摆放位置，使其反面朝上，单击软件界面上的"开始"按钮，在弹出的下拉列表中选择【红光】，如图 2-19 所示。然后按下扫描仪上的扫描键，开始扫描，如图 2-25 所示。

5）反面扫描完成后，按下扫描仪上的扫描键停止扫描，再单击扫描软件上的"停止"

按钮。使用套索工具选中扫描得到的无关数据，按下键盘上的 < Delete > 键将其删除，如图 2-26 所示。

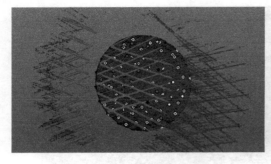 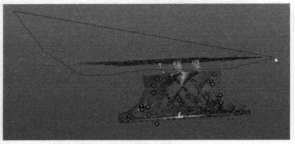

图 2-25　扫描涡轮叶轮的反面　　　　　　图 2-26　删除无关的扫描数据

6）单击【工程】选项卡中的【生成网格】命令，系统开始生成网格，并显示进度条。生成的涡轮叶轮网格模型如图 2-27 所示。

7）单击【保存】命令，在弹出的下拉列表中选择【网格文件（*.STL）】，如图 2-28 所示。弹出【另存为】对话框，选择保存路径并输入文件名后，单击【保存】按钮。

图 2-27　生成的涡轮叶轮网格模型　　　　图 2-28　保存为 STL 文件

任务 2　数据处理

1）打开软件。单击"开始"菜单中的 Geomagic Design X 2016 程序，或双击桌面上的图标 ，启动 Geomagic Design X 2016 应用软件。

2）导入模型。单击界面左上方的"导入"按钮 ，如图 2-29 所示。选择扫描生成的模型数据后单击【仅导入】，如图 2-30 所示。涡轮叶轮模型成功导入 Geomagic Design X 2016 软件，如图 2-31 所示。

3）单击【多边形】模块中的【修补精灵】命令，弹出【修补精灵】对话框。软件会自动检索面片模型中存在的各种缺陷，如非流形顶点、重叠单元面、悬挂的单元面、交差单元面等，如图 2-32 所示。单击"OK"按钮 ，软件自动修复检索到的缺陷。

项目 2　涡轮叶轮

图 2-29　单击"导入"按钮　　　　　　　　　图 2-30　【导入】对话框

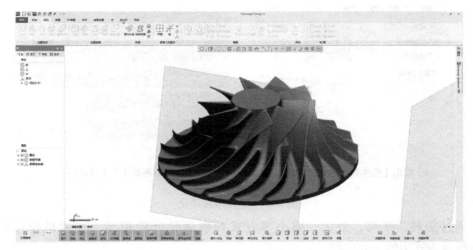

图 2-31　导入涡轮叶轮模型

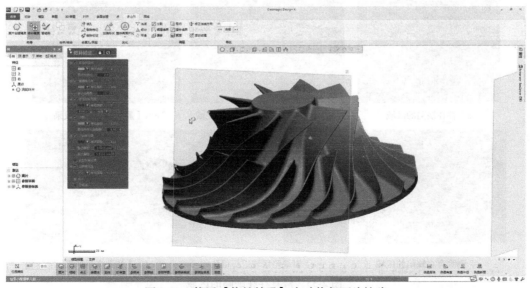

图 2-32　使用【修补精灵】自动修复面片缺陷

4）单击【多边形】模块中的【加强形状】命令，弹出图 2-33 所示对话框。【锐度】表示设置执行锐化的尖锐区域范围。【整体平滑】表示设置执行平滑的圆角区域范围。【加强水平】表示设置执行操作的迭代次数。这里将这 3 个选项都保持默认数值，然后单击"OK"按钮☑完成操作。加强形状用于锐化面片上的尖锐区域（棱角），同时平滑平面或圆柱面区域，以提高面片的质量。

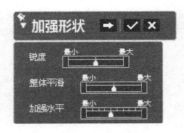

图 2-33　【加强形状】对话框

5）单击【多边形】模块中的【整体再面片化】命令，弹出图 2-34b 所示的对话框。保持默认的参数设置，单击"OK"按钮☑完成操作。使用【整体再面片化】命令，系统会重新计算整体面片并提高面片质量。

a）单击【整体再面片化】按钮　　　b）【整体再面片化】对话框

图 2-34　整体再面片化

6）通过旋转、平移、放大、缩小等方式观察涡轮叶轮模型，查找是否存在缺陷。图 2-35a 所示为其中的一处缺陷。

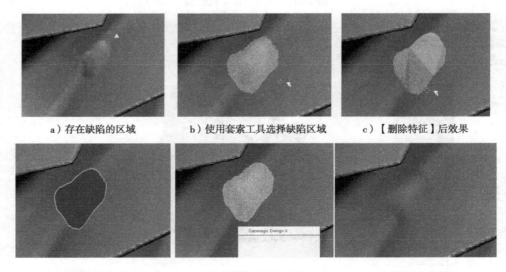

a）存在缺陷的区域　　b）使用套索工具选择缺陷区域　　c）【删除特征】后效果

d）删除存在缺陷的区域　　e）对缺陷区域进行填孔　　f）【填孔】后效果

图 2-35　修复存在缺陷的区域

首先尝试用【删除特征】命令修复该处缺陷，该命令会移除所选单元面并通过智能填孔修复该区域。使用图 2-36 所示的套索工具选择缺陷区域后，单击【多边形】模块中的【删除特征】命令，弹出图 2-37 所示的对话框。参数设置保持默认，单击"OK"按钮☑完成操作，效果如图 2-35c 所示。由于修复效果不好，按下键盘上的 <Ctrl+Z> 撤销删除特征操作。

图 2-36 选择套索工具并选择仅可见

接下来尝试用【填孔】命令修复该处缺陷。使用套索工具选择该处缺陷区域并按下键盘上的 <Delete> 键将其删除。单击【多边形】模块中的【填孔】命令，弹出图 2-38 所示的【填孔】对话框。选择图 2-35d 所示的孔，参数设置保持默认，单击"OK"按钮☑完成操作，效果如图 2-35f 所示。采用删除缺陷区域后再填孔的修复方法效果较好。

继续查找涡轮叶轮模型中是否还存在缺陷，若存在缺陷，继续用删除后填孔的方法进行修复。

图 2-37 【删除特征】对话框　　　　图 2-38 【填孔】对话框

7）单击【模型】模块中的【平面】命令，弹出【追加平面】对话框。如图 2-39 所示，单击【方法】右侧下拉箭头，在弹出的下拉列表中选择【选择多个点】。如图 2-40 所示，在涡轮叶轮底面选择 4 个点，单击"OK"按钮☑。3 个点可以创建一个平面，选择 4 个点创建的平面拟合精度更高。如图 2-41 所示，刚创建的"平面 1"默认处于选中状态，按下键盘上的 <Esc> 键取消选中。

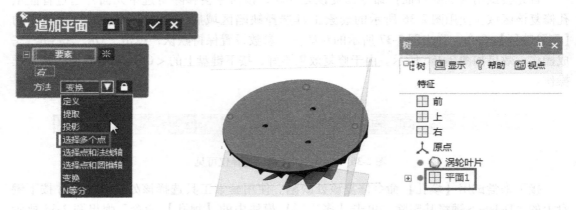

图 2-39 【追加平面】对话框   图 2-40 选择 4 个点创建平面   图 2-41 创建得到的平面 1

8)单击【模型】模块中的【线】命令,弹出【添加线】对话框。如图 2-42 所示,单击【方法】右侧下拉箭头,在弹出的下拉列表中选择【检索圆柱轴】。如图 2-43 所示,使用套索工具选择涡轮叶轮底部圆柱面上的区域,注意不要选择该圆柱面上质量不好的区域,同时所选的区域大致均布在该圆柱面的圆周上,单击 "OK" 按钮 ✓ 后,创建得到"线 1"。

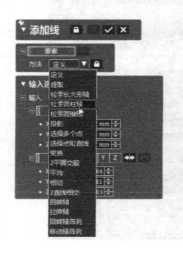

图 2-42 【添加线】对话框   图 2-43 使用套索工具选择涡轮叶轮底部圆柱面上的区域

9)单击【对齐】模块中的【手动对齐】命令,弹出图 2-44a 所示的对话框。单击 "下一步" 按钮 ➡ 后,对话框如图 2-44b 所示。选择【X-Y-Z】选项,再单击对话框中的【位置】,然后选择 "线 1" 和 "平面 1",可以在图形窗口中选择,也可以在特征树中选择,位置就是 "线 1" 和 "平面 1" 的交点位置。单击对话框中的【Z轴】,然后选择 "平面 1",Z 轴就是 "平面 1" 的法线方向。检查坐标系的 Z 轴是否是由涡轮的底部指向头部,若不是,则双击 Z 轴箭头使其反向。单击 "OK" 按钮 ✓ 后,创建得到的坐标系如图 2-45 所示。

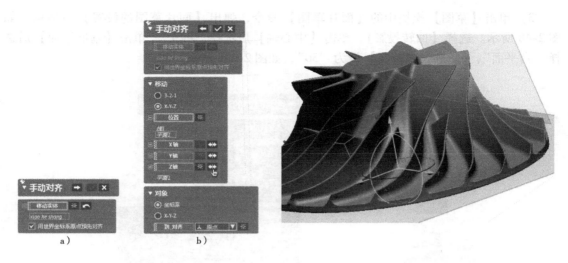

图 2-44　【手动对齐】对话框　　　　　　　图 2-45　创建坐标系

任务 3　逆向建模

2.3.1　主体建模

1）选择特征树中的"平面 1"和"线 1"，单击【删除】，在弹出的对话框中单击【是】，如图 2-46 所示。

2）单击【模型】模块中的【线】命令，弹出【添加线】对话框。如图 2-47 所示，单击下拉箭头，在下拉列表中选择【2 平面交差】。选择【上】平面和【右】平面，单击"OK"按钮✓后，创建得到【上】平面和【右】平面的交线，即"线 1"。

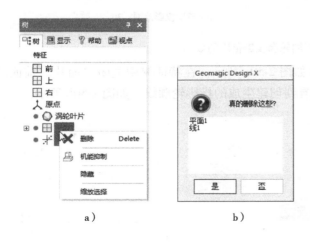

图 2-46　删除平面 1 和线 1

图 2-47　【添加线】对话框

3）单击【草图】模块中的【面片草图】命令，弹出【面片草图的设置】对话框，如图 2-48 所示。选择【回转投影】，单击【中心轴】后选择"线 1"，单击【基准平面】后选择"右平面"，【轮廓投影范围】设为"30"，如图 2-49b 所示。

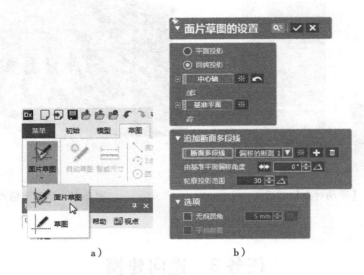

图 2-48　面片草图

a）轮廓投影范围：0

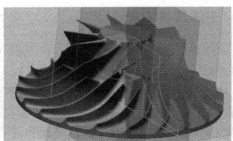

b）轮廓投影范围：30

图 2-49　不同轮廓投影范围的效果

单击"OK"按钮☑后，投影轮廓线如图 2-50a 所示。在特征树中关闭"面片"前面的眼睛图标◉，隐藏面片，可以更清楚地看到创建生成的投影轮廓线，如图 2-50b 所示。

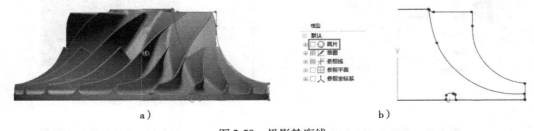

图 2-50　投影轮廓线

4)单击【草图】模块中的【直线】命令,弹出【直线】对话框。选择坐标系的原点作为直线的起点,向右拖动,系统自动捕捉到水平约束——,在超过投影轮廓线的位置单击选择直线的终点,如图2-51所示。创建直线过程中按下键盘上的<Esc>键可以取消连续直线的创建。用同样的方法创建第2条直线,可以左右拖动该直线,使其通过投影轮廓线的端点,如图2-52所示。用同样的方法继续创建其余直线,如图2-53所示。

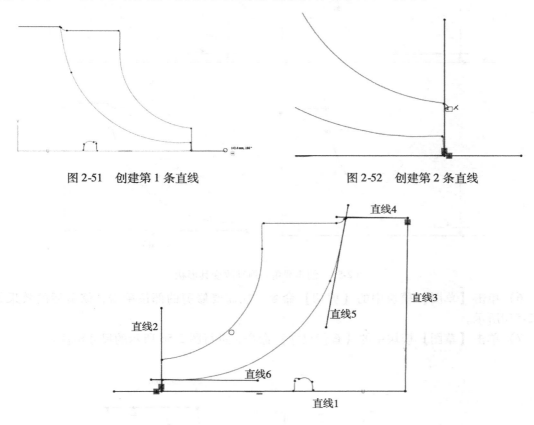

图2-51 创建第1条直线　　　　　　图2-52 创建第2条直线

图2-53 继续创建直线

5)单击【草图】模块中的"圆角"命令,如图2-54所示。依次选择图2-53所示的直线6和直线5,继续按住鼠标左键不放,拖动改变圆角的大小,使其与轮廓线重合,如图2-55a、b、c所示。释放鼠标左键后,会自动创建该圆角的半径尺寸。双击该半径尺寸,并输入"80.8"。

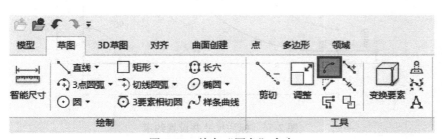

图2-54 单击"圆角"命令

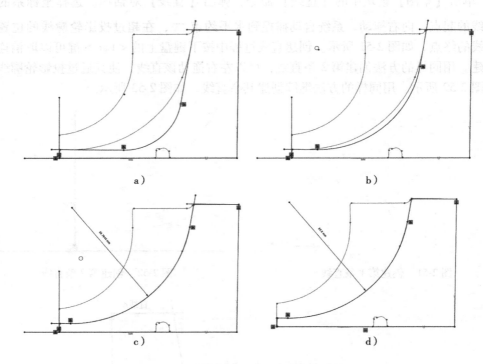

图 2-55 创建圆角并拖动改变其形状

6)单击【草图】模块中的【剪切】命令,在需要修剪的部位单击,修剪后的效果如图 2-55d 所示。

7)单击【草图】模块中的【智能尺寸】命令,进行图 2-56 所示的尺寸标注。

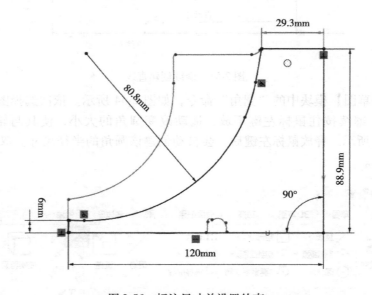

图 2-56 标注尺寸并设置约束

8)当需要添加草图要素间的相互约束条件时,可以先选择第一个需要约束的草图要素,再按住<Shift>键双击另一个草图要素,此时弹出关于两者约束条件的操作窗口,如图2-57所示。在【共同的约束条件】中显示的是两个要素之间的相互约束条件,在【独立的约束条件】中显示的是两个要素各自的约束条件。

在图2-57中存在过约束,过约束用红色表示。直线3和直线4的垂直约束,直线1和直线3的90°角度约束,以及直线1和直线4之间的平行约束,这三者中去掉一个就能解决过约束的问题。如果要删除约束条件,可以在【约束条件】对话框中选择要删除的约束,然后单击下方的【移除约束】按钮。

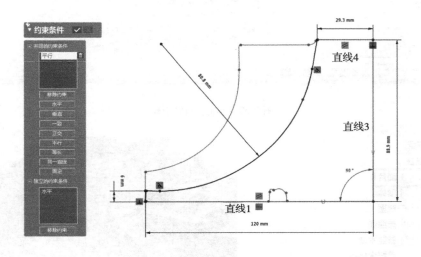

图2-57 设置约束条件

9)单击界面左上角的"退出"按钮，或界面右下角的"退出"按钮，退出草图。

10)单击【模型】模块中【创建实体】组中的【回转】命令,如图2-58所示,自动生成回转实体,如图2-59所示。

图2-58 单击【回转】命令

11)单击模型特征树中"面片"前面的方框,出现眼睛图标，使面片可见,如图2-60所示。观察面片和回转实体的接近程度,可以发现在涡轮的上端部分,回转实体与面片不够接近,如图2-61所示。

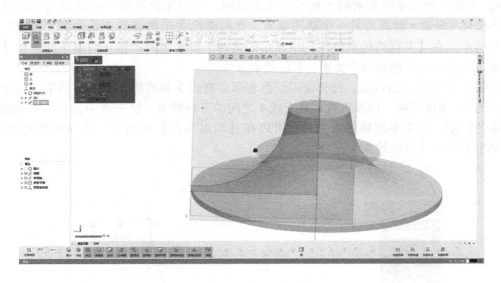

图 2-59　生成回转实体

图 2-60　使面片可见　　　　图 2-61　上端部分回转实体与面片不够接近

12）双击特征树中的"草图 1（面片）"，如图 2-62 所示，进入草图模块。双击图 2-63 所示的尺寸，将其由原来的"29.3"修改至"29.5"，然后退出草图。可以看到修改后回转实体与面片相接近，如图 2-64 所示。

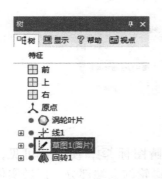
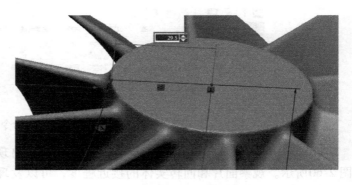

图 2-62　双击"草图 1（面片）"　　　　图 2-63　将尺寸值修改至 29.5

13）涡轮叶轮的主体部分建模完成，如图2-65所示。

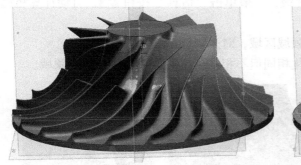

图2-64 修改后回转实体与面片相接近　　　　图2-65 涡轮叶轮的主体

2.3.2 叶片建模

1）检查涡轮叶片的拟合精度。单击界面右侧的【Accuracy Analyzer（TM）】，如图2-66a所示，在弹出的对话框中选择【体偏差】。偏差图如图2-66b所示，从偏差图中可以看出拟合精度符合要求。

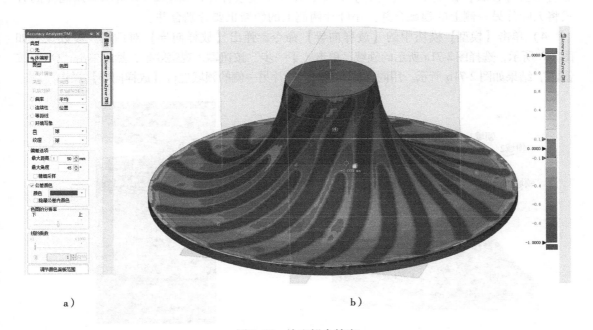

a)　　　　　　　　　　　　　　　　b)

图2-66 检查拟合精度

2）单击【领域】模块中的【自动分割】命令，弹出图2-67所示的对话框。保持默认的参数设置，单击"OK"按钮✓后，结果如图2-68所示，不同的颜色表示不同的领域。

【自动分割】对话框中主要选项说明如下。

【敏感度】：指曲率敏感度，敏感度值越低，划分的领域数量越少，反之，划分的领域数量越多，选择范围是0~100。

【面片的粗糙度】:指当前多边形模型的粗糙度情况。用于计算曲率时,忽略粗糙度对领域划分的影响。自平滑至粗糙分为4个等级,一般单击"估算"按钮,自动计算粗糙度情况。

【保持当前领域】:指不改动已划分的领域区域,对未划分区域进行领域划分。

【合并相同的原始形状】:指将曲率变化相同但不相互连接的领域合并为同一领域。

图 2-67 【自动分割】对话框　　　　　图 2-68 自动分割后的领域

3)查找质量较好且相邻的 1 个大叶片和 1 个小叶片。选择大叶片其中一侧上的领域,然后单击【领域】模块中的【合并】命令,将 3 个领域合并,如图 2-69 所示。用同样的方式将大叶片另一侧上的领域合并。小叶片两侧上的领域也要分别合并。

4)单击【模型】模块中的【放样向导】命令,弹出【放样向导】对话框。参数设置如图 2-70 所示,选择图 2-71a 所示的领域,单击"下一步"按钮,观察效果,然后单击"OK"按钮,结果如图 2-71b 所示。用同样的方式对大叶片另一侧的领域进行【放样向导】操作。

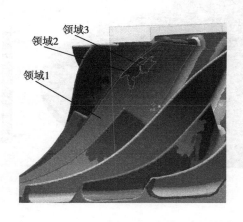

图 2-69 大叶片其中一侧上的 3 个领域　　　图 2-70 【放样向导】对话框

5)单击界面右侧的【Accuracy Analyzer(TM)】,在弹出的对话框中选择【体偏差】,偏差分析如图 2-72 所示。从偏差图中可以看出放样面的绝大部分都是绿色的,虽然有一部分是红色,表示偏差较大,但是该部分以后会被裁剪掉。

6)用同样的方式对小叶片两侧的领域分别进行【放样向导】操作,结果如图 2-73 所示。

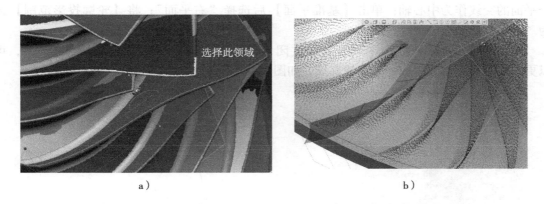

图 2-71 从领域中提取放样对象

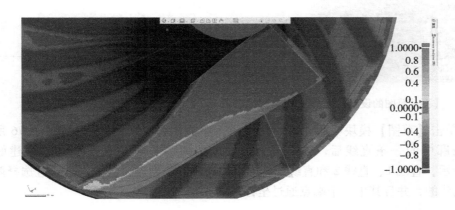

图 2-72 放样面的偏差分析

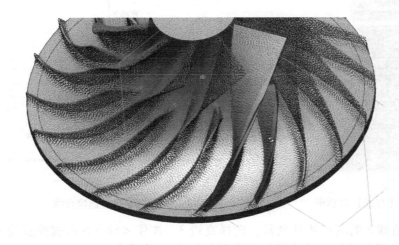

图 2-73 创建的 4 个放样面

7) 单击【草图】模块中的【面片草图】命令,弹出【面片草图的设置】对话框,如图 2-74 所示。选择【回转投影】;单击【中心轴】后选择"上平面"和"右平面",以这两

个平面的交线作为中心轴;单击【基准平面】后选择"右平面";将【轮廓投影范围】设为"30"。

单击"OK"按钮后,在特征树中关闭"面片"前面的眼睛图标,隐藏面片,可以更清楚地看到创建生成的投影轮廓线,如图 2-75 所示。

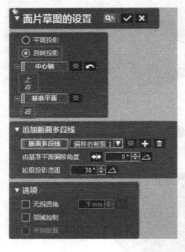

图 2-74 【面片草图的设置】对话框

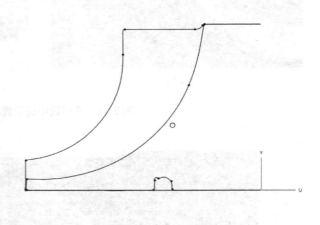

图 2-75 投影轮廓线

8)单击【草图】模块中的【直线】按钮,弹出【直线】对话框,如图 2-76 所示。选择投影轮廓中的一条直线后,单击对话框中的按钮,一条直线就自动创建好了。如图 2-77 所示,直线 1、直线 2 和直线 3 就是采用这种方法创建,直线 4 是通过选择两个端点的方式创建的,并且其中一个端点通过坐标系的原点。

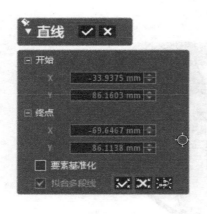

图 2-76 【直线】对话框

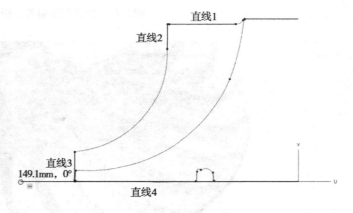

图 2-77 创建直线

9)拖动直线 3 的端点使其延长。选择直线 3,按住 <Shift> 键选择直线 4,此时弹出【约束条件】对话框。单击【共同的约束条件】中的【垂直】,单击"OK"按钮,为直线 3 和直线 4 添加垂直约束,如图 2-78 所示。

10)单击【草图】模块中的【3 点圆弧】命令,选择图 2-77 中直线 2 和直线 3 之间的圆弧,单击"OK"按钮,圆弧就自动创建完成了。

项目 2 涡轮叶轮

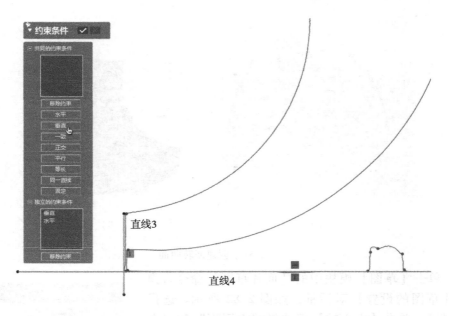

图 2-78 添加垂直约束

11)拖动直线的端点使其延长,然后使用【草图】模块中的【剪切】命令进行修剪。

12)单击【草图】模块中的【智能尺寸】命令,进行图 2-79 所示的尺寸标注,其中下方的尺寸值为 120mm,右侧的尺寸值为 85.8mm,并设置相切等约束条件。单击界面左上角的"退出"按钮，或界面右下角的"退出"按钮，退出草图。

13)单击【模型】模块中【创建曲面】组中的【回转】命令,弹出图 2-80 所示的【回转】对话框。设置【方法】为【两方向】,拖动图 2-81a 所示的 2 个箭头,使创建的回转曲面能够与之前创建的 4 个放样面相交。单击【回转】对话框中的"OK"按钮，结果如图 2-81b 所示。

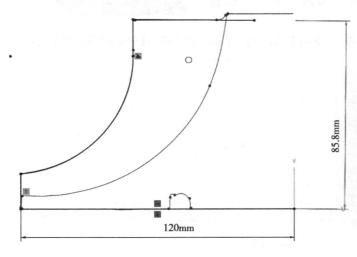

图 2-79 调整后的草图曲线

图 2-80 【回转】对话框

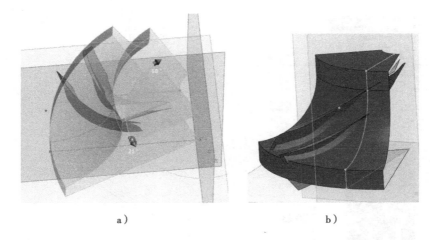

图 2-81 创建回转曲面

14)单击【草图】模块中的【面片草图】命令,弹出【面片草图的设置】对话框,如图 2-82 所示。选择【回转投影】,单击【中心轴】后选择"上平面"和"右平面",以这两个平面的交线作为中心轴,单击【基准平面】后选择"右平面",调整【轮廓投影范围】和【由基准平面偏移角度】这两个参数的值,直至小叶片上端的截面线投影到草图平面上,如图 2-83a 所示。

15)单击【草图】模块中的【直线】命令,弹出【直线】对话框。选择图 2-83b 中箭头所指的直线后,单击对话框中的按钮,直线就自动创建好了。拖动这条直线的两个端点使其延长,如图 2-83c 所示。单击界面左上角的"退出"按钮,或界面右下角的"退出"按钮,退出草图。

16)单击【模型】模块中【创建曲面】组中的【回

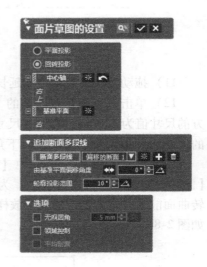

图 2-82 【面片草图的设置】对话框

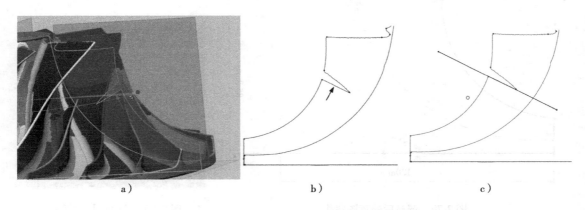

图 2-83 创建小叶片上端的截面线

转】命令，弹出图 2-80 所示的【回转】对话框。设置【方法】为【两方向】，拖动图 2-84a 所示的 2 个箭头，使创建的回转曲面能够与小叶片相交。为便于观察，单击图 2-85 所示模型特征树中"曲面体"前面的方框，出现眼睛图标 👁，使之前创建的曲面可见，如图 2-84b 所示。单击【回转】对话框中的"OK"按钮 ✓，结果如图 2-84c 所示。

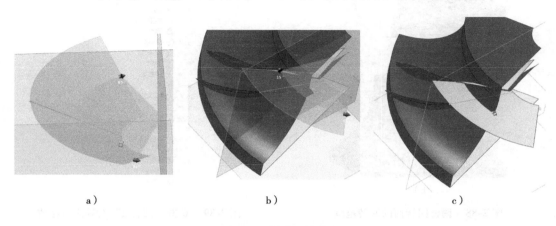

图 2-84　创建回转曲面

17）单击【模型】模块中的【曲面偏移】命令，弹出图 2-86 所示的【曲面偏移】对话框。输入【偏移距离】为"0mm"，选择涡轮主体上的 3 个面，单击"OK"按钮 ✓，结果如图 2-87 所示。

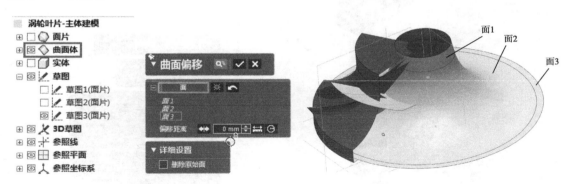

图 2-85　模型特征树　　图 2-86　【曲面偏移】对话框　　图 2-87　抽取涡轮主体上的 3 个面

18）观察模型后发现，步骤 13）创建的回转曲面并未完全与放样曲面相交，如图 2-88a 所示。因此双击图 2-89 所示特征树中的"回转 2"，拖动箭头对回转角度进行修改，修改后结果如图 2-88b 所示。

19）观察模型后发现，之前创建的 4 个放样面并未与涡轮主体完全相交，如图 2-90a 所示，因此需要对这 4 个放样面进行延长。单击【模型】模块中的【延长曲面】命令，出现图 2-91 所示的对话框。设置【距离】为"10mm"，选择图 2-90a 中箭头所指的放样面，出现图 2-91b 所示的预览效果，此时放样面与涡轮主体完全相交，单击"OK"按钮 ✓。用同样的方法延长其余 3 个放样面。

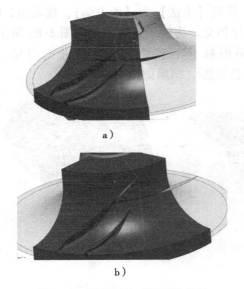

图 2-88 调整回转面的回转范围　　　　图 2-89 双击"回转 2"对其进行修改

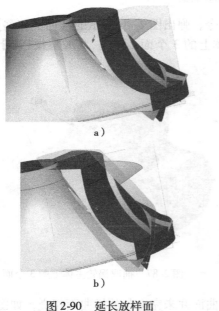

图 2-90 延长放样面　　　　图 2-91 【延长曲面】对话框

20）单击【模型】模块中的【剪切曲面】命令，出现图 2-92a 所示的对话框。单击【工具要素】后选择图 2-93a 所示的 3 个面作为工具，单击【对象体】后选择 4 个放样面作为对象，单击"下一步"按钮，出现图 2-92b 所示的对话框。在图形窗口中选择要保留的曲面，如图 2-93b 所示。单击"OK"按钮，完成对 4 个放样面的修剪。

21）单击【模型】模块中的【剪切曲面】命令，出现图 2-94a 所示的对话框。单击【工具要素】，选择剪切后的 4 个放样面、回转 3 和曲面偏移 1 作为工具，单击【对象体】，选择回转 2 作为对象，单击"下一步"按钮，出现图 2-94b 所示的对话框。在图形窗口中

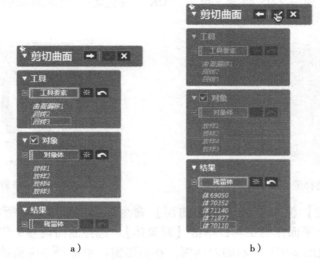

图 2-92 【剪切曲面】对话框

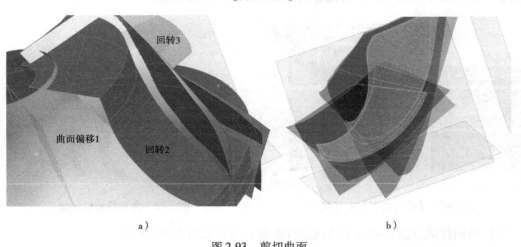

图 2-93 剪切曲面

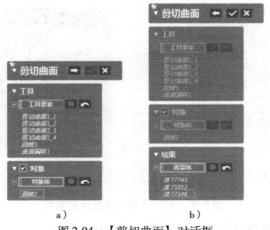

图 2-94 【剪切曲面】对话框

选择要保留的曲面，如图 2-95 所示。单击"OK"按钮☑，完成对回转 2 的修剪，结果如图 2-96 所示。

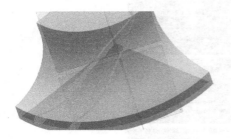 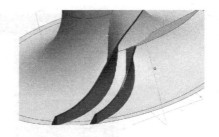

图 2-95　选择要保留的曲面　　　　　　　　图 2-96　曲面修剪后的效果

22）单击【模型】模块中的【剪切曲面】命令，出现图 2-97a 所示的对话框。单击【工具要素】，选择上平面作为工具；单击【对象体】，选择曲面偏移 1 作为对象；单击"下一步"按钮➡，出现图 2-97b 所示的对话框。在图形窗口中选择要保留的曲面，单击"OK"按钮☑，从主体上抽取的 3 个面被剪切掉一半，如图 2-98 所示。

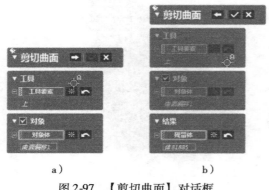 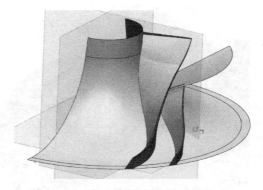

　　　　a）　　　　　　　　b）

图 2-97　【剪切曲面】对话框　　　　　　　图 2-98　曲面修剪后的效果

23）用同样的方法对回转 3 曲面进行修剪，结果如图 2-99b 所示。

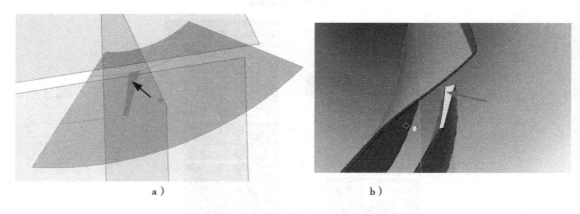

a）　　　　　　　　　　　　　　b）

图 2-99　回转 3 剪切曲面后的效果

24）用同样的方法继续剪切曲面，结果如图 2-100 所示。

项目 2　涡轮叶轮

a）剪切曲面前　　　　　　　　　　　　b）剪切曲面后

图 2-100　曲面剪切后的效果

25）单击【模型】模块中的【缝合】命令，出现图 2-101a 所示的对话框。选择组成小叶片的所有面片，单击"下一步"按钮，出现图 2-101b 所示的对话框。单击"OK"按钮，组成小叶片的面片被缝合成一个实体。用同样的方法缝合大叶片，缝合后的结果如图 2-102 所示。

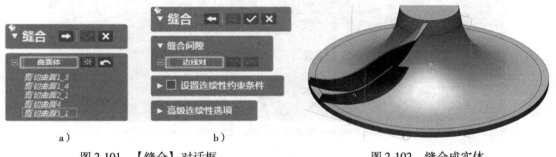

a）　　　　　　　b）

图 2-101　【缝合】对话框　　　　　　　图 2-102　缝合成实体

26）单击【模型】模块中的"圆形阵列"命令，如图 2-103 所示。单击【体】，选择缝合后的小叶片和大叶片，单击【回转轴】，选择线 1，设置【要素数】为"10"，如图 2-104 所示。单击"OK"按钮，结果如图 2-105 所示。

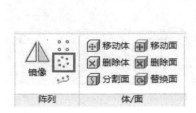

图 2-103　"圆形阵列"命令

图 2-104　【圆形阵列】对话框

27）单击【模型】模块中的【布尔运算】命令，弹出图 2-106 所示的对话框。选择【操作方法】为【合并】，选择所有实体，单击"OK"按钮。

28）单击【圆角】命令，选择【可变圆角】。叶轮大叶片一侧内部圆角半径为 3mm，外部圆角半径为 1mm。另一侧外部圆角半径为 0.3mm，内部圆角半径为 1.5mm，如图 2-107 所示。

图 2-105 阵列后效果

图 2-106 【布尔运算】对话框

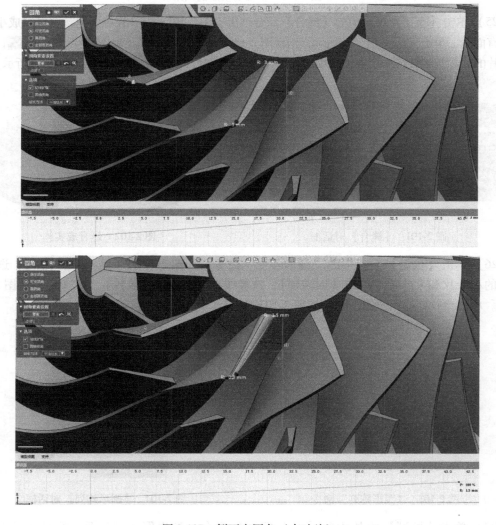
图 2-107 倒可变圆角（大叶片）

29）对大叶片根部进行倒圆角处理，圆角半径为 4mm，如图 2-108 所示。

30）单击【圆角】命令，选择【可变圆角】。叶轮小叶片一侧内部圆角半径为 2mm，外部圆角半径为 1mm。另一侧外部圆角半径为 0.5mm，内部圆角半径为 1mm，如图 2-109 所示。

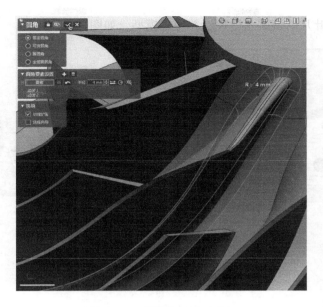

图 2-108　倒圆角（大叶片）

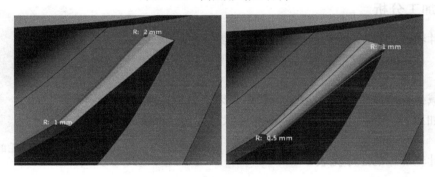

图 2-109　倒可变圆角（小叶片）

31）对小叶片根部进行倒圆角处理，圆角半径为 4mm，如图 2-110 所示。

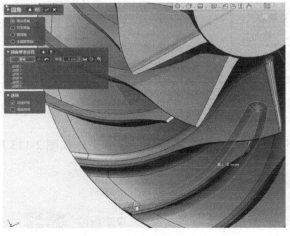

图 2-110　倒圆角（小叶片）

32）完成所有叶片的倒圆角处理，并导出模型，如图2-111所示。

图2-111 倒圆角完成

任务4 加工编程

2.4.1 加工分析

涡轮叶轮零件形状比较复杂，加工精度要求高，属于薄壁零件，加工时容易产生变形，而且加工叶片时容易产生干涉。

为了减少五轴铣削的加工量，在数控车床上先对毛坯进行车削加工，得到叶轮回转体的基本形状，如图2-112所示。

涡轮叶轮加工的工序见表2-1。

图2-112 涡轮叶轮的毛坯

表2-1 涡轮叶轮加工工序

序号	工序	类型	工序子类型	刀具
1	叶片粗加工	多叶片铣削 mill_multi_blade	多叶片粗加工 multi_blade_rough	D8R4
2	主叶片（大叶片）精加工		叶片精加工 blade_finish	D8R4
3	分流叶片（小叶片）精加工		叶片精加工 blade_finish	D8R4
4	轮毂精加工		轮毂精加工 hub_finish	D8R4
5	圆角精加工		圆角精加工 blend_finish	D6R3

2.4.2 编程准备

在创建涡轮叶轮的加工程序之前，需要做的准备工作如图2-113所示。

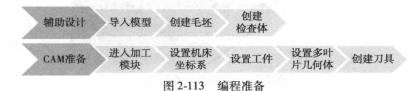

图2-113 编程准备

1. 辅助设计

1) 导入模型。使用【文件】/【导入】/【Parasolid】命令，导入 x_t 格式的涡轮叶轮模型，如图 2-114 所示。

图 2-114　导入模型

2) 创建毛坯几何体。通过【旋转】命令，选择叶轮边线，矢量选择叶轮轴线。创建的包覆面如图 2-115 所示。

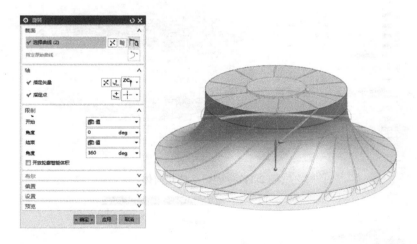

图 2-115　创建包覆面

通过【拉伸】命令，选择边缘线进行拉伸。拉伸的结束选择【直至延伸部分】，选择叶轮顶面，布尔运算选择【求和】，使其与创建的包覆面求和，如图 2-116 所示。

同样通过【拉伸】命令，选择下边缘线向下进行拉伸。拉伸的结束距离设为 25mm，布尔运算选择【求和】，如图 2-117 所示。至此，完成毛坯几何体的创建。

为便于观察，可通过单击【编辑对象显示】命令，或者直接采用快捷键 <Ctrl + J>，适当更改毛坯几何体的颜色及透明度，如图 2-118 所示。

3) 创建检查体。通过【拉伸】命令，将毛坯几何体的下边缘线拉伸 250mm，作为支撑柱（也可将其视为夹具），如图 2-119 所示。

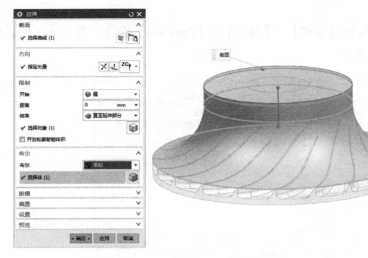

图 2-116 拉伸

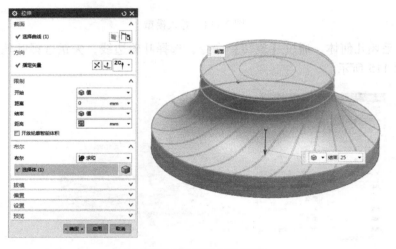

图 2-117 创建毛坯几何体

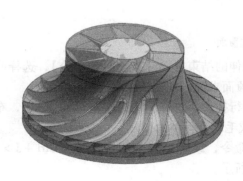

图 2-118 对对象进行调整

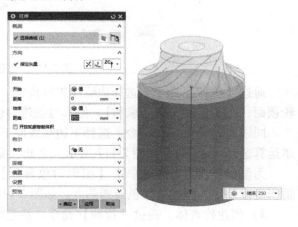

图 2-119 创建检查体

2. CAM 准备

1) 进入加工模块。在弹出的加工环境设置中选择【cam_general】和【mill_multi_blade】（多轴铣叶轮模板）。

2) 设置机床坐标系。进入加工模块后切换至【几何视图】（功能区环境下位于导航器上部，经典工具条环境下位于导航器下方），如图 2-120 所示。

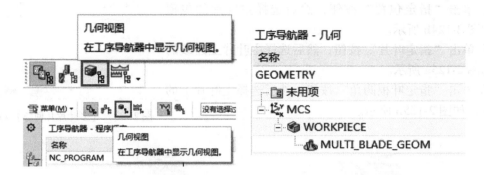

图 2-120　几何视图

双击【MCS】，弹出【MCS】对话框。单击对话框中的"CSYS 对话框"按钮，然后选择毛坯顶面中心作为坐标原点，并使 ZM 轴朝上，如图 2-121 所示。

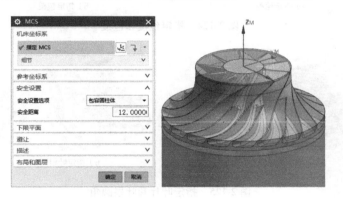

图 2-121　MCS 设置

3) 设置工件。双击【WORKPIECE】，依次指定部件、毛坯和检查体，如图 2-122 所示。

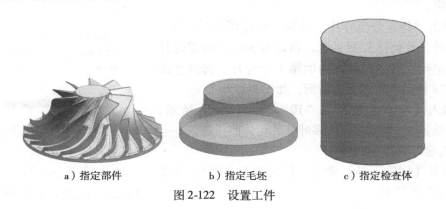

a) 指定部件　　　b) 指定毛坯　　　c) 指定检查体

图 2-122　设置工件

4）设置多叶片几何体。双击工序导航器几何视图中的【MULTI_BLADE_GEOM】，弹出【多叶片几何体】对话框，如图2-123所示。

① 单击"指定轮毂"按钮，然后选择内部的大面，如图2-124a所示。

② 单击"指定包覆"按钮，然后选择主叶片的包覆面，如图2-124b所示。

③ 单击"指定叶片"按钮，然后选择主叶片上连续的面，如图2-125a所示。

④ 单击"指定叶根圆角"按钮，然后选择主叶片上的圆角面，如图2-125b所示。

图2-123 【多叶片几何体】对话框

a）指定轮毂　　　　　　　　b）指定包覆

图2-124 指定包覆和包覆

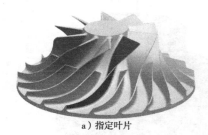

a）指定叶片　　　　　　　　b）指定叶根圆角

图2-125 指定叶片和叶根圆角

注意：包覆、叶片和叶根圆角需要在同一个主叶片上指定。

⑤ 单击"指定分流叶片"按钮，弹出图2-126所示的【分流叶片几何体】对话框。指定分流叶片时需选择之前所指定叶片沿逆时针方向的第1个叶片。选择壁面如图2-127a所示，再选择圆角面，如图2-127b所示。

⑥ 输入【叶片总数】为"10"，如图2-128所示，再单击【确定】。至此，完成多叶片几何体的指定。

5）创建刀具。

① 单击【创建刀具】命令，创建刀具D8R4，参数设置如图2-129所示。

图2-126 【分流叶片几何体】对话框

② 单击【创建刀具】命令，创建刀具 D6R3，参数设置如图 2-130 所示。

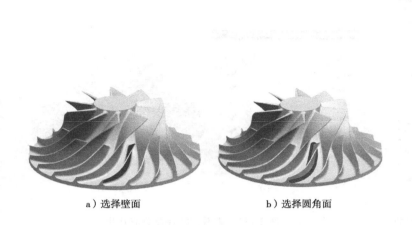

a）选择壁面　　　　　　b）选择圆角面

图 2-127　指定分流叶片　　　　　　　　　图 2-128　叶片总数

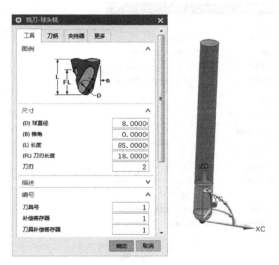

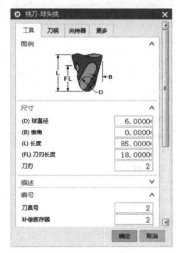

图 2-129　创建刀具 D8R4　　　　　　　图 2-130　创建刀具 D6R3

2.4.3　多叶片粗加工

1）单击【创建工序】命令打开对话框。选择【类型】为"mill_multi_blade"，选择【工序子类型】为"多叶片粗加工"，其余参数设置如图 2-131 所示，再单击【确定】。

2）弹出图 2-132a 所示的【多叶片粗加工】对话框。几何体已经设置完成，单击"显示"按钮，可以查看对应的轮毂、包覆、叶片等几何体，如图 2-132b 所示。

3）在【多叶片粗加工】对话框中单击【驱动方法】组中的"叶片粗加工"按钮，如图 2-133a 所示，弹出【叶片粗加工驱动方法】对话框，参数设置如图 2-133b 所示。单击【预览】组中的"显示"按钮，可以预览驱动方法，如图 2-133b 所示，再单击【确定】。

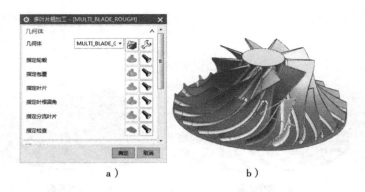

图 2-131 【创建工序】对话框　　　　　图 2-132 查看主叶片及分流叶片

图 2-133 驱动方法

4) 在【多叶片粗加工】对话框中单击【刀轨设置】组中的"切削层"按钮，弹出【切削层】对话框。选择【深度模式】为【从包覆至插补轮毂】，【距离】为恒定"1mm"，如图 2-134 所示。

5) 在【多叶片粗加工】对话框中单击【刀轨设置】组中的"切削参数"按钮，弹出【切削参数】对话框。输入【叶片余量】为"0.6"，【轮毂余量】为"0.4"，【检查余量】为"0.6"，如图 2-135 所示。

图 2-134 【切削层】对话框

图 2-135 【切削参数】对话框

6）在【多叶片粗加工】对话框中单击【刀轨设置】组中的"进给率和速度"按钮，然后输入主轴速度为"8000"，输入切削进给率为"4000"，如图 2-136a 所示。单击"主轴速度"后面的"基于此值计算进给和速度"按钮，计算得到表面速度和每齿进给量，如图 2-136b 所示。

a)   b)

图 2-136 【进给率和速度】对话框

7）生成刀轨。在【多叶片粗加工】对话框中单击【操作】组中的"生成"按钮，叶片粗加工的刀轨如图 2-137a 所示。单击"确认"按钮，弹出【刀轨可视化】对话框；单击【2D 动态】选项卡，再单击"播放"按钮，开始切削加工仿真，仿真结果如图 2-137b所示。

8）单击【多叶片粗加工】对话框中的【确定】按钮后，在工序导航器的程序顺序视图中会出现【MULTI_BLADE_ROUGH】程序。

图 2-137 生成刀轨和 2D 动态仿真

2.4.4 主叶片精加工

1）单击【创建工序】命令，然后选择【类型】为【mill_multi_blade】，选择【工序子类型】为"叶片精加工"，其余参数设置如图 2-138 所示，再单击【确定】。

2）弹出【叶片精加工】对话框，几何体已经设置完成，此处不必设置。

3）在【叶片精加工】对话框中单击【驱动方法】组中的"叶片精加工"按钮，如图 2-139a 所示，弹出【叶片精加工驱动方法】对话框。选择【要精加工的几何体】为【叶片】，选择【要切削的面】为【左面、右面、前缘】，选择【切削模式】为【单向】，选择【切削方向】为【顺铣】，其余参数设置如图 2-139b 所示。

图 2-138 创建工序（叶片精加工）

4）在【叶片精加工】对话框中单击【刀轨设置】组中的"切削层"按钮，弹出【切削层】对话框。选择【深度模式】为【从包覆插补至

图 2-139 驱动方法

轮毂】，选择【每刀切削深度】为【残余高度】，输入【残余高度】为"0.02"，其余参数设置如图2-140所示。

5) 在【叶片精加工】对话框中单击【刀轨设置】组中的"进给率和速度"按钮，然后输入主轴速度为"8000"，输入切削进给率为"4000"，再单击"基于此值计算进给和速度"按钮，计算得到表面速度和每齿进给量。

6) 在【叶片精加工】对话框中单击【操作】组中的"生成"按钮，可以看到主叶片精加工的刀轨，如图2-141a所示，刀轨放大后如图2-141b所示。

图2-140 【切削层】对话框

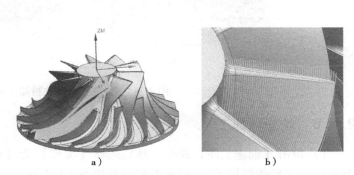

图2-141 生成主叶片精加工刀轨

2.4.5 分流叶片精加工

分流叶片的精加工工序与主叶片的精加工工序参数一致。可以通过复制的方法进行创建。

1) 选择工序导航器的程序顺序视图中的【BLADE_FINISH】程序，单击右键，在弹出的右键菜单中选择【复制】。再次单击右键，在弹出的右键菜单中选择【粘贴】，此时会出现【BLADE_FINISH_COPY】程序。

2) 双击【BLADE_FINISH_COPY】程序，弹出图2-142a所示的【叶片精加工】对话框。单击【驱动方法】组中的按钮，弹出【叶片精加工驱动方法】对话框。选择【要精

图2-142 修改驱动方法

加工的几何体】为【分流叶片1】，单击【确定】，如图2-142b所示。

3）在【叶片精加工】对话框中单击【操作】组中的"生成"按钮，可以看到分流叶片精加工的刀轨，如图2-143a所示，刀轨放大后如图2-143b所示。

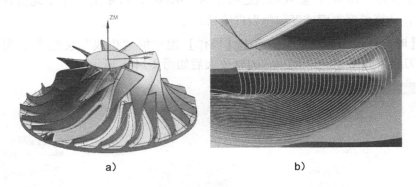

图2-143　生成分流叶片精加工刀轨

2.4.6　轮毂精加工

1）单击【创建工序】命令，然后选择【类型】为【mill_multi_blade】，选择【工序子类型】为"轮毂精加工"，其余参数设置如图2-144所示，再单击【确定】。

2）弹出【轮毂精加工】对话框，几何体已经设置完成，此处不必设置。

3）在【轮毂精加工】对话框中单击【驱动方法】组中的"轮毂精加工"按钮，弹出【轮毂精加工驱动方法】对话框。将前缘的【径向延伸】设为"50"，其余参数设置如图2-145a所示。前缘径向延伸"0"和"50"的效果分别如图2-145b、c所示。

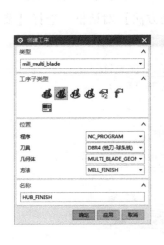

图2-144　创建工序（轮毂精加工）

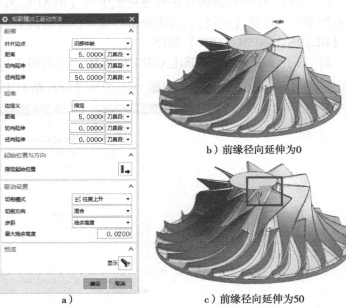

图2-145　轮毂精加工的驱动方法

4)在【轮毂精加工】对话框中单击【刀轨设置】组中的"进给率和速度"按钮,输入主轴速度为"8000",输入切削进给率为"4000"。单击"基于此值计算进给和速度"按钮,计算得到表面速度和每齿进给量。

5)在【轮毂精加工】对话框中单击【操作】组中的"生成"按钮,可以看到轮毂精加工的刀轨,如图 2-146 所示。

6)单击【轮毂精加工】对话框中的【确定】按钮后,工序导航器的程序顺序视图中会出现【HUB_FINISH】程序。

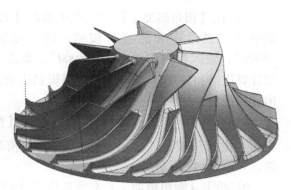

图 2-146　生成轮毂精加工刀轨

2.4.7　圆角加工

圆角加工包括主叶片的圆角精加工和分流叶片的圆角精加工。首先创建主叶片的圆角精加工程序,然后通过复制粘贴的方式创建分流叶片的圆角精加工程序。具体步骤如下:

1)单击【创建工序】命令,然后选择【类型】为【mill_multi_blade】,选择【工序子类型】为"圆角精加工",其余参数设置如图 2-147 所示,再单击【确定】。

2)弹出【圆角精加工】对话框,几何体已经设置完成,此处不必设置。

3)在【圆角精加工】对话框中单击【驱动方法】组中的"编辑"按钮,弹出【圆角精加工驱动方法】对话框。选择【要精加工的几何体】为【叶根圆角】,选择【要切削的面】为【左面、右面、前缘】,其余参数设置如图 2-148 所示。

图 2-147　创建圆角精加工工序

图 2-148　圆角精加工驱动方法

4）在【圆角精加工】对话框中单击【刀轨设置】组中的"进给率和速度"按钮；输入主轴速度为"8000"，输入切削进给率为"4000"，单击"基于此值计算进给和速度"按钮，计算得到表面速度和每齿进给量。

5）在【圆角精加工】对话框中单击【操作】组中的"生成"按钮，生成主叶片圆角精加工的刀轨，如图2-149所示。

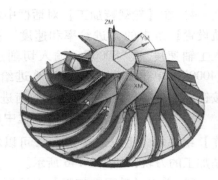

图2-149　生成主叶片圆角精加工刀轨

6）单击【圆角精加工】对话框中的【确定】按钮后，工序导航器的程序顺序视图中会出现【BLEND_FINISH】程序。

7）在工序导航器的程序顺序视图中选择【BLEND_FINISH】程序，单击右键，并在弹出的右键菜单中选择【复制】。再次单击右键，在弹出的右键菜单中选择【粘贴】，此时会出现【BLEND_FINISH_COPY】程序。

8）双击【BLEND_FINISH_COPY】程序，弹出图2-150a所示的【圆角精加工】对话框。单击【驱动方法】组中的按钮，弹出【圆角精加工驱动方法】对话框。选择【要精加工的几何体】为【分流叶片1倒圆】，单击【确定】，如图2-150b所示。

9）单击【圆角精加工】对话框【操作】组中的"生成"按钮，可以看到分流叶片圆角精加工的刀轨，如图2-151所示。

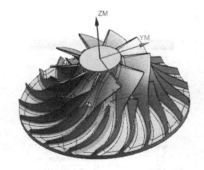

a）　　　　　　　　　b）

图2-150　分流叶片1倒圆　　　　　图2-151　生成刀轨

2.4.8　复制程序

1）单击图2-152所示的【创建程序】命令，弹出【创建程序】对话框，参数设置如图2-153a所示。单击【确定】后，弹出图2-153b所示的【程序】对话框。单击【确定】，

此时工序导航器中会出现 ROUGH。将【MULTI_BLADE_ROUGH】程序拖至【ROUGH】下面。

2）用同样的方法创建【BLADE】、【HUB】和【BLEND】程序，将【BLADE_FINISH】和【BLADE_FINISH_COPY】程序拖至【BLADE】下面，将【HUB_FINISH】程序拖至【HUB】下面，将【BLEND_FINISH】和【BLEND_FINISH_COPY】程序拖至【BLEND】下面，如图 2-154 所示。

图 2-152 【创建程序】命令

a）

b）

图 2-153 创建程序

图 2-154 工序导航器-程序顺序视图

3）选择【MULTI_BLADE_ROUGH】程序（叶片粗加工），单击右键，并在弹出的右键菜单中选择【对象】/【变换】命令。在【变换】对话框中选择【类型】为【绕直线旋转】，选择【直线方法】为【点和矢量】，指定点为叶轮的圆心，指定矢量为 +ZC，其余参数设置如图 2-155 所示。单击【显示结果】按钮，可以预览复制的刀轨，再单击【确定】按钮。同样的方法完成其他刀轨的复制，如图 2-156 所示。

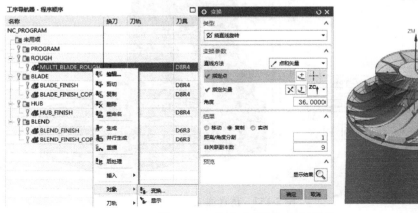

图 2-155 通过【变换】复制程序

图 2-156　复制刀轨

2.4.9　仿真加工

在 NX 软件中，仿真加工的方式有两种，一种是调入仿真机床进行仿真加工，另一种是直接将生成的程序直接转换成 2D 或 3D 形式进行仿真加工。仿真加工如图 2-157 所示。

在 NX 中利用加工模块分别对叶轮流道面和叶片生成相应的刀轨，利用五轴加工中心仿真环境，通过模拟数控机床的实际运动，减少实际的切削验证，解决了刀具、工件与机床部件和夹具的碰撞问题，验证了数控加工程序和后处理器的正确性，从而缩短生产周期，降低成本。

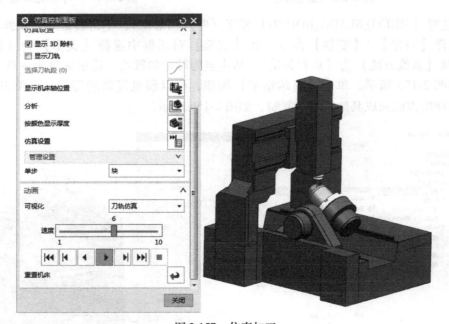

图 2-157　仿真加工

项目 3　挤牙膏器

本项目以图 3-1 所示的挤牙膏器为载体，使用手持激光扫描仪 BYSCAN 510 进行数据采集，使用 Geomagic Design X 2016 软件进行数据处理，使用 Siemens NX 10 软件进行产品逆向建模，最后利用 Geomagic Control X 2018 软件进行产品分析比对并输出报告。项目实施流程如图 3-2 所示。

图 3-1　挤牙膏器实物

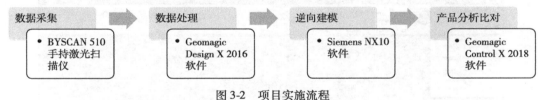

图 3-2　项目实施流程

任务 1　数据采集

本项目数据采集使用手持激光扫描仪 BYSCAN 510，配套扫描软件 ScanViewer。具体扫描步骤与过程可参考涡轮叶轮项目。

3.1.1　贴标记点

开始扫描前，先在挤牙膏器的正、反面都贴上标记点，如图 3-3 所示。因其内部有许多筋板，所以在给挤牙膏器贴标记点时需注意不要使标记点太接近筋板或边缘。

图 3-3　贴标记点

3.1.2 扫描标记点

1) 打开手持激光扫描仪 BYSCAN 510 的配套扫描软件 ScanViewer。

2) 在扫描控制面板中,将【扫描解析度设置】设为 1mm,将【曝光参数设置】设为 1ms,选中【标记点】选项后单击【开始】按钮,如图 3-4 所示。

3) 如图 3-5 所示,将扫描仪正对挤牙膏器,按下扫描仪上的扫描键,开始扫描。扫描过程中的软件界面如图 3-6 所示。扫描完成后保存数据。

图 3-4 参数设置

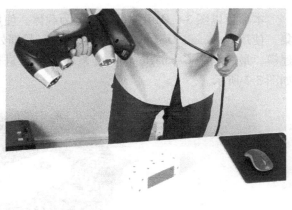
图 3-5 扫描挤牙膏器

图 3-6 扫描标记点时的软件界面

3.1.3 扫描激光点

1) 在扫描控制面板中,将【扫描解析度设置】设为 0.3mm,将【曝光参数设置】设为 2.1ms,单击【开始】按钮,开始扫描,如图 3-7 所示。

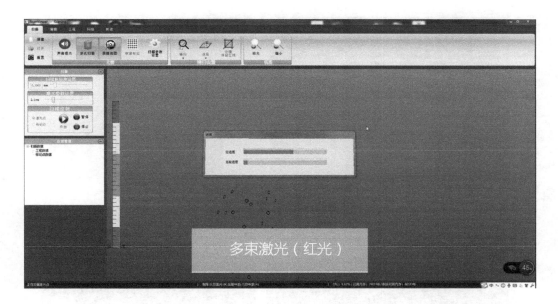

图 3-7　扫描激光点的参数设置

2）按下扫描仪上的扫描键，开始扫描。扫描过程中的软件界面如图 3-8 所示。

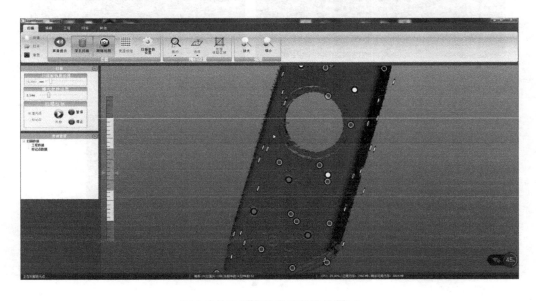

图 3-8　扫描激光点时的软件界面

由于工件较小，可以拿在手中，并可切换到单束激光扫描的模式。单束激光和多束激光交替使用，有助于扫描细节区域，如图 3-9 和图 3-10 所示。

3）扫描完成后，按下扫描仪上的扫描键停止扫描，再单击扫描软件中的【暂停】按钮。使用套索工具选中扫描的无关数据，按下键盘上的 < Delete > 键将其删除，如图 3-11 所示。

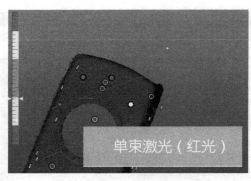 

图 3-9　单束激光扫描

 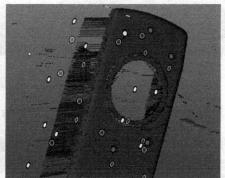

图 3-10　多束激光与单束激光扫描

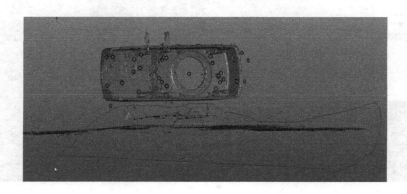

图 3-11　选中并删除无关数据

4）扫描完成后，单击【工程】选项卡中的【生成网格】命令，系统开始生成网格，并显示进度条，如图 3-12 所示。

5）单击【网格】选项卡中的【保存】命令，在弹出的下拉列表中选择【网格文件（*.STL）】，如图 3-13 所示，弹出【另存为】对话框。选择保存路径并输入文件名后，单击【保存】按钮。

项目3 挤牙膏器

图 3-12 生成挤牙膏器网格模型　　　　　　图 3-13 保存为 STL 文件

6）下盖的扫描与上盖相似，扫描完成后保存数据，如图 3-14 和图 3-15 所示。

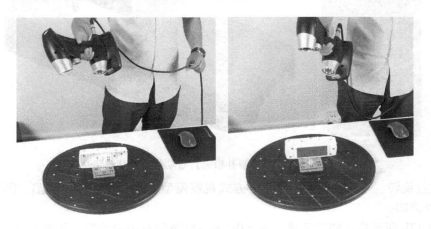

图 3-14 挤牙膏器下盖实际扫描

图 3-15 挤牙膏器下盖扫描软件界面

任务 2　数据处理

3.2.1　导入模型

1）打开软件。单击"开始"菜单中的 Geomagic Design X 2016 程序，或双击桌面上的图标，启动 Geomagic Design X 2016 应用软件。

2）导入模型数据。单击界面左上方的"导入"按钮，选择扫描生成的模型数据导入到 Geomagic Design X 2016 软件。

3.2.2 上盖扫描数据处理

1）单击【多边形】模块中的【修补精灵】命令，弹出【修补精灵】对话框。软件会自动检索面片模型中存在的各种缺陷，如非流形顶点、重叠单元面、悬挂的单元面、交差单元面等，如图 3-16 所示。单击"OK"按钮，软件自动修复检索到的缺陷。

图 3-16　使用【修补精灵】自动修复面片缺陷

2）通过旋转、平移、放大、缩小等方式观察模型，查找是否存在缺陷。图 3-17a 所示为其中一处缺陷。

如图 3-17b 所示选择缺陷区域，按 <Delete> 键删除选中的数据，如图 3-17c 所示。单击【多边形】模块中的【填孔】命令，弹出图 3-18 所示的对话框。选择图 3-17d 所示的孔，参数设置保持默认，单击"OK"按钮完成操作，效果如图 3-17e 所示。

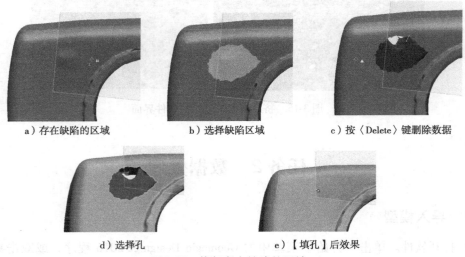

a）存在缺陷的区域　　　　b）选择缺陷区域　　　　c）按〈Delete〉键删除数据

d）选择孔　　　　e）【填孔】后效果

图 3-17　修复存在缺陷的区域

通过【填孔】命令对其他缺陷处的孔进行填充,如图 3-19 所示。

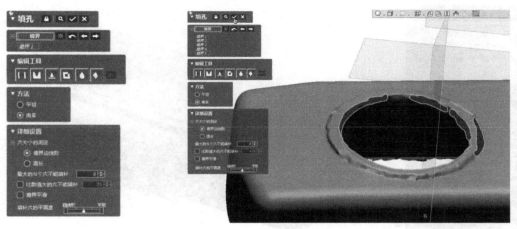

图 3-18 【填孔】对话框　　　　　图 3-19 填孔

3）单击【多边形】模块中的【加强形状】命令,弹出图 3-20 所示对话框。对话框中的 3 个选项都保持默认数值,然后单击"OK"按钮☑完成操作。加强形状用于锐化面片上的尖锐区域(棱角),同时平滑平面或圆柱面区域,提高面片的质量。

4）选中挤牙膏器上盖,选择【菜单】/【文件】/【输出】命令,弹出【输出】对话框,如图 3-21a 所示。单击"OK"按钮☑进行输出,如图 3-21b 所示。

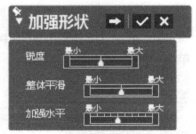

图 3-20 【加强形状】对话框

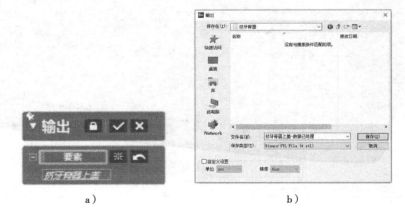

图 3-21 输出挤牙膏器上盖数据

3.2.3 下盖扫描数据处理

1）单击【多边形】模块中的【修补精灵】命令,单击"OK"按钮☑,软件自动修复检索到的缺陷,如图 3-22 所示。

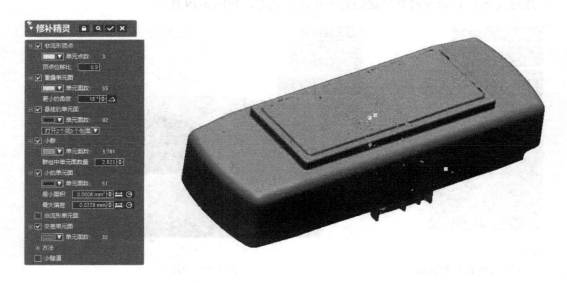

图 3-22　使用【修补精灵】自动修复面片缺陷（下盖）

2）单击【多边形】模块中的【填孔】命令，选择所需填充的孔，参数设置保持默认，单击"OK"按钮✓完成操作，如图 3-23 所示。

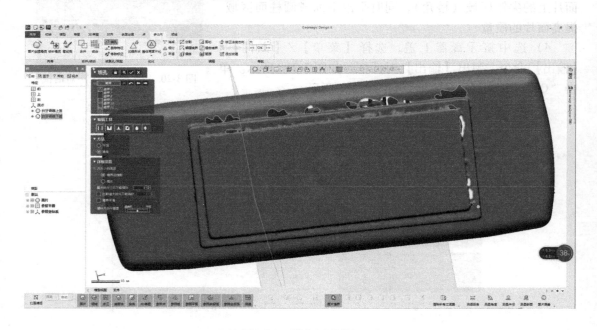

图 3-23　填孔（下盖）

3）单击【多边形】模块中的【加强形状】命令，参数保持默认数值，进行加强形状。

4）选中挤牙膏器下盖，选择【菜单】/【文件】/【输出】命令，弹出【输出】对话框，如图 3-24a 所示。单击"OK"按钮✓进行输出，如图 3-24b 所示。

项目 3 挤牙膏器

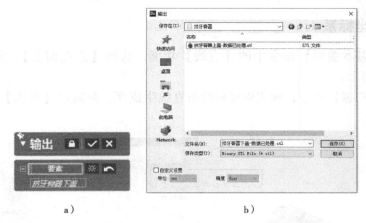

图 3-24 输出挤牙膏器下盖数据

任务 3　逆向建模

3.3.1　导入模型

1）分别导入挤牙膏器上盖、下盖的扫描文件。更改一个扫描文件的颜色，以便于区分上盖和下盖，如图 3-25 所示。

图 3-25　导入扫描文件

2）根据个人习惯，可将片体颜色进行更改，使后续创建的片体为蓝色，便于观察，如图 3-26 所示。

图 3-26　更改片体颜色

3.3.2 确定坐标系

1）通过【基本曲线】命令中的【直线】功能，选择【点在面上】，分别创建两条直线，如图3-27所示。

2）通过【扫掠】命令，利用创建的两条直线生成面，并通过【扩大】命令进行调整，如图3-28所示。

图 3-27　创建直线　　　　　　　　　　图 3-28　扫掠后调整面大小

3）暂定创建的平面为水平面。通过【格式】/【WCS】/【定向】命令，类型选择【自动判断】，捕捉点注意点选【面上的点】，单击3个大致成90°的点，如图3-29所示。

4）通过【基本曲线】命令中的【直线】功能，在下盖侧面绘制一条直线。完成后通过【投影曲线】命令将直线投影至XC-YC平面，如图3-30所示。

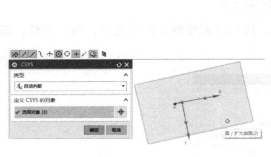 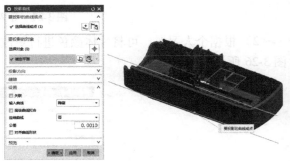

图 3-29　定向　　　　　　　　　　　　图 3-30　绘制直线并投影

5）通过【格式】/【WCS】/【定向】命令，首先单击X轴，再单击投影直线，使X轴的方向与投影直线的方向呈平行状态，如图3-31所示。

6）通过【拉伸】命令，拉伸之前已经创建的直线，并测量其与另一侧的距离，如图3-32所示。

项目3 挤牙膏器

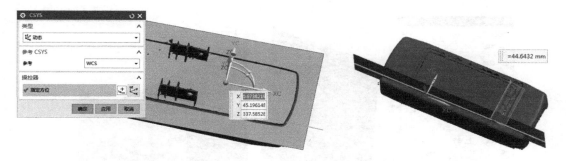

图 3-31　X 轴定向　　　　　　　　　图 3-32　拉伸并测量距离

7）通过【偏置曲面】命令将面偏置测量距离的一半。可通过【变换】命令，将面镜像至另一侧以观察结果，如图 3-33 和图 3-34 所示。

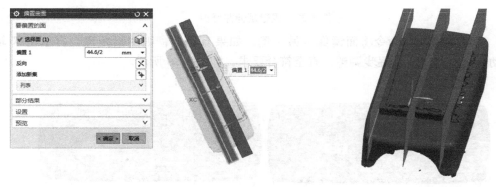

图 3-33　偏置曲面　　　　　　　　　图 3-34　镜像面

8）通过【基本曲线】命令中的【圆弧】功能，选择面上的点创建圆弧，并通过【曲线长度】命令将圆弧延长，如图 3-35 所示。

9）通过【规律延伸】命令，进行拉伸，并用 XC-ZC 平面进行修剪，如图 3-36 所示。

图 3-35　创建曲线　　　　　　　　　图 3-36　规律延伸并修剪

10）将面镜像至另一侧。通过【基本曲线】命令中的【圆弧】功能，选择两端点作为起点和终点，重新绘制圆弧，以保证其关于 XC-ZC 平面对称。再次通过【规律延伸】命令进行拉伸，并适当调整曲面大小，如图 3-37 所示。

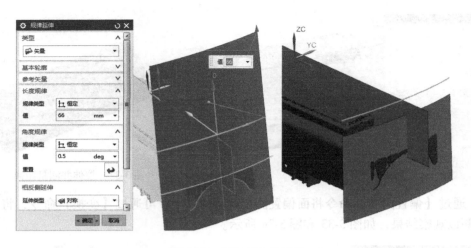

图 3-37　规律延伸并修剪

11）通过【变换】命令将面镜像至另一侧。如果有偏差，则测量距离，再单击 X 轴，通过移动测量距离的一半逐步调整，直至符合要求，如图 3-38 所示。

图 3-38　确定对称面

12）至此零件的中心确定完成，可以通过保存坐标系或者添加方块等方式，也可根据个人习惯进行标记创建，用于标记坐标系，以防止坐标系丢失，如图 3-39 所示。

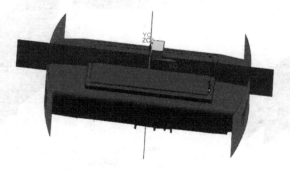

图 3-39　标记坐标系

3.3.3　下盖逆向建模

1. 挤牙膏器下盖主体部分

1）通过【基本曲线】命令中的【直线】功能，在侧面绘制直线，并通过【规律延伸】

命令进行拉伸，并对两个片体进行修剪，如图 3-40 所示。

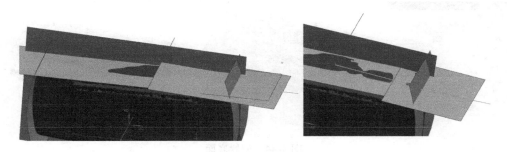

图 3-40　创建片体并修剪

2) 通过【变换】命令将片体镜像并观察，若发现存在一定偏差，将其进行均分，如图 3-41 所示。

3) 将原始面通过【偏置曲面】功能向内偏置 0.15mm，再次进行镜像并观察结果，如图 3-42 和图 3-43 所示。

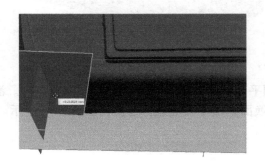

图 3-41　观察发现片体存在一定偏差

图 3-42　偏置曲面

图 3-43　重新镜像面

4) 观察顶面部分面与底面为平行关系，大致测量距离后，通过【偏置曲面】命令，将底面向上偏置一定距离，如图 3-44 所示。

5) 通过【基本曲线】命令中的【直线】功能，在两斜面处分别创建两条平行于 YC 方向的直线，并通过【规律延伸】命令进行拉伸，如图 3-45 所示。

6) 通过【修剪体】、【修剪片体】命令进行修剪。如果长度不够，可通过【延伸片体】命令进行延伸，直至最终修剪完成，并将其通过【缝合】命令缝合成实体，如图 3-46 所示。

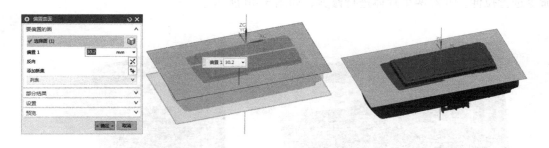

图 3-44 偏置底面

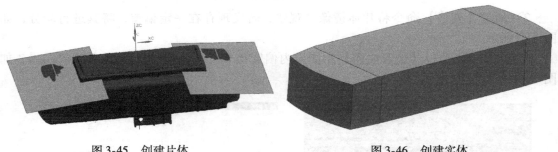

图 3-45 创建片体　　　　　　　　图 3-46 创建实体

7）观察实际产品，判断其应对称。将右侧平面扩大后镜像至左侧，观察贴合情况。测量发现偏差约为 0.1mm，符合要求。通过【替换面】命令将面进行替换，如图 3-47 所示。

图 3-47 替换面

8）接下来绘制凸台特征。通过【基本曲线】命令中的【直线】功能，绘制一条平行于 YC 轴方向的直线，并进行拉伸，如图 3-48 所示。

9）将面镜像至另一侧，此处为挤牙膏器实际使用时用于贴在墙上，因此不是关于坐标系对称。通过【偏置曲面】命令进行偏置调整，如图 3-49 所示。

10）测量凸台顶部距离后，通过【偏置曲面】命令进行偏置，如图 3-50 所示。

11）通过【加厚】命令进行加厚，再通过【替换面】命令进行替换，如图 3-51 所示。

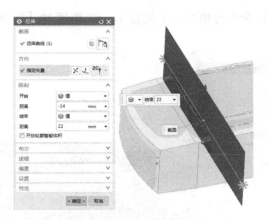

图 3-48　创建直线并拉伸

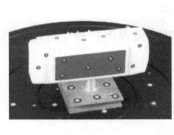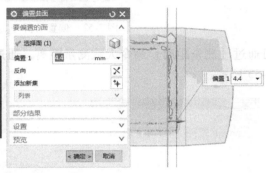

图 3-49　偏置曲面

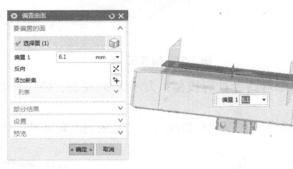

图 3-50　测量距离后偏置曲面

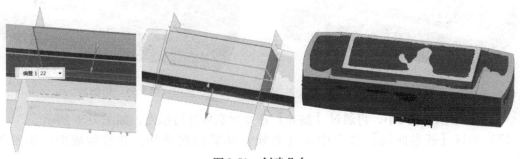

图 3-51　创建凸台

12）通过【偏置面】命令将顶面向下偏置1mm，再通过【拆分体】命令对实体进行拆分，如图3-52所示。

图 3-52　偏置后拆分实体

13）可通过【测量距离】命令大致测量距离，通过【偏置面】命令对四周的面进行偏置，如图3-53所示。

图 3-53　偏置面

14）通过【拔模】命令对凸台进行拔模，大小为3°左右，便于产品脱模，如图3-54所示。

15）通过【边倒圆】命令，根据贴合情况进行产品的倒圆角处理，如图3-55所示。

图 3-54　拔模　　　　　　　　　　图 3-55　边倒圆

16）分析壁厚的大小，再通过【抽壳】命令对实体进行抽壳，如图3-56所示。

17）通过【基本曲线】命令中的【直线】功能创建直线，并拉伸成面，如图3-57所示。

项目3 挤牙膏器

图 3-56 分析壁厚并抽壳

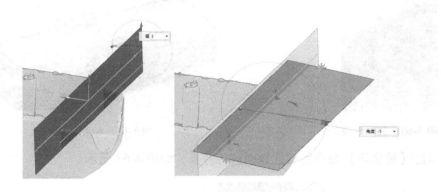

图 3-57 创建面

18）通过【修剪体】命令，将面修剪后进行缝合。通过【拆分体】命令将实体进行拆分，并倒 1.5mm 圆角，如图 3-58 所示。

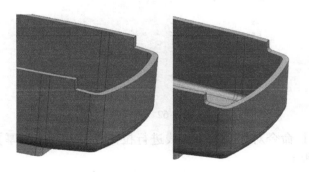

图 3-58 拆分体并倒圆角

2. 挤牙膏器下盖内部特征

1）通过【基本曲线】命令中的【圆弧】功能，取消勾选【线串模式】，勾选【整圆】，创建圆。双击创建的圆，观察圆的直径大小，大概为 6.5mm，如图 3-59 所示。

2）通过【基本曲线】命令中的【直线】功能，捕捉圆弧圆心，绘制一条过圆心且平行于 Z 轴方向的直线。再通过【管道】命令，创建直径为 6.5mm 的管道，如图 3-60 所示。

3）通过【扩大】命令创建底面的平面。测量分析后，通过【偏置曲面】命令将面向下偏置 1mm，如图 3-61 所示。

图 3-59　创建圆

图 3-60　创建管道

图 3-61　偏置面

4）通过【替换面】命令将圆柱顶面进行替换，如图 3-62 所示。

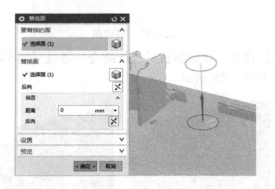

图 3-62　替换面

5）通过【拉伸】命令对圆柱的中心线进行拉伸，再通过【加厚】命令向两边各加厚 0.5mm，如图 3-63 所示。

图 3-63　拉伸并加厚

6）通过【替换面】命令，进行多次替换。替换完成后，添加0.3°的脱模斜度，如图3-64和图3-65所示。

图3-64 替换面

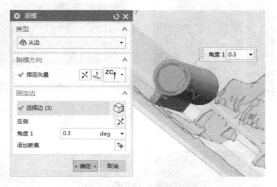

图3-65 添加脱模斜度

7）通过【管道】命令，创建直径为4mm的圆柱；然后偏置一个距顶面2mm的面，用其对圆柱体进行修剪。之后对两个圆柱体进行求差操作，并添加一定的脱模斜度，如图3-66所示。

8）通过【移动对象】命令，将特征进行复制移动，再通过【变换】命令将特征镜像至另一侧，如图3-67所示。

9）通过【基本曲线】命令中的【直线】功能，创建一条直线，并将直线沿Z轴延伸，

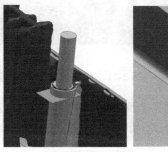

图3-66 特征创建

角度为0.1°。再对延伸片体进行加厚处理，使其顶部接近扫描数据，并对另一侧也添加0.1°的脱模斜度，如图3-68所示。

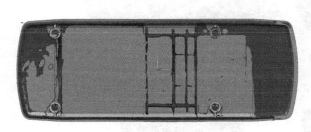

图3-67 镜像特征

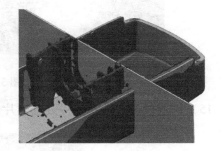

图3-68 加厚

10）创建其他位置贴合筋板的面，并进行修剪、求差，最后进行倒圆角处理，如图3-69所示。

11）通过【移动对象】命令将筋板移动复制到合适位置，如图3-70所示。

12）根据模型数据创建平面，并对中间两块筋板进行修剪，如图3-71所示。

13）与之前筋板的创建方法相同，通过【规律延伸】、【加厚】、【替换面】等命令创建剩余筋板，如图3-72所示。

 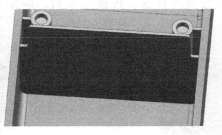 

图 3-69 筋板　　　　　　　　　　　　　图 3-70 移动对象

 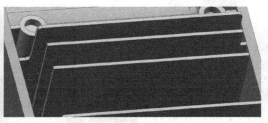 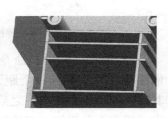

图 3-71 修剪体（中间筋板）　　　　　　　图 3-72 创建剩余筋板

14）先创建 3 个面，然后通过【面倒圆】命令进行片体倒圆。片体缝合后，通过【修剪体】命令修剪三个筋板，如图 3-73 所示。

 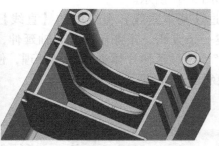

图 3-73 修剪三个筋板

15）以同样的方法在另一处创建片体，并对另一个筋板进行修剪，如图 3-74 所示。

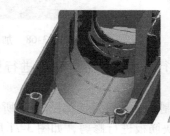 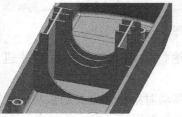

图 3-74 修剪另一个筋板

16）利用【基本曲线】命令中的【圆弧】功能，通过 3 点创建整圆。查看圆的大小，然后捕捉该圆弧的圆心，通过圆心创建一条平行于 X 轴方向的直线，并根据之前的分析测量，创建直径为 4mm 的管道。最后进行求差操作，如图 3-75 所示。

图 3-75 制作孔

17）通过【拉伸】命令将圆柱的轴线沿 Y 轴方向进行拉伸，创建两个平面。通过【相交曲线】命令求取交线，如图 3-76 所示。

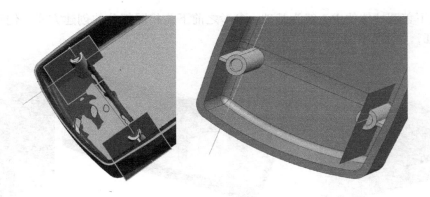

图 3-76 拉伸并求取交线

18）通过【直纹】命令创建一个面，再通过【加厚】命令对面进行加厚，并添加 0.3°的脱模斜度，如图 3-77 所示。

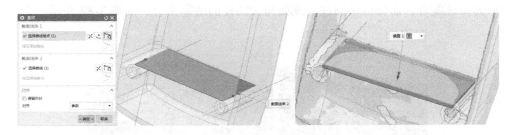

图 3-77 创建面并加厚

19）创建片体，片体缝合后对加厚的面（体）进行修剪，如图 3-78 所示。

20）至此，挤牙膏器下盖逆向造型基本完成，包含部分特征制作，如图 3-79 所示。

图 3-78　修剪最后一个筋板　　　　　　　图 3-79　挤牙膏器下盖基本完成

3.3.4　上盖逆向建模

1）将下盖主体部分镜像至另一侧。通过观察发现与扫描数据较为贴合，如图 3-80 所示。

2）以下盖圆柱体的中心线为基础，参考之前下盖相同部分的创建方式，创建上盖圆柱及筋板特征，如图 3-81 所示。

 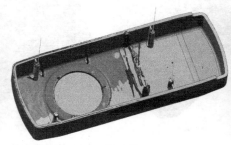 

图 3-80　观察下盖镜像后　　　　　　　图 3-81　创建上盖圆柱及筋板特征
　　　　与上盖贴合情况

3）为保证配合，需保证小圆柱的顶面与大圆柱的底部贴合，如图 3-82 所示。

 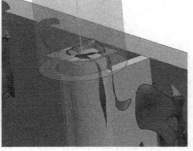

图 3-82　上、下面需贴合

4）通过【抽取几何特征】命令，抽取 3 个面，如图 3-83 所示。

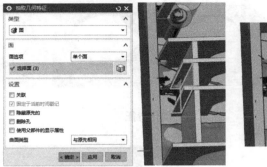

图 3-83　抽取面

5）该部分的形状较为简单，无复杂曲面，创建方法与之前下盖特征的创建方法相似。完成上盖该部分内部特征的创建，如图 3-84 所示。

6）通过【基本曲线】命令中的【圆弧】功能捕捉圆心，通过圆心绘制一条平行于 Z 轴方向的直线。该轴线并不能保证在 YC-ZC 平面上，需进一步处理，如图 3-85 所示。

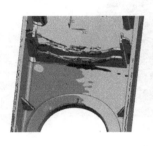

图 3-84　创建内部特征

7）通过【相交曲线】或【投影曲线】等方式，在 YC-ZC 平面上生成一条直线，如图 3-86 所示。

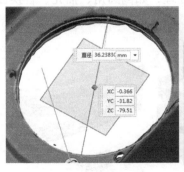

图 3-85　创建轴线　　　　　　　　　图 3-86　创建直线

8）接下来细化该处的特征，如图 3-87 所示。

9）创建一个圆弧，将其拉伸后对半切除，然后镜像至另一侧。重新通过 3 点（起点和终点选择两端点）绘制圆弧，并进行拉伸，如图 3-88 所示。

10）将拉伸的片体镜像至另一侧，对上、下实体均进行修剪，并倒圆角，如图 3-89 所示。

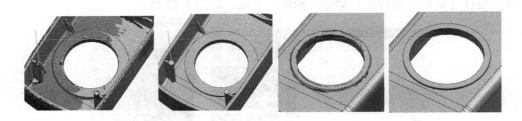

图 3-87 细化特征

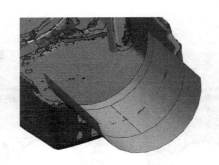

图 3-88 创建圆弧

图 3-89 处理挤牙膏器顶部特征

3.3.5 细节处理

挤牙膏器上、下盖大致建模完成后,对模型进行细化处理,如倒圆角、抽壳、替换面等操作,并制作出配合卡扣,如图 3-90 和图 3-91 所示。

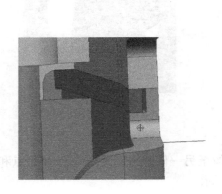

图 3-90 卡扣处理

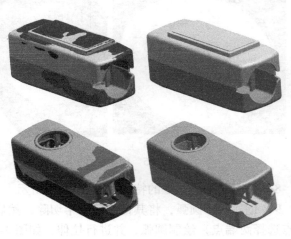

图 3-91 挤牙膏器逆向造型完成

任务 4　产品分析比对

3.4.1　数据的导入

模型和扫描数据可直接拖至 Geomagic Control X 软件窗口或通过工具条中的【导入】功能导入，如图 3-92 所示。

图 3-92　导入数据

3.4.2　初始模型对齐

单击【初始对齐】按钮，然后勾选【利用特征识别提高对齐精度】，如图 3-93a 所示，再单击"OK"按钮，使扫描数据与逆向建模模型通过特征自动对齐，如图 3-93b 所示。

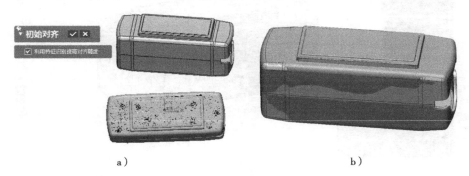

a)　　　　　　　　　　　　　　b)

图 3-93　初始对齐

3.4.3　最佳拟合对齐

初始对齐是通过零件的特征自动对齐的，不是最佳的拟合状态。单击【最佳拟合对齐】命令，再单击"OK"按钮，让扫描数据与逆向建模模型对齐到最佳状态，如图 3-94 所示。

图 3-94　最佳拟合对齐

3.4.4 3D 比较

单击【3D 比较】命令，进行 3D 数据比对，如图 3-95 所示。

图 3-95　3D 比较

单击"下一阶段"按钮，进入下一个阶段，如图 3-96 所示。在【3D 比较】对话框中，修改【使用指定公差】为"±0.1"，再单击"OK"按钮确定。

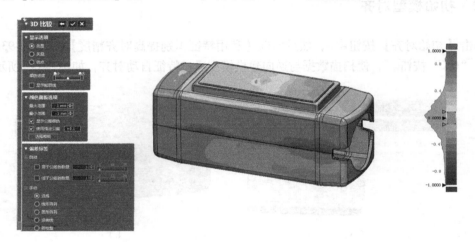

图 3-96　3D 比较颜色显示

可通过单击 3D 比较属性中的【直方图】按钮 直方图 ▶ 查看更详细的数据，如图 3-97 所示。

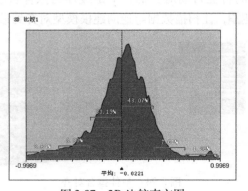

图 3-97　3D 比较直方图

3.4.5　2D 比较

单击【平面】命令，创建中心面，如图 3-98 所示。
单击【2D 比较】命令，弹出图 3-99 所示对话框。

　　图 3-98　创建中心面　　　　　　　　　　图 3-99　2D 比较

以创建的基准平面截取横截面，单击"下一阶段"按钮➡。勾选对话框中的【使用指定公差】选项。鼠标左键单击线框边线，将自动显示数模与扫描数据横截面之间的偏差值，如图 3-100 所示。完成之后单击"OK"按钮✓。

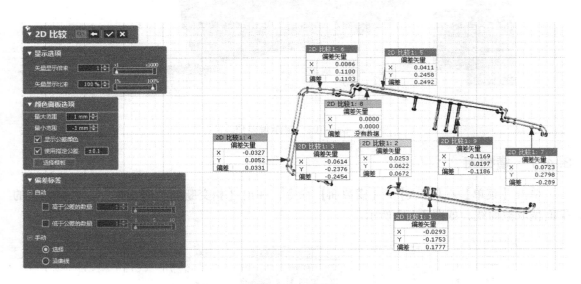

图 3-100　偏差显示

3.4.6　比较点

单击【比较点】命令，弹出如图 3-101 所示对话框，选择默认参数。
鼠标左键单击颜色偏离较深的区域，将自动显示偏差值，如图 3-102 所示。确认完成之后单击"OK"按钮✓。

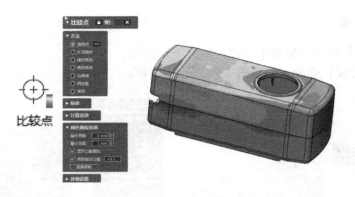

图 3-101　比较点

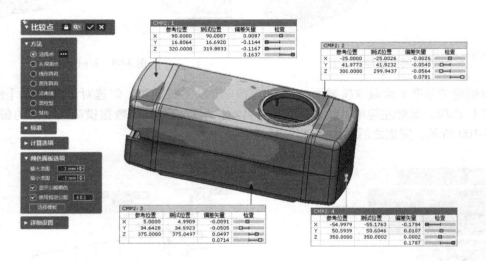

图 3-102　显示偏差值

3.4.7　横截面

单击【菜单】/【插入】/【横截面】命令，弹出【相交断面】对话框，以之前创建的平面截取截面线，如图 3-103 所示。

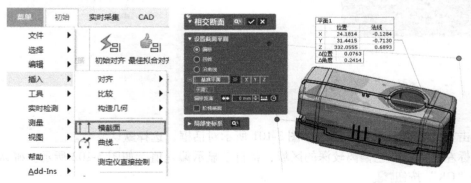

图 3-103　横截面

单击"OK"按钮☑,将自动跳转至尺寸标注界面,如图3-104所示。

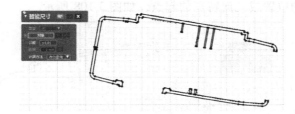

图 3-104　尺寸标注界面

3.4.8　标注功能

1. 智能尺寸
可创建长度、半径、椭圆或角度尺寸,如图3-105所示。

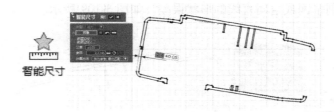

图 3-105　智能尺寸

2. 长度尺寸
测量所选目标实体之间的长度尺寸,如图3-106所示。

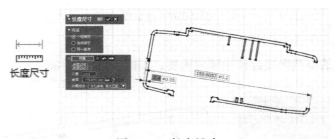

图 3-106　长度尺寸

3. 角度尺寸
测量所选目标实体之间的角度尺寸,如图3-107所示。

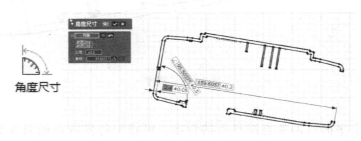

图 3-107　角度尺寸

4. 基准

基准是建立特征的位置或几何特征的起点，如图 3-108 所示。

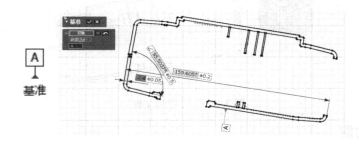

图 3-108　基准

5. 直线度

直线度用于控制平面或空间直线的形状误差，如图 3-109 所示。

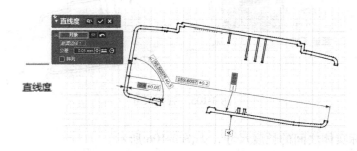

图 3-109　直线度

6. 圆度

圆度是指零件的横截面接近理论圆的程度，如图 3-110 所示。

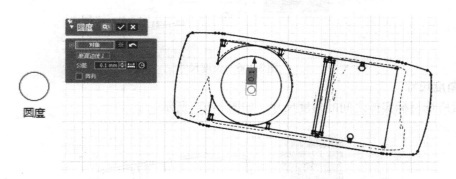

图 3-110　圆度

7. 平行度

平行度指两个平面或两条直线平行的程度。可将平行度视为倾斜度的特殊情况，如图 3-111 所示。

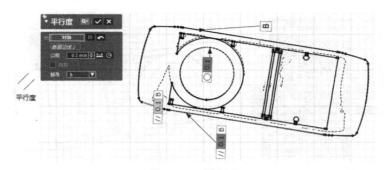

图 3-111 平行度

8. 垂直度

垂直度用于评价直线之间、平面之间或直线与平面之间的垂直状态，如图 3-112 所示。

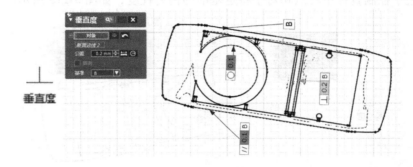

图 3-112 垂直度

9. 倾斜度

根据指定参考系，倾斜度用于控制曲面、轴或平面相对于非 90°的理论正确角度之间的偏差，如图 3-113 所示。

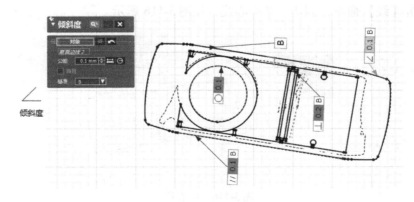

图 3-113 倾斜度

10. 位置度

位置度用于描述特征相对于参考基准或其他特征的准确位置，如图 3-114 所示。

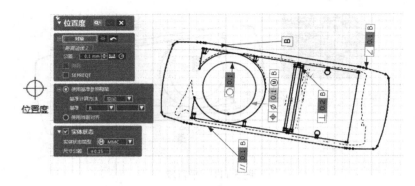

图 3-114 位置度

11. 同心度

同心度用于控制直径中心点相对于特定基准的偏离程度,如图 3-115 所示。

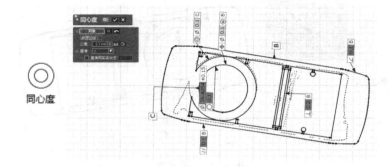

图 3-115 同心度

12. 对称度

对称度用于控制两个特征的中点相对于特定基准的偏离程度。

首先单击【线】命令,创建一条中心线,如图 3-116 所示。

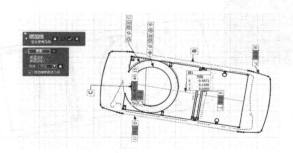

图 3-116 中心线

将中心线设为基准,如图 3-117 所示。

选择线 1、线 2 为对象,以 D 为基准分析两条线的对称度,如图 3-118 所示。

操作完成后,单击屏幕右下角两个按钮中的左侧按钮退出操作界面,如图 3-119 所示。

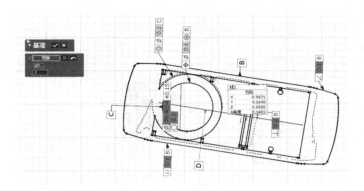

图 3-117　将中心线设为基准

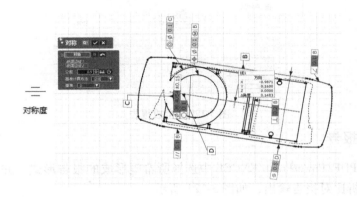

图 3-118　对称度

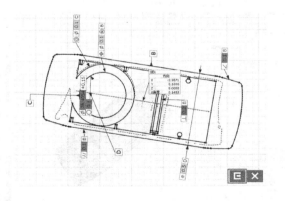

图 3-119　完成后退出

3.4.9　生成报告

单击【生成报告】命令，弹出图 3-120 所示对话框，单击【生成】。

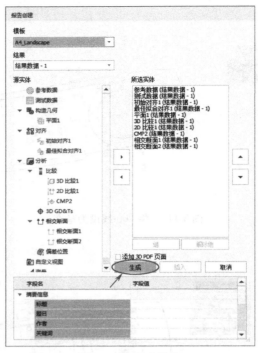

图 3-120　生成报告

3.4.10　输出报告

从左上角的 PDF/PowerPoint/EXCEL 中选择所希望形成的报告形式，并保存到指定目录下，就完成了分析比对报告输出，如图 3-121 所示。

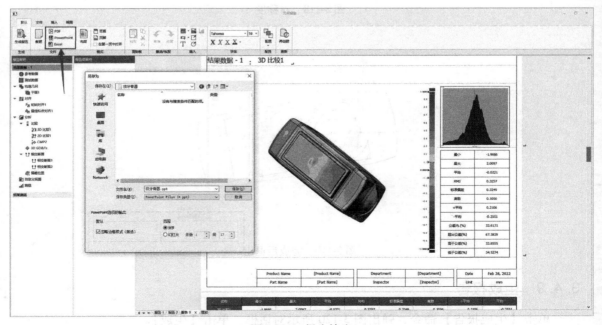

图 3-121　报告输出

项目 4　游戏手柄

本项目以图 4-1 所示的摇杆游戏手柄为载体，使用 Win3DD 单目三维扫描仪进行数据采集，使用 Geomagic Wrap 2017 软件进行数据处理，最后在 Siemens NX 10 软件中完成产品逆向建模和数控加工仿真。项目实施流程如图 4-2 所示。

图 4-1　摇杆游戏手柄

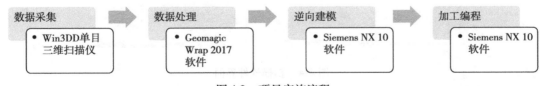

图 4-2　项目实施流程

任务 1　数据采集

采集游戏手柄的模型数据时，使用的是 Win3DD 单目三维扫描仪。Win3DD 系列产品在延续经典的双目系列技术优势的基础上，对外观设计、软件功能和附件配置等均进行大幅提升，除具有高精度外，还具有易学、易用等特点。Win3DD 单目三维扫描仪的硬件系统由扫描头、云台和三脚架 3 部分组成，如图 4-3 所示，其配套的扫描软件界面如图 4-4 所示。

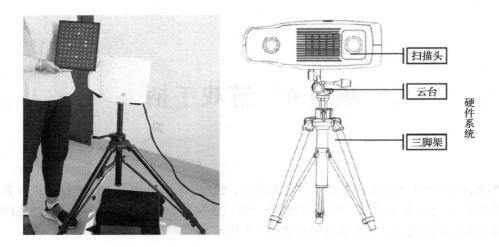

图 4-3 Win3DD 单目三维扫描仪

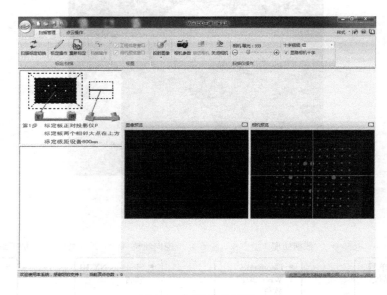

图 4-4 扫描软件界面

4.1.1 扫描仪标定

开始扫描前,应先标定系统。标定的精度将直接影响扫描精度。如果在使用过程中已经标定过系统,在系统未发生任何变动的情况下,进行下一次扫描时可以不用再标定。一般遇到以下情况时需要进行标定:

1)扫描仪初次使用,或长时间放置后使用。
2)扫描仪使用过程中发生碰撞,导致相机位置偏移。
3)扫描仪在运输过程中发生严重振动。
4)扫描过程中频繁出现拼接错误、拼接失败等现象。
5)扫描过程中,扫描数据不完整,数据质量严重下降。

Win3DD 单目三维扫描仪的系统标定步骤如下所述。

1）启动 Win3DD 扫描系统，使扫描系统预热 5~10min，以保证标定状态与扫描状态尽可能相近。打开配套扫描软件，单击【扫描标定切换】按钮，进入扫描标定界面，然后根据界面左上角的标定指示进行标定操作。

2）将标定板放在扫描区域内，如图 4-5 所示。标定板上一共有 99 个点，其中有 5 个大点，大点在标定过程中起定位作用，如图 4-6 所示。

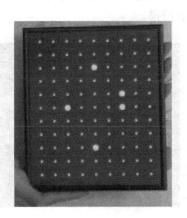

图 4-5　将标定板放在扫描区域内　　　　　图 4-6　标定板

3）调整云台的高度以及俯仰角，如图 4-7 所示，使投影至标定板上的黑色垂直线与软件界面中的白色垂直线对齐，如图 4-8 所示。调整完成后锁紧云台（注意：后续每次调整完都需锁紧云台）。

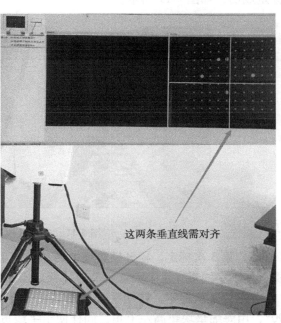

图 4-7　调整云台位置　　　　　图 4-8　两条垂直线需对齐

4）第 1 次标定。再次调整标定板的位置，使黑色垂直线贯穿 4 个大点，如图 4-9 所示。然后单击【标定操作】按钮进行扫描。在扫描过程中，需注意不能有物体遮挡住扫描光源。扫描后，屏幕上会显示扫描结果，如图 4-10 所示。左图为扫描结果，绿色的十字线表示圆心。每次扫描需超过 88 个点，才能认定为有效扫描。

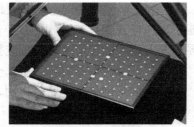

图 4-9　调整标定板位置进行第 1 次标定

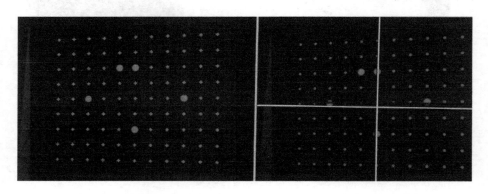

图 4-10　第 1 次标定结果

5）第 2 次标定。松开锁紧处，记住手柄的位置，并转动手柄 3 圈，使云台向上移动。调整后，十字线可能会有位置的移动，需重新调整标定板的位置，使十字线贯穿 4 个大点（后续每次调整均需十字线贯穿大点）。然后单击【标定操作】按钮，软件界面左侧会显示标定成功，如图 4-11 所示。

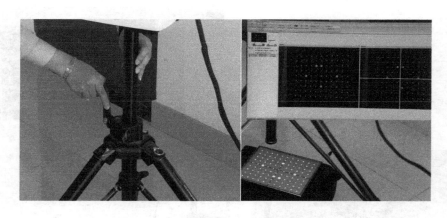

图 4-11　升高云台后进行第 2 次标定

6）第 3 次标定。松开锁紧处，转动手柄 6 圈，使云台向下移动。再次调整标定板的位置，使十字线贯穿 4 个大点。单击【标定操作】按钮进行第 3 次标定，如图 4-12 所示。

7）第 4 次标定。转动手柄 3 圈，使云台向上移动，调整至最初位置。将标定板旋转 90°，用海绵垫块垫高左上角，同样使十字线贯穿 4 个大点。单击【标定操作】按钮进行第 4 次标定，如图 4-13 所示。

项目 4 游戏手柄

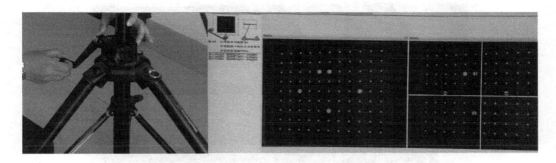

图 4-12 下降云台后进行第 3 次标定

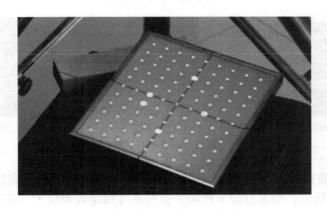

图 4-13 垫高左上角进行第 4 次标定

8) 第 5 次标定。将标定板逆时针旋转 90°，单击【标定操作】按钮进行第 5 次标定，如图 4-14 所示。

9) 第 6 次标定。再次将标定板逆时针旋转 90°，单击【标定操作】按钮进行第 6 次标定，如图 4-15 所示。

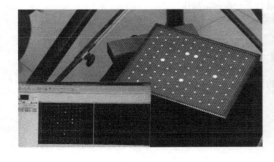 

图 4-14 第 5 次标定　　　　　　　　　图 4-15 第 6 次标定

10) 第 7 次标定。再次将标定板逆时针旋转 90°，标定板回到初始位置。将海绵垫块垫在标定板的右侧，然后单击【标定操作】按钮，如图 4-16 所示。

11) 第 8~10 次标定。每逆时针旋转标定板 90°，单击一次【标定操作】按钮进行标定，共重复 3 次，如图 4-17 所示。

图 4-16　第 7 次标定

　　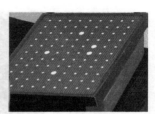

a）第8次标定　　　　　　　　b）第9次标定　　　　　　　　c）第10次标定

图 4-17　第 8～10 次标定

12）至此，共进行了 10 次标定，在标定信息显示区提示【标定完成】，并在下方有精度显示，如图 4-18 所示。如果标定不成功，会提示【标定误差较大，请重新标定】。

```
第1次标定成功，请按照提示进行下一步标定操作！
第2次标定成功，请按照提示进行下一步标定操作！
第3次标定成功，请按照提示进行下一步标定操作！
第4次标定成功，请按照提示进行下一步标定操作！
第5次标定成功，请按照提示进行下一步标定操作！
第6次标定成功，请按照提示进行下一步标定操作！
第7次标定成功，请按照提示进行下一步标定操作！
第8次标定成功，请按照提示进行下一步标定操作！
第9次标定成功，请按照提示进行下一步标定操作！
第10次标定成功，请按照提示进行下一步标定操作！
标定完成
标定结果平均误差：0.027
点击标定操作按钮可进行新的一轮标定流程！
```

图 4-18　标定后的信息提示

4.1.2　贴标记点

为完整地扫描三维物体，通常需要在被扫描物体表面贴上标记点，并且要求标记点粘贴牢固、平整。贴点前，需喷涂显影剂以增强扫描效果。根据游戏手柄的尺寸及其对应的扫描范围，选择内径为 3mm 的标记点，如图 4-19 所示。

在粘贴标记点时应注意以下几点。

1）贴点尽可能随意一些。避免人为分组，

图 4-19　内径为 3mm 的标记点

如图 4-20 所示。

2）避免标记点呈直线或等边、等腰三角形排布，如图 4-21 所示。

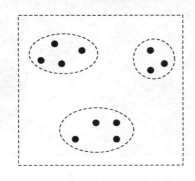 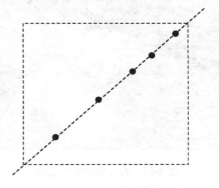

图 4-20　（错误）人为分组标记点　　　　图 4-21　（错误）标记点直线排布

3）标记点尽可能不要贴在面与面的交接处，尽量贴在平坦处，如图 4-22 所示。
4）为保证扫描质量，在扫描转盘上也需要贴上相应的标记点，如图 4-23 所示。

图 4-22　标记点需避免在交接处　　　　图 4-23　在转盘上贴点

为便于贴点，用棉签擦拭掉游戏手柄表面的部分显影剂，留出贴点位置，然后再粘贴标记点，如图 4-24 所示。游戏手柄的正、反面都需要扫描，因此都需要粘贴标记点，如图 4-25 和图 4-26 所示。

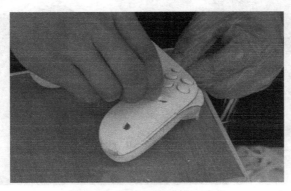

图 4-24　用棉签擦拭便于贴点

图 4-25　在正面粘贴标记点

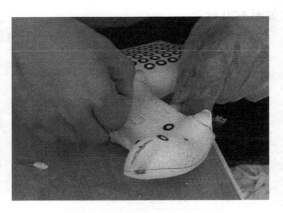
图 4-26　在背面粘贴标记点

4.1.3　数据扫描

为防止扫描过程中扫描件与转盘产生相对位移，可用橡皮泥进行固定。如果产生相对位移，扫描得到的数据会出现分层现象，导致扫描精度下降，甚至导致该扫描数据无法使用。数据扫描的具体步骤如下所述。

1）新建工程，命名为"游戏手柄1"（可自由命名，但为区分扫描内容，建议命名与扫描件相关），如图 4-27 所示。

2）调整转盘使扫描件对准扫描仪，单击【扫描操作】按钮，进行第 1 次扫描，如图 4-28 所示。

图 4-27　新建工程

图 4-28　第 1 次扫描结果

3）轻轻旋转转盘，防止在转动过程中扫描件与转盘的相对位置发生变化，再次单击【扫描操作】按钮进行扫描。旋转转盘进行多次扫描，直至正面数据扫描完整，周边数据完全。与背面数据手动拼合时需要用到周边数据，如图4-29所示。

图4-29　反复扫描并观察结果

4）确保数据无缺损后，将其保存为"点云文件（*.asc）"，如图4-30所示。

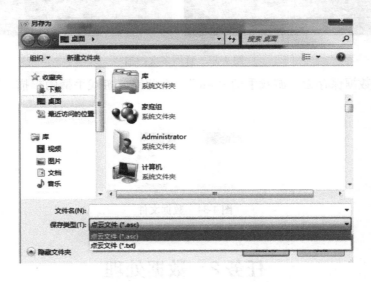

图4-30　保存游戏手柄正面数据

5）翻转游戏手柄，取下之前粘贴的橡皮泥，再选择合适的固定不动的位置进行固定，避免粘贴在活动位置处，例如游戏手柄的摇杆，以免对扫描数据产生影响，如图4-31所示。

6）同样，新建一个工程，命名为"游戏手柄2"。旋转转盘，选择标记点较多的面进行第1次扫描，如图4-32所示。

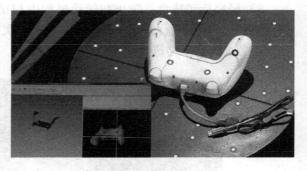

图 4-31　（错误位置）避免固定在晃动处　　图 4-32　背面第 1 次扫描

7）与正面的扫描方式相同，旋转转盘进行多次扫描，直至背面数据完整，周边数据完全，如图 4-33 所示。

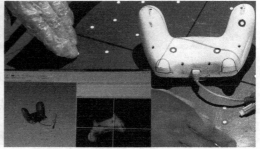

图 4-33　背面多次扫描

8）将所得数据保存为"游戏手柄 2. asc"，至此完成游戏手柄的扫描。2 份点数据文件如图 4-34 所示。

图 4-34　数据文件

任务 2　数据处理

Geomagic Wrap 2017 软件所提供的强大的工具箱，能够以易用、低成本、快速而精确的方式，将 3D 扫描数据直接转换为 3D 多边形和曲面模型，用于制造、艺术和工业设计等。这里采用 Geomagic Wrap 2017 软件进行游戏手柄正、反面扫描数据的手动拼接及数据封装。

4.2.1　手动拼接

1）导入扫描数据文件"游戏手柄 1. asc"和"游戏手柄 2. asc"，如图 4-35 所示。

图 4-35　导入扫描数据

2）为便于观察，使用【着色点】命令进行着色处理，如图 4-36 所示。

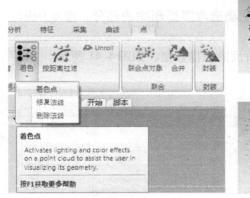

a）选择【着色点】命令　　b）着色前　　c）着色后

图 4-36　着色处理

3）隐藏"游戏手柄 2"，对"游戏手柄 1"的数据进行处理，如图 4-37 所示。

4）使用【非连接项】命令，如图 4-38 所示，单击【确定】后，系统自动选中外部点

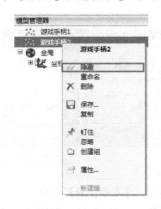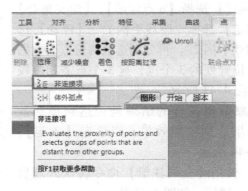

图 4-37　隐藏"游戏手柄 2"数据　　　　图 4-38　使用【非连接项】命令

并显示为红色；然后按<Delete>键或单击软件中的【删除】按钮将这些红色的点数据删除，如图4-39所示。

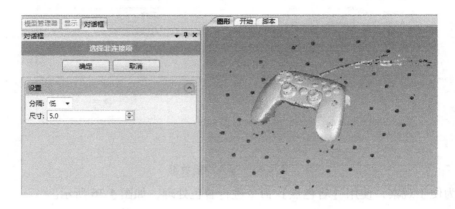

图4-39 删除外部点

5）接下来，通过按住鼠标左键手动选取需要删除的部分，按<Delete>键将其删除，如图4-40所示。

6）用同样的方法处理游戏手柄背面的扫描数据，如图4-41所示。

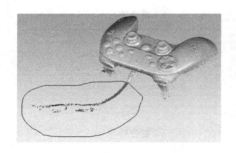 

图4-40 手动选取需去除点　　　图4-41 处理后的游戏手柄背面扫描数据

7）在模型管理器选中2份点云数据，使用【对齐】/【手动注册】命令进行拼合，如图4-42所示。

【模式】选择【n点注册】，【定义集合】中【固定】选择"游戏手柄1"，【浮动】选择"游戏手柄2"。再分别选取两份点云数据上相近位置处的3个点，这3个点需相距较远，如图4-43所示。

观察数据拼合的结果是否达到预期效果，如果达到预期效果，则单击【确定】，如图4-44所示。

8）使用【全局注册】命令再次处理后完成数据拼接，如图4-45所示。

图4-42 【手动注册】命令

项目 4　游戏手柄

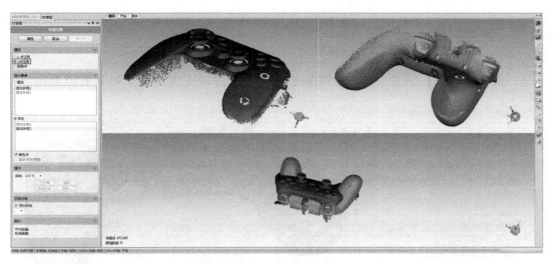

图 4-43　选择 3 个点进行拼合

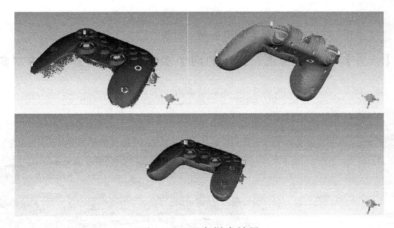

图 4-44　观察拼合效果

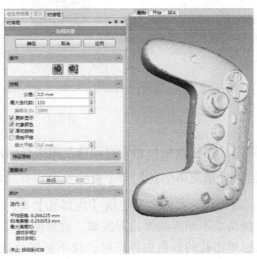

图 4-45　全局注册

4.2.2 数据封装

1)使用【封装】 命令对点云数据进行计算,得到多边形模型,如图4-46所示。

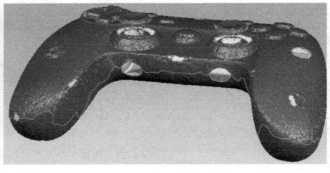

图4-46 封装

2)使用【合并】命令,将正、反面的两个多边形对象合并为单个对象,如图4-47所示。

图4-47 合并片体

3)使用【网格医生】命令自动修复多边形网格内的缺陷,如图4-48所示。

4)使用【填充单个孔】命令将孔补全,有【曲率】 、【切线】 和【平面】 3种方式。可以填充孔的类型包括【内部孔】 、【边界孔】 ,并可以以【搭桥】 的方式连接两个不相连的多边形区域。这里使用【曲率】的方式填充【内部孔】。填充单个孔时,鼠标单击孔的绿色边界即可;找不到绿色边界时可以单击"下一个"按钮,可自动跳转至下一个需要修补的孔,如图4-49所示。

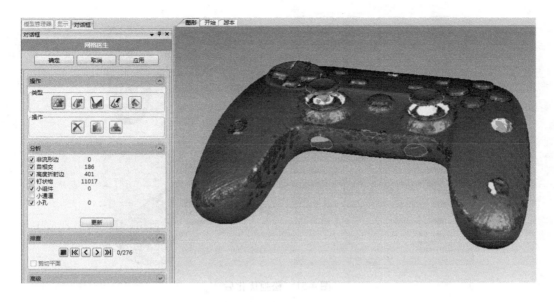

图 4-48 【网格医生】自动修复

5）如果三角面片数量较多，可通过【简化】命令，输入想要消减到的百分比数量进行简化，如图 4-50 所示。

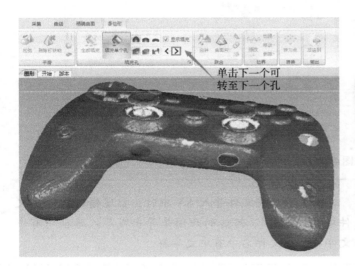

图 4-49 填充孔

图 4-50 减少面片数量

6）孔填充后，观察发现表面瑕疵较多，可通过【松弛】命令，根据情况调整参数，再单击【应用】对面片进行处理，最大限度地减少单独多边形之间的角度，使多边形网格更平滑，如图 4-51 所示。

7）将生成的三角面片保存为 STL 文件，作为 NX 中逆向建模的基础。建议以英文命名并存于英文路径，如图 4-52 所示。

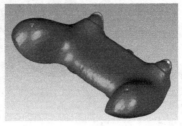

图 4-51 松弛优化后

图 4-52 保存为 STL 文件

如果后续将文件导入 NX 中时，出现错误提示"文件中解析错误"，可能的原因是导出或导入路径中有中文字符，或文件名中有中文字符。

任务 3　逆向建模

在后续测量平面与导入小平面体的距离时，可采用如下方法：按下"启用捕捉点"按钮，不勾选【面上的点】，选择逆向建模时创建的面；再勾选【面上的点】，选择导入小平面体上的点，如图 4-53 所示。

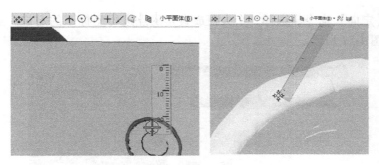

图 4-53 测量距离的一种方法

4.3.1 确定坐标系

1)导入游戏手柄的 STL 文件,如图 4-54 所示。

2)游戏手柄的主体结构对称,首先观察游戏手柄,以便确定坐标轴及镜像面,如图 4-55 所示。

图 4-54 导入 STL 文件

图 4-55 观察游戏手柄

3)通过【基本曲线】命令中的"直线"功能,勾选【无界】(屏幕界面多大线就多长),通过【点构造器】,【类型】选择为【点在面上】,然后在合适处选择两点,完成直线的创建。用同样的方法绘制一条竖直的线,并通过【扫掠】命令形成一个面。可以通过测量距离进行分析,如图 4-56 所示。

图 4-56 创建一个面

4）通过【定向】命令，选择 3 个大概成直角的点进行坐标系的定位。至此，Z 轴方向确定完成，如图 4-57 所示。

图 4-57　确定 Z 轴方向

5）接下来需确定 X、Y 轴的方向。一般选择左右两圆圆心的连线来进行确定，如图 4-58 所示。

6）通过【基本曲线】中的"圆弧"功能，勾选【整圆】，再通过【点构造器】类型选择为【点在面上】；然后根据实物想象，大致选择在圆角根部的 3 个点，绘制左、右两边的两个圆，如图 4-59 所示。

图 4-58　需确定 X、Y 轴方向

图 4-59　绘制两个圆

7）通过【基本曲线】中的"直线"功能，选择两圆心进行直线的绘制，并将直线投影至以 Z 轴为法向量的平面上。建议在设置中将输入曲线隐藏，便于新手辨别哪条曲线为新生成的曲线，如图 4-60 所示。

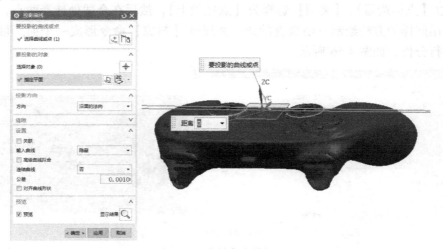

图 4-60　绘制直线并投影

8）通过【定向】命令，选择 X 轴，再选择投影的直线，完成 X 轴的初步定向，如图 4-61 所示。

9）初步坐标系确定后，根据个人习惯，可在原点创建一个小方块或者保存坐标系，防止坐标系丢失，如图 4-62 所示。

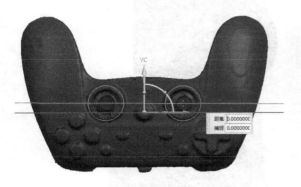 

图 4-61　X 轴初步定向　　　　　　　　　　图 4-62　创建小方块

4.3.2　铺面

1）通过【插入】/【基准/点】/【点】命令，选择【点在面上】，然后在面上合适处选择点，注意避免在圆角处选择点，如图 4-63 所示。

2）选择所创建的点，在任意一点上单击右键，选择【新建组】（此处为 NX 10.0 版本与之前版本区别处，需特别注意），如图 4-64 所示。

3）通过【拟合曲面】命令，选择组中的任意一点进行曲面创建。如效果不理想，可通过修改【参数化】中的内容进

图 4-63　创建点

图 4-64　创建组

行大致调整。需注意，阶次越高，其光顺性就会越差。可通过【面分析】进行观察，并通过【扩大】命令将面适当扩大，如图4-65所示。

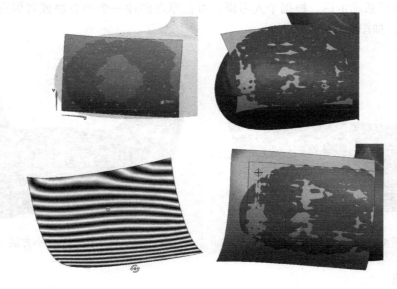

图 4-65　拟合曲面

4）通过【基本曲线】中的"圆弧"功能先绘制两段圆弧，再分别以这两个圆弧的两个端点作为起点和终点，并在大致成一条直线处选择圆弧的另一点绘制两段圆弧。利用【通过曲线网格】命令创建曲面，如图4-66所示。

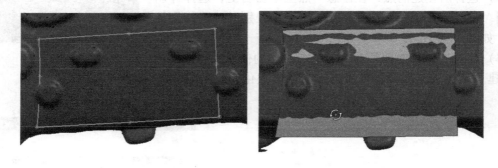

图 4-66　创建曲面并扩大

5）通过【基本曲线】中的"圆弧"功能创建两段圆弧，然后通过【扫掠】命令创建曲面并扩大，如图4-67所示。

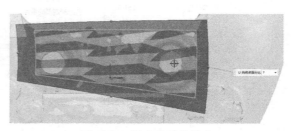

图 4-67　创建曲面并扩大

6)通过【拟合曲面】命令进行曲面拟合,并适当扩大,如图 4-68 所示。

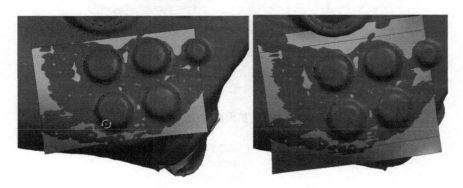

图 4-68 拟合曲面并扩大

7)通过【基本曲线】中的"圆弧"功能绘制 4 段圆弧,再通过【通过曲线网格】命令创建曲面,并适当扩大,如图 4-69 所示。

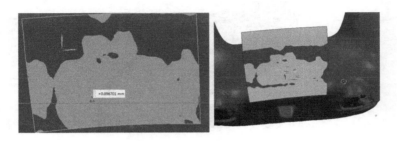

图 4-69 创建曲面并扩大

8)在表面上创建点后,通过【拟合曲面】命令拟合曲面。如拟合曲面不理想,也可通过【X 型】命令进行调整,如图 4-70 所示。

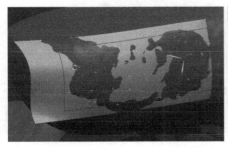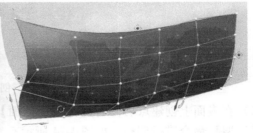

图 4-70 创建曲面并调整

9)通过【扩大】命令将曲面扩大,并通过【X 型】命令进行逐步调整,如图 4-71 所示。

10)通过【基本曲线】中的"圆弧"功能绘制 4 段圆弧,再利用【通过曲线网格】命令创建曲面,并适当扩大,如图 4-72 所示。

11)通过【基本曲线】中的"圆弧"功能绘制 4 段圆弧,再利用【通过曲线网格】命令创建曲面,并适当扩大。然后采用【X 型】命令进行调整,如图 4-73 所示。

图 4-71　扩大曲面并调整

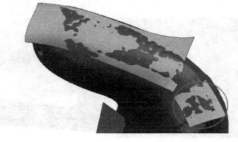

图 4-72　创建曲面

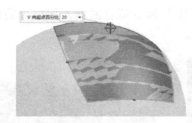

图 4-73　创建曲面

12）在表面上创建点后，通过【拟合曲面】命令拟合曲面。如拟合曲面不理想，也可通过【X 型】命令进行调整，如图 4-74 所示。

图 4-74　创建曲面

13）通过【基本曲线】中的"圆弧"功能绘制 4 段圆弧，再利用【通过曲线网格】命令创建曲面，并适当扩大。如曲面不理想，也可通过【X 型】命令进行调整，如图 4-75 所示。

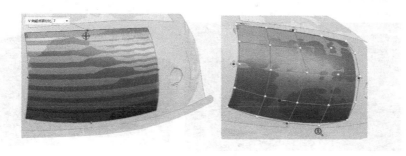

图 4-75　创建曲面

14）通过【基本曲线】中的"圆弧"功能绘制圆弧，并通过【规律延伸】命令拉伸成曲面，完成后将其进行扩大，如图 4-76 所示。

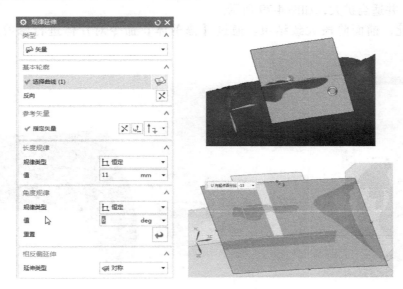

图 4-76　创建曲面

15）通过【基本曲线】中的"圆弧"功能绘制 4 段圆弧，再利用【通过曲线网格】命令创建曲面，并适当扩大。如曲面不理想，可通过【X 型】命令进行调整，如图 4-77 所示。

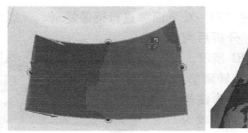

图 4-77　创建曲面

16）通过【扩大】命令将曲面扩大，并通过【X 型】命令进行逐渐调整，如图 4-78 所示。

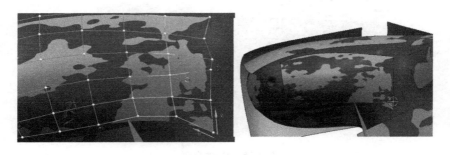

图 4-78 扩大曲面

17）通过【基本曲线】中的"圆弧"功能绘制圆弧，并通过【规律延伸】命令将曲线拉伸成曲面，并适当扩大，如图 4-79 所示。

18）至此，铺面阶段大致结束。通过【修剪体】命令对片体进行修剪，如图 4-80 所示。

图 4-79 规律延伸　　　　　　　　图 4-80 修剪面片

4.3.3 绘制大致形状

1）通过【变换】命令，将片体通过"XC-ZC 平面"进行镜像复制。观察发现另一侧面有一些偏差，需对坐标系进行调整。分析后判断，需绕 Y 轴适当旋转坐标系，如图 4-81 所示。

2）通过【定向】命令将坐标系进行旋转调整，直至测量数据大致符合要求。由于该产品为注塑件，在实际生产中可能会产生变形，因此大面较为贴合，部分面偏差在允许范围内即可。坐标系调整后，重新镜像，并重新创建原点标记（创建方块或

图 4-81 镜像面

保存坐标系），如图 4-82 所示。

3）接下来需对中间镜像拼接处进行处理，保证中间光顺连接。通过【基本曲线】中的"直线"功能，创建一条过坐标原点且平行于 Y 轴的直线，并拉伸成平面，如图 4-83 所示。

图 4-82 调整坐标系后重新镜像

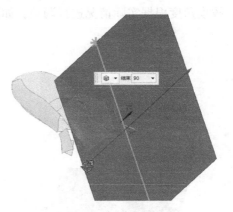
图 4-83 创建平面

4）通过【偏置表面】命令，将平面向左、右两侧各偏置一定距离，如图 4-84 所示。

5）利用偏置的面，对面进行修剪，如图 4-85 所示。

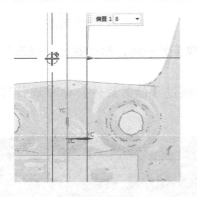
图 4-84 偏置面

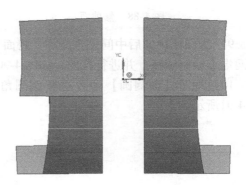
图 4-85 修剪面

6）通过【桥接曲线】命令进行桥接，并利用【通过曲线网格】命令创建曲面。用同样的方法完成其他面的创建，如图 4-86 和图 4-87 所示。

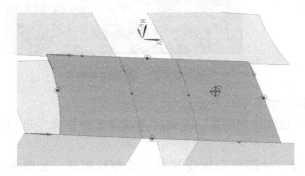
图 4-86 创建曲面

图 4-87 其他位置创建曲面

7)至此,坐标系已经完全确定。修剪一侧的面,仅保留一侧,如图4-88所示,完成后镜像至另一侧即可。

8)通过【基本曲线】中的"圆弧"功能绘制圆弧,并通过【规律延伸】命令进行拉伸,拔模角度根据实际情况进行调整,如图4-89所示。

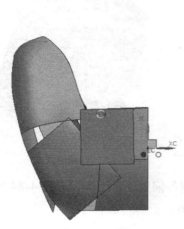

图4-88 修剪面

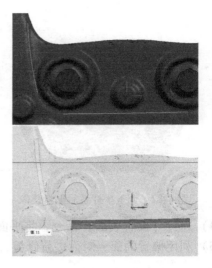

图4-89 规律延伸

9)为保证镜像后中间完全光顺,将面切除后镜像至另一侧,然后通过"圆弧"功能捕捉两端点绘制圆弧,并进行拉伸,如图4-90所示。

10)通过【面倒圆】命令进行倒圆角处理,并通过【修剪片体】命令进行修剪,如图4-91所示。

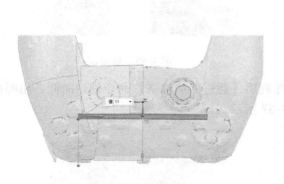

图4-90 规律延伸

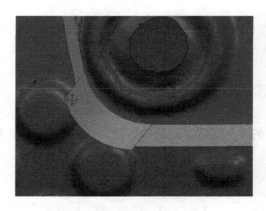

图4-91 面倒圆并修剪

11)通过【修剪片体】命令进行修剪,如图4-92所示。

12)通过【基本曲线】中的"圆弧"功能绘制圆弧,并通过【规律延伸】命令进行拉伸,如图4-93所示。

13)通过【桥接曲线】命令进行桥接,并调整参数使其接近模型轮廓线,如图4-94所示。

图 4-92 修剪片体

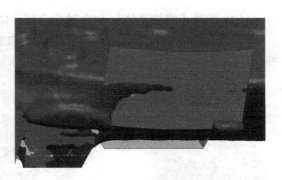
图 4-93 规律延伸

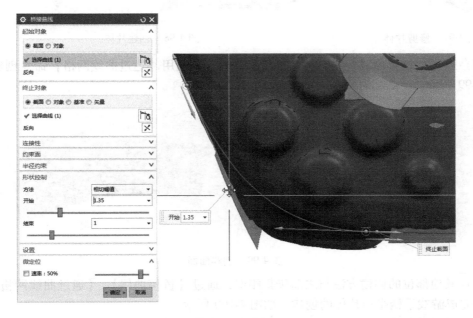
图 4-94 桥接曲线

14）利用【在面上偏置曲线】命令进行曲线偏置，并将曲线桥接。利用【通过曲线网格】命令创建曲面，如图 4-95 所示。

15）对面进行适当修剪及桥接后，利用【通过曲线网格】命令创建曲面，如图 4-96 所示。

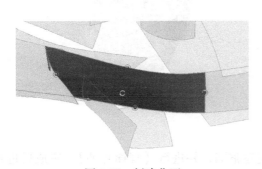
图 4-95 创建曲面

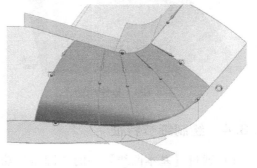
图 4-96 创建曲面

16）通过【修剪片体】命令对多余的片体进行修剪，如图4-97所示。

17）通过【基本曲线】中的"圆弧"功能创建圆弧，并通过【规律延伸】命令进行拉伸，如图4-98所示。

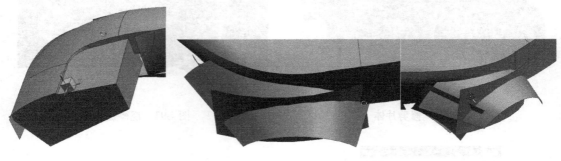

图4-97　修剪片体　　　　　　　　　图4-98　创建片体

18）通过【桥接曲线】进行桥接；适当修剪后，利用【通过曲线网格】命令创建曲面，如图4-99所示。

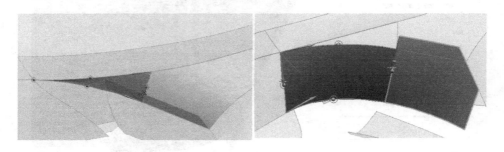

图4-99　创建曲面

19）其他部位的创建方法与之前步骤相似，通过【桥接曲线】、【通过曲线网格】等命令最终完成游戏手柄半边片体的创建，如图4-100所示。

20）通过【缝合】命令将片体缝合成实体，如图4-101所示。

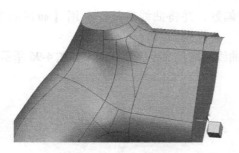 

图4-100　创建曲面　　　　　　　　　图4-101　缝合成实体

4.3.4　绘制细节特征

1）通过【基本曲线】中的"圆弧"功能创建整圆，并进行【规律延伸】，完成后进行修剪，如图4-102所示。

2）接下来创建一个球，其直径可通过 3 点绘制整圆大致判断。通过【球】命令，以之前的圆心作为球心创建球，然后向下移动一定距离直至合适位置，如图 4-103 所示。

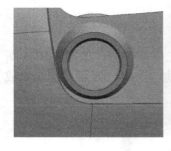 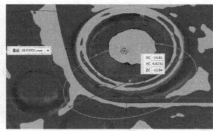 

图 4-102　创建特征　　　　　　　　　　　图 4-103　创建球

3）创建中间圆台特征，如图 4-104 所示。

4）创建 4 个圆柱特征，并通过游戏手柄表面偏置一定的距离作为修剪面进行修剪，再进行拔模及倒圆角处理，如图 4-105 所示。

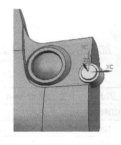   

图 4-104　创建圆台　　　　　　　　　　　图 4-105　创建按钮特征

5）继续进行细节特征的创建，直至完成所有特征的创建，如图 4-106 所示。

6）对模型进行倒圆角处理，完成游戏手柄逆向建模，如图 4-107 所示。

图 4-106　游戏手柄　　　　　　　　　　　图 4-107　完成游戏手柄逆向建模

任务4 加工编程

4.4.1 工艺分析

考虑到游戏手柄反面的余量较大，为保证工件刚性首先加工反面部分，如图4-108a所示。然后在方框内浇注石灰，再装夹反面部分，进行正面部分的加工。正、反面都加工完成后，最后对接缝处进行打磨。

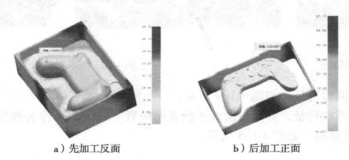

a）先加工反面　　　　　　b）后加工正面

图4-108　加工方案

为节约加工时间，先使用三轴进行粗加工，然后用五轴加工局部特征，加工工序见表4-1。

表4-1　游戏手柄的加工工序

工序	类型	工序子类型		刀具	加工余量/mm
反面粗加工	mill_contour	型腔铣	CAVITY_MILL	D20R0.8	0.3
	mill_contour	拐角粗加工	CORNER_ROUGH	D8	0.3
反面半精加工	mill_contour	区域轮廓铣	CONTOUR_AREA	R3	0.2
	mill_contour	流线	STREAMLINE	R3	0.2
反面精加	mill_contour	区域轮廓铣	CONTOUR_AREA	R3	0
	mill_contour	流线	STREAMLINE	R3	0
正面粗加工	mill_contour	型腔铣	CAVITY_MILL	D20R0.8	0.3
				D6R0.5	0.3
				D3R0.5	0.3
				D1	0.3
正面半精加工	mill_contour	区域轮廓铣	CONTOUR_AREA	R3	0.2
	mill_contour	区域轮廓铣	CONTOUR_AREA	D6R0.5	0
	mill_contour	深度轮廓加工	ZLEVEL_PROFILE	D6R0.5	0
	mill_contour	区域轮廓铣	CONTOUR_AREA	D3R0.5	0
正面精加工	mill_multi-axis	可变轮廓铣	VARIABLE_CONTOUR	R3	0
	mill_contour	区域轮廓铣	CONTOUR_AREA	D6R0.5	0
	mill_multi-axis	可变流线铣	VARIABLE_STREAMLINE	D2	0
	mill_multi-axis	外形轮廓铣	CONTOUR_PROFILE	D8	0
				D2	0
	mill_contour	清根参考刀具	FLOWCUT_REF_TOOL	R2	0.07
				R1	0
				R0.5	0
	mill_contour	区域轮廓铣	CONTOUR_AREA	R1	0

4.4.2 编程准备

在创建游戏手柄的加工程序之前，需要进行图 4-109 所示的准备工作。

图 4-109 编程准备

1. 辅助设计

1）导入数据。使用【文件】/【导入】/【Parasolid】命令，导入 x_t 格式的游戏手柄模型。

2）模型合并。将图 41-110a 所示的 3 个实体移动至其他图层，然后使用【合并】命令将其余实体进行合并，如图 4-110b 所示。为便于后期刀轨的观察，将模型颜色改为灰色。

a）导入的模型　　　　　　　　　b）处理后的模型

图 4-110 处理模型数据

3）模型摆正。由于游戏手柄模型的摆放位置与绝对坐标系并不对齐，所以需要将其摆正。创建图 4-111 所示的平面、直线和圆弧。

单击【插入】/【基准/点】/【基准 CSYS】命令，弹出【基准 CSYS】对话框。【类型】选择【平面，X 轴，点】，再依次选择之前创建的平面、直线和圆弧的圆心，创建图 4-112 所示的基准坐标系。使用【格式】/【WCS】/【保存】命令保存基准坐标系。

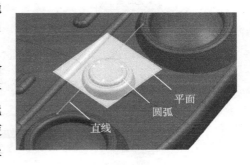

图 4-111 创建平面、直线和圆弧

单击【编辑】/【移动对象】命令，然后选择所有的模型数据，【运动】选择【CSYS 到 CSYS】，设置【指定起始 CSYS】为刚创建的基准坐标系，【指定目标 CSYS】为"绝对 CSYS"，结果如图 4-113 所示，此时模型位置已经摆正。

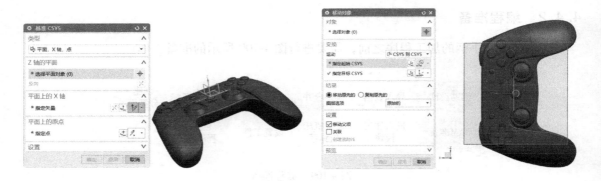

图 4-112　创建基准坐标系　　　　　图 4-113　模型摆正

4）创建毛坯。选择【应用模块】中的【电极设计】，然后选择【创建方块】命令，如图 4-114 所示。

图 4-114　【创建方块】命令

考虑到粗加工刀具的直径为 20mm，毛坯四周需留出 5mm 左右的余量，所以输入【间隙】值为"25mm"，如图 4-115 所示。

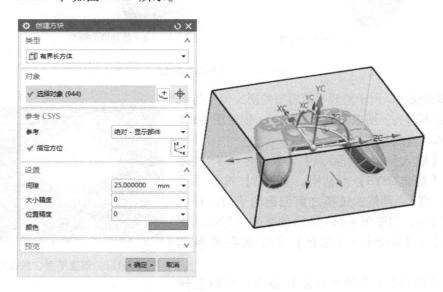

图 4-115　【创建方块】对话框

上、下两个面不需要留出太多余量，保留 5mm 左右即可，所以使用【偏置面】命令将其向内偏置"20mm"，如图 4-116 所示。

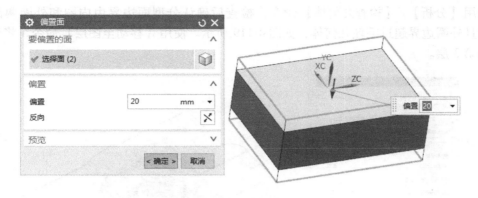

图 4-116　偏置面

5）创建分型面。使用【分析】/【形状】/【斜率】命令，观察后判断分型面的大致位置，如图 4-117 所示。

图 4-117　斜率分析

使用【拉伸】命令，选择位于分型面处的边进行拉伸，拉伸长度可以适当加大，保证其超过毛坯。由于该产品为对称件，创建其中一半分型面后，使用【变换】命令将其镜像复制到另一侧，并使用【缝合】命令进行缝合，如图 4-118 所示。

图 4-118　创建分型面

使用【分析】/【检查几何体】命令，检查后确认分型面边界由内圈和外圈两部分组成，并且外圈边界超过毛坯几何体，如图4-119所示。使用【移动至图层】命令，将分型面移动到第3层。

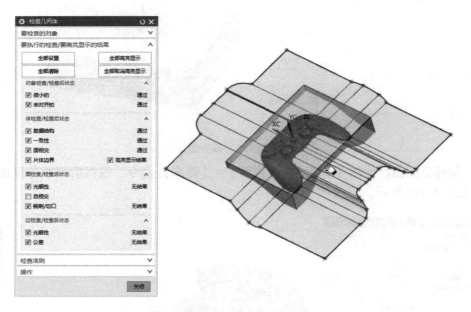

图4-119　检查几何体

2. CAM 准备

1）进入【加工】模块，弹出【加工环境】对话框，然后选择【cam_general】和【mill_contour】，如图4-120所示。

2）设置机床坐标系。切换到工序导航器几何视图，将【MCS_MILL】重命名为【MCS_MILL-F】，将【WORKPIECE】重命名为【WORKPIECE-F】（F表示反面），如图4-121所示。

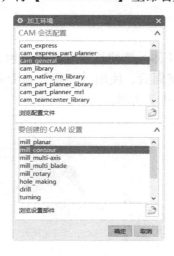

图4-120　【加工环境】对话框　　　　　　图4-121　几何视图

双击【MCS_MILL-F】，弹出【MCS 铣削】对话框，单击对话框中的"CSYS 对话框"按钮弹出【CSYS】对话框。选择毛坯顶面中心作为坐标原点，并调整坐标轴方向，使 ZM 轴朝上，使 XM 轴与毛坯的长边对齐，如图 4-122 所示。

图 4-122　调整机床坐标系

在【MCS 铣削】对话框中，【安全设置选项】选择【刨】，然后选择毛坯顶面并将其向上偏置 50mm，将偏置得到的平面指定为安全平面，如图 4-123 所示。

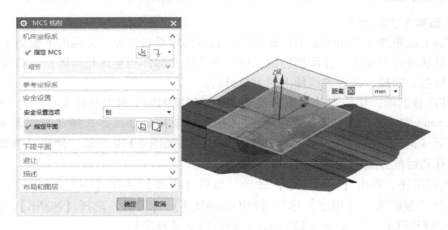

图 4-123　安全平面设置

3）设置工件。双击图 4-121 所示的【WORKPIECE-F】，弹出【工件】对话框，然后依次指定部件和毛坯，如图 4-124 所示。

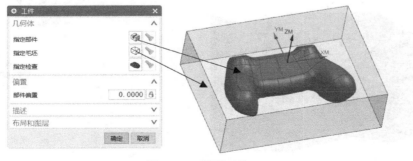

图 4-124　设置工件

4）创建程序组。切换到程序顺序视图，选择【PROGRAM】后单击鼠标右键，选择【插入】/【程序组】命令，创建图 4-125 所示的两个程序组。

a) b)

图 4-125 创建程序组

4.4.3 反面粗加工

1. 反面第 1 次粗加工

反面第 1 次粗加工采用的是型腔铣工序，通过移除垂直于固定刀轴的平面切削层中的材料对轮廓形状进行粗加工。使用型腔铣工序，必须定义部件和毛坯几何体，一般用于移除模具型腔与型芯、凹模、铸件和锻件上的大量材料。

1）移动分型面。切换到【建模】模块。考虑到反面加工时刀具最大半径为 3mm，并且需要留 0.3mm 左右的加工余量，所以使用【移动对象】命令，并选择【复制原先的】，将分型面向下（向游戏手柄的正面方向）移动 3.3mm，如图 4-126 所示。使用【移动至图层】命令，将移动后的分型面放置到第 4 层。

2）创建工序。单击【创建工序】命令，然后【类型】选择【mill_contour】，【工序子类型】选择"型腔铣"，【程序】选择【PROGRAM_F】，【刀具】选择【NONE】，【几何体】选择【WORKPIECE-F】，如图 4-127 所示，最后单击【确定】。

图 4-126 移动分型面

图 4-127 创建型腔铣工序

3)设置检查几何体。弹出图 4-128 所示的【型腔铣】对话框后,单击"指定检查"按钮,选择步骤 1)中移动后的分型面作为检查几何体。

4)创建刀具。单击【型腔铣】对话框中【工具】组中的"新建"按钮,如图 4-129 所示,弹出【新建刀具】对话框。输入刀具名称后单击【确定】,弹出【铣刀-5 参数】对话框。输入图 4-130 所示的参数后单击【确定】。

图 4-128 【型腔铣】对话框　　　　　　　图 4-129 新建刀具

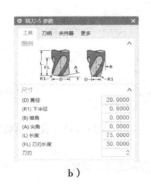

a)　　　　　　　　b)　　　　　　　　c)

图 4-130 创建刀具 D20R0.8

5)设置刀轨。【切削模式】设置为【跟随周边】,【平面直径百分比】为"70",【最大距离】为"0.5"(对于钢件,一般为 0.3~0.5),如图 4-131 所示。

6)设置切削参数。【切削顺序】设置为【深度优先】,勾选【岛清根】,部件侧面余量设为"0.3",如图 4-132 所示。

7)设置非切削移动。【进刀类型】设为【沿形状斜进刀】,【斜坡角】设为"3",【高度】设为"1",其余参数设置如图 4-133 所示。

8)设置进给率和主轴速度。输入主轴转速和切削进给率之后,单击"基于此值计算进给和速度"按钮,如图 4-134 所示。

图 4-131 设置刀轨

图 4-132 设置切削参数

图 4-133 设置非切削移动

图 4-134 设置进给率和主轴速度

9)生成刀轨。单击【型腔铣】对话框中【操作】组中的"生成"按钮,生成图 4-135 所示的刀轨。该刀轨超出了毛坯范围,把应该留下的一圈边框也铣削掉了,所以需要修改。

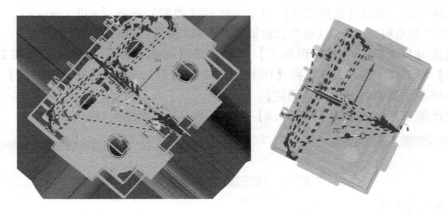

图 4-135 初次生成的刀轨

10)编辑刀轨。双击工序导航器程序顺序视图中的【CAVITY_MILL】程序,在弹出的【型腔铣】对话框中单击"指定修剪边界"按钮,弹出【修剪边界】对话框。选择毛坯顶面,然后【修剪侧】选择【外部】,勾选【余量】并输入"14",单击【确定】后重新生成刀轨。通过加工仿真验证重新生成后的刀轨是否正确,如图 4-136 所示。

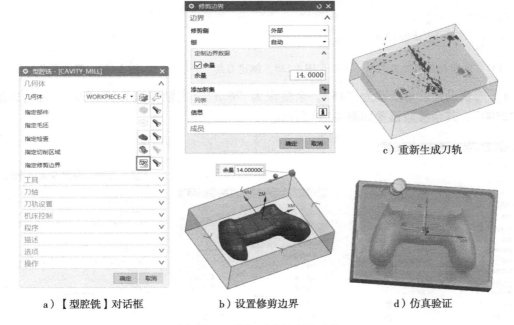

a)【型腔铣】对话框　　b)设置修剪边界　　c)重新生成刀轨　　d)仿真验证

图 4-136 编辑后重新生成刀轨

2. 反面第 2 次粗加工

第 1 次粗加工后,游戏手柄的拐角处可能加工不到,需要再创建一个工序进行第 2 次粗加工。反面第 2 次粗加工使用拐角粗加工工序,对反面第 1 次粗加工时拐角中刀具处理不到

的遗留材料进行粗加工。使用拐角粗加工工序，必须定义部件和毛坯几何体，并将之前粗加工工序中使用的刀具指定为"参考刀具"，以确定切削区域。

1）创建工序。单击【创建工序】命令，然后【类型】选择【mill_contour】，【工序子类型】选择"拐角粗加工"，其余参数设置如图4-137所示，最后单击【确定】。

2）创建刀具。单击【拐角粗加工】对话框中【工具】组中的"新建"按钮，弹出【新建刀具】对话框。输入刀具名称【D8】后单击【确定】，弹出【铣刀-5 参数】对话框。输入图4-138所示的参数，单击【确定】。

3）设置参考刀具。选择【D20R0.8】作为参考刀具，如图4-139所示。

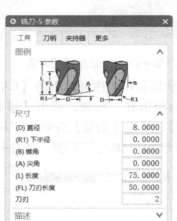

图4-137 创建拐角粗加工工序　　　图4-138 创建刀具　　　图4-139 设置参考刀具

4）刀轨设置。角度设为"45"，余量设为"0.3"，主轴转速为"3000"，切削进给率为"2500"，其余参数设置如图4-140所示。

a)　　　　　　　　　　　b)　　　　　　　　　　　c)

图4-140 刀轨设置

5)生成刀轨。单击【操作】组中的"生成"按钮,生成图 4-141 所示的刀轨,观察后有一部分刀轨不需要,所以需要进行修剪。

6)编辑刀轨。单击【拐角粗加工】对话框中【几何体】组中的"指定修剪边界"按钮,弹出【修剪边界】对话框。【修剪侧】选择【内部】,单击几个点形成修剪边界,将需要修剪的刀轨包含在内即可,如图 4-141 所示。需要注意的是,修剪边界创建在 XC-YC 平面上,所以根据情况调整坐标系的方位。编辑后重新生成的刀轨如图 4-142 所示。

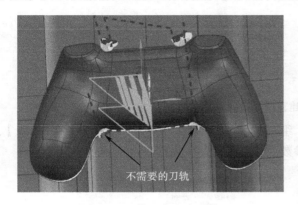
图 4-141 第 1 次生成刀轨

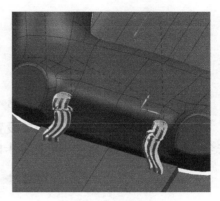
图 4-142 编辑后重新生成刀轨

4.4.4 反面半精加工

1. 区域轮廓铣

区域轮廓铣是使用区域铣削驱动方法来加工切削区域中的曲面,在加工过程中刀轴方向保持固定。需要指定部件几何体,选择面以指定切削区域,编辑驱动方法以指定切削模式。建议用于精加工特定区域。

1)创建工序。单击【创建工序】命令;然后【类型】选择【mill_contour】,【工序子类型】选择"区域轮廓铣",其余参数设置如图 4-143 所示,再单击【确定】。

2)指定切削区域。单击【区域轮廓铣】对话框中【几何体】组中的"指定切削区域"按钮,弹出【切削区域】对话框,选择图 4-144 所示的面作为切削区域。

图 4-143 创建区域轮廓铣工序

图 4-144 指定切削区域

3）创建刀具。创建刀具 R3，参数如图 4-145 所示。

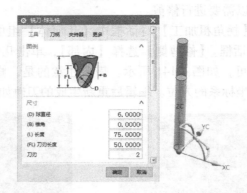

图 4-145　创建直径为 6mm 的球刀

4）设置切削参数。单击【刀轨设置】组中的"切削参数"按钮，【策略】、【余量】、【安全设置】和【更多】这 4 个选项卡的参数设置如图 4-146 所示。

图 4-146　设置切削参数

5）设置非切削移动。将进刀半径设为"20"，如图 4-147 所示。

6）设置进给率和速度。输入主轴速度为"6500"，输入切削进给率为"1500"，然后单击主轴转速后面的"基于此值计算进给和速度"按钮，如图 4-148 所示。

7）生成刀轨。单击【操作】组中的"生成"按钮，生成图 4-149 所示的刀轨，经检查确认该刀轨不会穿透分型面。

  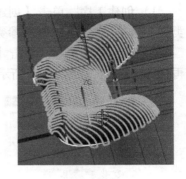

图 4-147　非切削移动　　　　　图 4-148　进给率和速度　　　　图 4-149　观察刀轨

8）编辑刀轨。单击【驱动方法】组中的"编辑"按钮，在弹出的【区域铣削驱动方法】对话框中设置【最大距离】为"0.3"，如图 4-150 所示。重新生成的刀轨如图 4-151 所示。观察刀轨后发现，游戏手柄有两个位置没有加工到，需添加刀轨对其进行加工。

图 4-150　区域铣削驱动方法

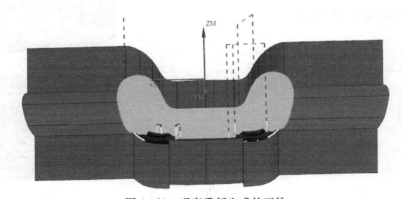

图 4-151　观察重新生成的刀轨

2. 流线

流线是指使用流曲线和交叉曲线来引导切削模式并遵照驱动几何体形状的固定轴曲面轮廓铣工序。使用流线时要控制光顺切削模式的流和方向。

1）创建工序。单击【创建工序】命令，然后【类型】选择【mill_contour】，【工序子类型】选择"流线"，其余参数设置如图 4-152 所示，最后单击【确定】。

2）指定切削区域。单击【流线】对话框中【几何体】组中的"指定切削区域"按钮，再选择区域 1，如图 4-153 所示。

图 4-152 创建【流线】工序

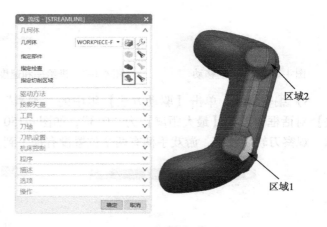

图 4-153 设置切削区域

3）设置流线驱动方法。单击【驱动方法】组中的"编辑"按钮，将【选择方法】设为【自动】。单击"指定切削方向"按钮，然后选择图 4-154c 所示圆圈内的一个箭头，表示切削方式为从上到下。【刀具位置】设为【相切】，【切削模式】设为【往复】，步距的数量设为"30"。【切削步长】设为【公差】，如图 4-154 所示。

a）

b）

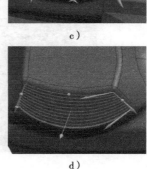

c）

d）

图 4-154 设置流线驱动方法

4）设置投影矢量。【矢量】选择【垂直于驱动体】，如图 4-155 所示。

5）设置刀具。【刀具】选择球头铣刀【R3】，如图 4-155 所示。

6）设置刀轴。【轴】选择【动态】，然后将坐标轴倾斜 5°，如图 4-155 所示。

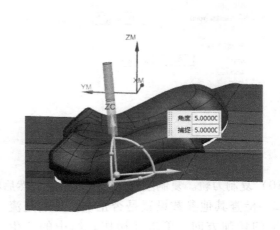

图 4-155　设置投影矢量和刀轴

7）设置切削参数。单击【刀轨设置】组中的"切削参数"按钮；然后输入【部件余量】为"0.2"，输入【内公差】和【外公差】均为"0.01"，【过切时】设为【跳过】，勾选【优化刀轨】，如图 4-156 所示。

图 4-156　设置切削参数

8）设置进给率和速度。输入主轴速度为"6500"，输入切削进给率为"1500"，单击主轴转速后面的"基于此值计算进给和速度"按钮，如图 4-157 所示。

9）生成刀轨。单击【操作】组中的"生成"按钮，生成图 4-158 所示的刀轨。

图 4-157 设置进给率和速度

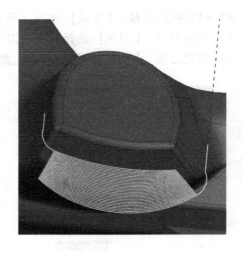

图 4-158 生成刀轨

10)复制刀轨。复制刚创建的流线刀轨,然后将切削区域由图 4-153 所示的区域 1 改为区域 2,检查其他参数设置是否正确,特别是流线驱动的切削方向。单击【操作】组中的"生成"按钮,重新生成刀轨。创建的两个流线程序的刀轨如图 4-159 所示。

11)干涉检查。切换到【建模】模块后,使用【拉伸】命令创建一个边框来模拟毛坯余量,以检验加工时刀具是否会与毛坯边框发生干涉,如图 4-160 和图 4-161 所示。

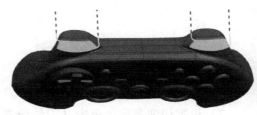

图 4-159 两个流线程序的刀轨

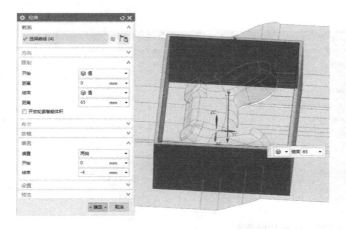

图 4-160 拉伸创建毛坯余量

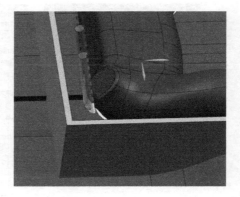

图 4-161 检查刀具与毛坯余量是否干涉

4.4.5 反面精加工

反面精加工是在反面半精加工的基础上进行的,具体操作步骤如下。

1)复制刀轨。复制反面半精加工中创建的 3 个刀轨,将其内部粘贴后得到 3 个刀轨,

分别是 CONTOUR_AREA_COPY、STREAMLINE_COPY_1 和 STREAMLINE_COPY_COPY，如图 4-162 所示。

图 4-162　复制刀轨后内部粘贴

2）修改区域轮廓铣刀轨。双击工序导航器中的【CONTOUR_AREA_COPY】。

单击【驱动方法】组中的"编辑"按钮，然后修改【最大距离】、【步距已应用】和【与 XC 的夹角】，如图 4-163 所示。

图 4-163　区域铣削驱动方法

在【刀轨设置】组中，将【部件余量】设为"0"；并将主轴转速设为"6500"，将切削进给率设为"1000"，再单击主轴转速后面的"基于此值计算进给和速度"按钮，如图 4-164 所示。

图 4-164　余量和切削进给率设置

单击【操作】组中的"生成"按钮,生成图 4-165 所示的刀轨。

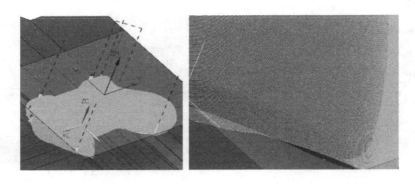

图 4-165 生成刀轨

3)修改流线刀轨。双击工序导航器中的【STREAMLINE_COPY_1】。

单击【驱动方法】组中的"编辑"按钮,将【步距数】设为"50",将【内公差】和【外公差】都设为"0.01",如图 4-166 所示。

在【刀轨设置】组中,将【部件余量】设为"0";将主轴转速设为"6500",将切削进给率设为"1000",再单击主轴转速后面的"基于此值计算进给和速度"按钮,如图 4-164 所示。

单击【操作】组中的"生成"按钮,生成图 4-167 所示的刀轨。

另一个流线刀轨【STREAMLINE_COPY_COPY】的修改方法相同,此处不再赘述。

图 4-166 流线驱动方法

图 4-167 修改后的流线刀轨

4)查看铣削效果。在工序导航器中选中最后一个刀轨【STREAMLINE_COPY_COPY】,单击右键后选择【工件】/【按颜色显示厚度】命令,查看铣削效果,如图 4-168 所示。

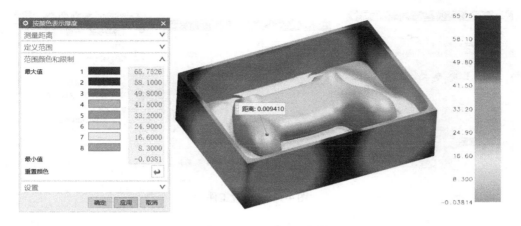

图 4-168　按颜色显示厚度

4.4.6　正面几何体设置

1）设置机床坐标系。切换至工序导航器的几何视图，然后单击【创建几何体】命令，【几何体子类型】选择"MCS"，输入名称为【MCS-Z】，如图 4-169 所示，最后单击【确定】。

在弹出的【MCS】对话框中，【安全设置选项】选择【刨】，然后选择毛坯顶面并将其向上偏置 50mm，将偏置得到的平面作为安全平面，如图 4-170 所示。

单击【机床坐标系】组中的"CSYS 对话框"按钮，然后选择毛坯顶面中心作为坐标原点，并调整坐标轴方向，使 ZM 轴朝上，使 XM 轴与毛坯的长边对齐。

图 4-169　创建几何体

 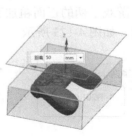
图 4-170　设置安全平面

2）设置工件。单击【创建几何体】命令，然后【几何体子类型】选择"WORK-PIECE"，输入名称为【WORKPIECE-Z】，如图 4-171 所示。单击【确定】后弹出【工件】对话框，依次选择游戏手柄模型作为【指定部件】，选择方块作为【指定毛坯】，再单击【确定】。

3）移动分型面。打开第 3 层中的初始分型面，然后使用【移动对象】命令，选择【复制原先的】，将其向下（往游戏手柄的反面方向）移动 3.3mm，如图 4-172 所示。使用【移动至图层】命令，将移动后的分型面放置到第 5 层。

图 4-171 设置工件

图 4-172 移动分型面

4.4.7 正面粗加工

游戏手柄的正面粗加工由 4 道工序组成,工序类型都是型腔铣,但是所使用的刀具各不相同,如图 4-173 所示。

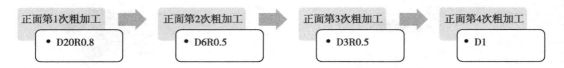

图 4-173 正面粗加工工序

1. 正面第 1 次粗加工

1)复制刀轨。选择反面粗加工刀轨【CAVITY_MILL】,将其复制后【内部粘贴】到【WORKPIECE-Z】中,得到刀轨【CAVITY_MILL_COPY】,如图 4-174 所示。

2)编辑刀轨。双击【CAVITY_MILL_COPY】,弹出【型腔铣】对话框。

单击"指定检查"按钮,选择移动后的分型面作为检查几何体。单击"指定修剪边界"按钮,在弹出的【修剪边界】对话框中取消选择【余量】,如图 4-175 所示。

单击"切削参数"按钮,将【刀路方向】设为【向内】,如图 4-176 所示。

单击"非切削移动"按钮,将开放区域中的【进刀类型】设为【圆弧】,并输入【半径】为刀具直径的 20%,如图 4-176 所示。

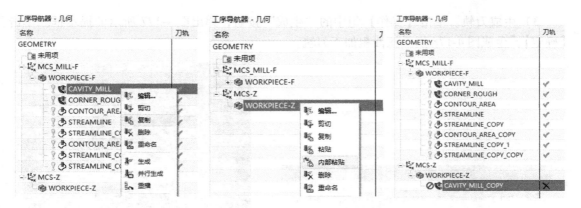

图 4-174　复制粗加工刀轨后将其内部粘贴

图 4-175　设置几何体

图 4-176　切削参数

3）生成刀轨。单击【操作】组中的"生成"按钮,弹出图4-177所示的提示框。单击【确定】,生成图4-178所示正面粗加工刀轨。

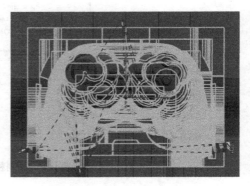

图4-177 【操作编辑】提示框　　　　　　图4-178 正面粗加工刀轨

2. 正面第2次粗加工

在第1次粗加工的基础上,用刀具D6R0.5对正面进行第2次粗加工。生成的刀轨如图4-179所示。

扫描二维码可观看操作过程。

图4-179 正面第2次粗加工刀轨　　　　　　视频:正面第2次粗加工

3. 正面第3次粗加工

在第2次粗加工的基础上,用刀具D3R0.5对正面进行第3次粗加工。生成的刀轨如图4-180所示。

扫描二维码可观看操作过程。

图4-180 正面第3次粗加工刀轨　　　　　　视频:正面第3次粗加工

4. 正面第 4 次粗加工

在第 3 次粗加工的基础上，用刀具 D1 对正面进行第 4 次粗加工。生成的刀轨如图 4-181 所示。

扫描二维码可观看操作过程。

图 4-181　正面第 4 次粗加工刀轨　　　　　　　视频：正面第 4 次粗加工

4.4.8　正面半精加工

1）复制刀轨。选择反面半精加工刀轨【CONTOUR_AREA】，将其复制后【内部粘贴】到【WORKPIECE-Z】中，得到刀轨【CONTOUR_AREA_COPY_1】。

2）指定检查。单击"指定检查"按钮，选择移动后的分型面作为检查几何体。

3）选择切削区域。单击"指定切削区域"按钮，重新指定切削区域，如图 4-182 所示。

4）生成刀轨。单击【操作】组中的"生成"按钮，生成图 4-183 所示正面半精加工刀轨。

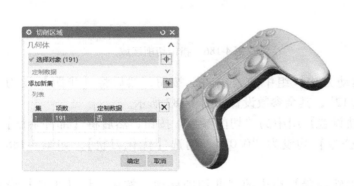
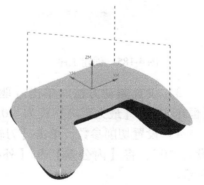

图 4-182　指定切削区域　　　　　　　　图 4-183　正面半精加工刀轨

4.4.9　正面精加工

游戏手柄正面精加工的流程如图 4-184 所示。

1. 手柄面精加工

1）创建工序。单击【创建工序】命令，然后【类型】选择【mill_contour】，【工序子

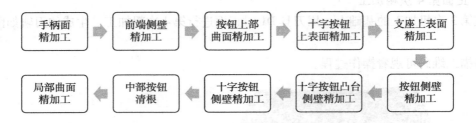

图 4-184 游戏手柄正面精加工的流程

类型】选择"区域轮廓铣",其余参数设置如图 4-185 所示,最后单击【确定】,打开【区域轮廓铣】对话框。

2)指定切削区域。单击"指定切削区域"按钮,指定图 4-186 所示的切削区域。

图 4-185 创建工序　　　　　　　　　图 4-186 指定切削区域

3)设置驱动方法。单击【驱动方法】组中的"编辑"按钮,然后将【步距】设为【恒定】,将【最大距离】设为"0.12",其余参数设置如图 4-187 所示。

4)设置切削参数。单击【刀轨设置】组中的"切削参数"按钮,然后将【部件余量】设为"0",将【内公差】和【外公差】均设为"0.01",勾选【优化刀轨】,如图 4-188 所示。

5)设置非切削移动。单击【刀轨设置】组中的"非切削移动"按钮,将【半径】设为刀具直径的 10%,如图 4-189 所示。

6)设置进给率和速度。单击【刀轨设置】组中的"进给率和速度"按钮,将主轴转速设为"8000",将切削进给率设为"1000",再单击主轴转速后面的"基于此值计算进给和速度"按钮,如图 4-190 所示。

7)生成刀轨。单击【操作】组中的"生成"按钮,生成图 4-191 所示的刀轨。

图 4-187 设置驱动方法

图 4-188 设置切削参数

图 4-189 设置非切削移动

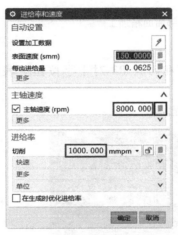

图 4-190 设置进给率和速度

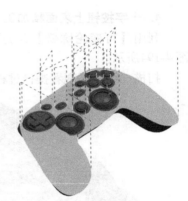

图 4-191 生成刀轨

2. 前端侧壁精加工

利用【深度轮廓加工】工序和刀具 D6R0.5 对前端侧壁进行精加工。生成的刀轨如图 4-192 所示。

扫描二维码可观看操作过程。

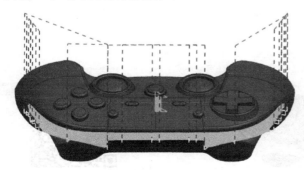

图 4-192 前端侧壁精加工

视频：前端侧壁精加工

3. 按钮上部曲面精加工

利用【区域轮廓铣】工序和刀具 D3R0.5 对按钮上部曲面进行精加工。生成的刀轨如图 4-193 所示。

扫描二维码可观看操作过程。

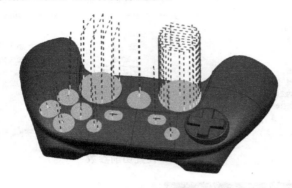

图 4-193　按钮上部曲面精加工

视频：按钮上部曲面精加工

4. 十字按钮上表面精加工

利用【可变轮廓铣】工序和刀具 R3 对十字按钮上表面进行精加工。生成的刀轨如图 4-194 所示。

扫描二维码可观看操作过程。

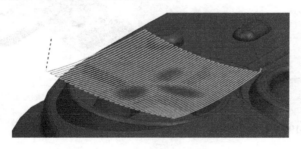

图 4-194　十字按钮上表面精加工

视频：十字按钮上表面精加工

5. 支座上表面精加工

利用【区域轮廓铣】工序和刀具 D6R0.5 对支座上表面进行精加工。生成的刀轨如图 4-195 所示。

扫描二维码可观看操作过程。

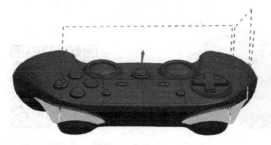

图 4-195　支座上表面精加工

视频：支座上表面精加工

6. 按钮侧壁精加工

利用【可变流线铣】工序和刀具 D2 对按钮侧壁进行精加工。生成的刀轨如图 4-196 所示。
扫描二维码可观看操作过程。

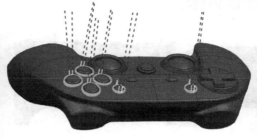

图 4-196 按钮侧壁精加工   视频：按钮侧壁精加工

7. 十字按钮凸台侧壁精加工

利用【外形轮廓铣】工序和刀具 D8 对十字按钮凸台侧壁进行精加工。生成的刀轨如图 4-197 所示。

扫描二维码可观看操作过程。

图 4-197 十字按钮凸台侧壁精加工   视频：十字按钮凸台侧壁精加工

8. 十字按钮侧壁精加工

利用【外形轮廓铣】工序和刀具 D2 对十字按钮侧壁进行精加工。生成的刀轨如图 4-198 所示。

扫描二维码可观看操作过程。

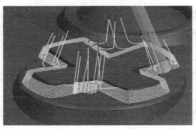

图 4-198 十字按钮侧壁精加工   视频：十字按钮侧壁精加工

9. 中部按钮清根

利用【清根参考刀具】工序和刀具 R2、R1、R0.5 对中部按钮进行 3 次清根。生成的刀轨如图 4-199 所示。

扫描二维码可观看操作过程。

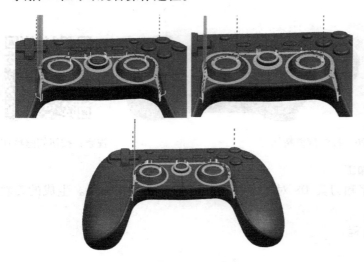

图 4-199　中部按钮清根　　　　　　　　　　　　　　　视频：中部按钮清根

10. 局部曲面精加工

1）检查铣削效果。在工序导航器中选择【FLOWCUT_REF_TOOL_COPY_COPY】，单击右键后选择【工件】/【按颜色显示厚度】命令。经检查发现十字按钮的壁面和两个圆形大按钮的侧面还有 0.2mm 左右的余量，如图 4-200 所示。

图 4-200　检查铣削效果

2）创建工序。单击【创建工序】命令，然后【类型】选择【mill_contour】，【工序子类型】选择"区域轮廓铣"，其余参数设置如图 4-201 所示，单击【确定】后打开【区域轮廓铣】对话框。

3）指定切削区域。单击【几何体】组中的"指定切削区域"按钮，指定图 4-202 所示的切削区域。

4）指定修剪边界。单击【几何体】组中的"指定修剪边界"按钮，然后将【修剪侧】设为【外部】，表示修剪边界外部的刀轨；通过【曲线】的方式指定十字按钮处的修剪边界，通过【点】的方式指定两个大按钮处的修剪边界，如图 4-203 所示。

项目4 游戏手柄

图 4-201 创建工序　　　　　　　　图 4-202 指定切削区域

图 4-203 指定修剪边界

5）设置驱动方法。将【非陡峭切削模式】设为【跟随周边】，将【步距】设为【恒定】并输入【最大距离】为"0.12mm"，如图 4-204a 所示。

6）设置余量。将【部件余量】设为"0"，如图 4-204b 所示。

7）设置进给率和速度。将主轴速度设为"12000"，将切削进给率设为"800"，如图 4-204c 所示。

a）　　　　　　　　　　b）　　　　　　　　　　c）

图 4-204 参数设置

173

8)生成刀轨。单击【操作】组中的"生成"按钮,生成图4-205所示的刀轨。

4.4.10 仿真加工

选择工序导航器程序顺序视图中的【NC_PROGRAM】,然后单击【确认刀轨】命令,弹出【刀轨可视化】对话框。单击【3D动态】选项卡中的"播放"按钮,可以查看切削效果,如图4-206所示。

单击【刀轨可视化】对话框中的【按颜色显示厚度】按钮,可以查看加工后的毛坯余量,如图4-207所示。

图4-205 生成刀轨

 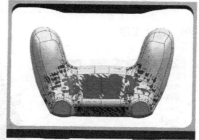

图4-206 加工仿真效果

 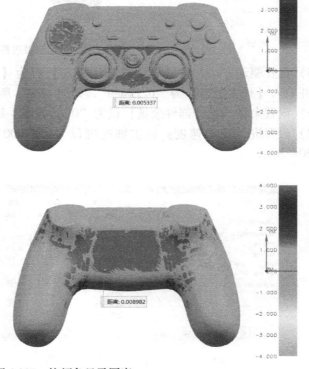

图4-207 按颜色显示厚度

参考文献

[1] 潘常春,李加文,卢骏. 逆向工程项目实践[M]. 杭州:浙江大学出版社,2014.

[2] 单岩,周文学,罗晓晔,等. UG NX6.0立体词典:产品建模[M]. 杭州:浙江大学出版社,2010.

[3] 王卫兵,林华钊,王志明. UG NX6.0立体词典:数控编程[M]. 杭州:浙江大学出版社,2010.

[4] 单岩,吴立军. Imageware逆向造型基础教程[M]. 3版. 北京:清华大学出版社,2018.

参考文献

[1] 陈香希, 李加文, 宋浩. 油田工程项目关键 [M]. 北京: 清华大学出版社, 2016.

[2] 申静, 赵方琨, 董召奎, 等. OC PKM 的系统构架与实际应用 [M]. 北京: 清华大学出版社, 2010.

[3] 王开民, 林春阳, 毛远洋. OC TXO1O 立体式力学验收标准 [M]. 北京: 清华大学出版社, 2010.

[4] 单青, 吴立波. Jumpware 视图及数据控制技术 [M]. 3 版. 北京: 清华大学出版社, 2018.

Figure 4-707. Spur flywheel by Quno.

Figure 4-708. Chuck ½ capacity, swing clip.

TEXTBOOKS FOR MACHINERY SPECIALITY OF HIGHER VOCATIONAL EDUCATION

高等职业教育机械类专业系列教材

3D Digital Design and Manufacturing

三维数字化设计与制造

Edited by Yudong Liang

梁宇栋 编

CHINA MACHINE PRESS

机械工业出版社

Based on typical products, this book mainly introduces the technical process of 3D digital design and manufacturing, including 3D reverse scanning, reverse mode selection, reverse modeling, as well as NC machining programming and simulation.

This book contains four projects, composed of 15 tasks. The project cases progress from simple ones to more complex ones, and the difficulty is gradually increased. The contents include data collection by virtue of scanning devices and supporting software, data processing based on Geomagic Design X or Geomagic Wrap, reverse modeling based on Geomagic Design X or Siemens NX, product comparison and analysis based on Geomagic Control X, as well as NC programming and machining based on Siemens NX.

In order to meet the needs of teaching, this book is provided with supporting teaching resources by SUNNYTECH and information-based teaching tools (Xuebei Classroom) offered by Xuebei Technology, which can improve the teaching efficiency and strengthen the teaching effect with richer contents, more diverse form and simpler teaching.

This book can be used as a textbook for 3D digital design and manufacturing course in higher vocational colleges and various kinds of skill trainings. It can also be used as a reference book for related engineers and technicians.

The accompanying electronic courseware is also available, which is free to download at www. cmpedu. com for teachers who choose this book as textbook. Hotline: 010-88379375.

Preface

3D digital design and manufacturing is a comprehensive technology, involving data modeling by virtue of data-based design software, reverse modeling, NC programming and machining. Therefore, it has been widely applied in various industrial fields such as aviation, aerospace, automobile, general machinery and electronics. Nowadays, there is a great demand for talents excelled at the 3D digital design and manufacturing technology.

This textbook, set out in a "project-driven, task-led" manner, introduces in detail the technical process of 3D digital design and manufacturing with typical products as teaching cases and in combination with related domestic competition projects and requirements in the higher vocational colleges. The contents include:

1) Data collection by scanning products with general scanning devices.

2) Data processing with Geomagic Design X or Geomagic Wrap to meet the needs of reverse modeling.

3) Reverse modeling with Geomagic Design X or Siemens NX based on the processed data.

4) Product analysis and comparison with Geomagic Control X between the processed data and the model out of reverse modeling, which ensures that the model be observed clearly and easily.

5) NC programming and machining with Siemens NX.

The author has been engaged in teaching and research of CAD/CAM/CAE for years and has rich working and teaching experience in 3D digital design and manufacturing. This book has been improved in structure and contents in its compiling process by reference to the experience of several senior modeling engineers from many companies, such as SUNNYTECH. Through 4 projects and 15 tasks, the knowledge, skills and methods of relevant specialties are organically combined, and the students can practice synchronously while learning, so as to improve their practical operation ability and apply what they have learned.

This book can be used as a textbook for 3D digital design and manufacturing course in higher vocational colleges and training materials for relevant competitions of vocational colleges, as well as a reference book for engineers of manufacturing enterprises in mould, machinery, household appliances and other industries, as well as industrial design companies.

This book is compiled by Yudong Liang of Tianjin Vocational College of Mechanics and Electricity. In the process of compiling, it has been helped by Professor Yan Shan of Hangzhou SUNNYTECH Development Co., Ltd.

There must be some insufficiency to be further corrected and improved in this book. Valuable opinions and advices from readers and professionals shall be highly appreciated and indispensable to its future enhancement.

Editor

Contents

Preface

Project 1   Model of the FIFA World Cup Trophy ·· 1

  Task 1   Data Collection ·· 1

    1.1.1   Scanner Calibration ··· 3

    1.1.2   Data Scanning ··· 6

  Task 2   Data Processing ·· 10

  Task 3   Machining Programming ··· 16

    1.3.1   Initial Setting ·· 16

    1.3.2   Roughing ·· 21

    1.3.3   Second Roughing ··· 26

    1.3.4   Semi-finishing ··· 28

    1.3.5   Finishing ·· 32

    1.3.6   Machining Simulation ··· 35

Project 2   Turbine Wheel ·· 38

  Task 1   Data Collection ··· 38

    2.1.1   Scanner Calibration ·· 38

    2.1.2   Stick Mark Points ·· 41

    2.1.3   Scan Mark Points ·· 41

    2.1.4   Scan Laser Points ·· 43

  Task 2   Data Processing ·· 46

  Task 3   Reverse Modeling ·· 51

    2.3.1   Main Body Modeling ·· 51

    2.3.2   Blade Modeling ··· 57

  Task 4   Machining Programming ··· 71

    2.4.1   Machining Analysis ·· 71

    2.4.2   Preparation for Programming ·· 71

    2.4.3   Multi-blade Roughing ··· 77

    2.4.4   Main Blade Finishing ··· 80

    2.4.5   Splitter Blade Finishing ··· 81

2. 4. 6	Hub Finishing	82
2. 4. 7	Blend Machining	83
2. 4. 8	Copy Program	85
2. 4. 9	Machining Simulation	87

Project 3　Toothpaste Dispenser 88

Task 1　Data Collection 88

3. 1. 1	Stick Mark Points	88
3. 1. 2	Scan Mark Points	89
3. 1. 3	Scan Laser Points	90

Task 2　Data Processing 93

3. 2. 1	Import Model	93
3. 2. 2	Scanning Data Processing of the Upper Cover	93
3. 2. 3	Scanning Data Processing of the Lower Cover	95

Task 3　Reverse Modeling 96

3. 3. 1	Import Model	96
3. 3. 2	Determine Coordinates	97
3. 3. 3	Reverse Modeling of the Lower Cover	100
3. 3. 4	Reverse Modeling of the Upper Cover	110
3. 3. 5	Detail Processing	112

Task 4　Product Analysis and Comparison 113

3. 4. 1	Data Import	113
3. 4. 2	Initial Model Alignment	113
3. 4. 3	Best Fit Alignment	114
3. 4. 4	3D Comparison	114
3. 4. 5	2D Comparison	116
3. 4. 6	Comparison Point	116
3. 4. 7	Cross Section	117
3. 4. 8	Marking Function	118
3. 4. 9	Generate Report	123
3. 4. 10	Export Report	124

Project 4　Gamepad 125

Task 1　Data Collection 125

4. 1. 1	Scanner Calibration	126
4. 1. 2	Paste the Mark Points	131
4. 1. 3	Data Scanning	132

Task 2　Data Processing 135

| 4. 2. 1 | Manual Stitching | 135 |

3D Digital Design and Manufacturing

4. 2. 2	Data Wrapping	138
Task 3	Reverse Modeling	141
4. 3. 1	Determine Coordinates	142
4. 3. 2	Pavement	144
4. 3. 3	Draw the Rough Shape	149
4. 3. 4	Draw Detail Features	154
Task 4	Machining Programming	155
4. 4. 1	Process Analysis	155
4. 4. 2	Preparation for Programming	156
4. 4. 3	Back Side Roughing	162
4. 4. 4	Back Side Semi-finishing	168
4. 4. 5	Back Side Finishing	174
4. 4. 6	Front Side Geometry Setting	177
4. 4. 7	Front Side Roughing	178
4. 4. 8	Front Side Semi-finishing	181
4. 4. 9	Front Side Finishing	182
4. 4. 10	Machining Simulation	188

Project 1　Model of the FIFA World Cup Trophy

The FIFA World Cup Trophy (also named Hercules Cup) represents the highest honors in football games. Designed by Italian artist Silvio Gazzaniga, the trophy looks like two Hercules lifting the Earth. The lines extrude upward from the base and rise in spiral curves. The images of the two athletes also emerge as Hercules. They extrude upward and lift the Earth. On this dynamic and compact trophy, the image of two athletes who are excited after victory is carved.

This project takes the model of the FIFA World Cup Trophy (Figure 1-1) as the carrier, using the desktop scanner EinScan-S to carry on the 3D scanning and Geomagic Design X 2016 as the data processing software,

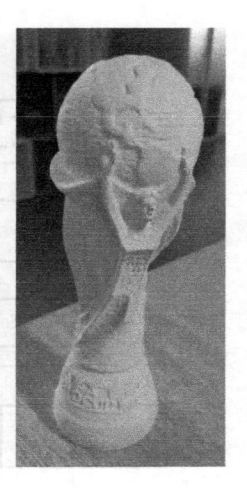

Figure 1-1　The physical and the model of the FIFA World Cup Trophy

and finally completing the numerical control machining simulation in the Siemens NX 10 software. Project implementation process is shown in Figure 1-2.

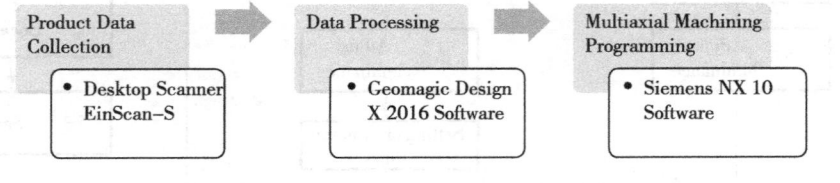

Figure 1-2　Project implementation process

Task 1　Data Collection

The model data is collected using EinScan-S, a high precision white light desktop 3D scanner. Its scanning accuracy is within 0.1mm. It has two scanning modes: turntable scanning and fixed scanning. In the turntable scanning mode, the 360° turntable takes only 2 minutes to scan a circle,

1 ◀

3D Digital Design and Manufacturing

and the maximum scanning range is 200mm × 200mm × 200mm. In the fixed scanning mode, the maximum scanning range is 700mm × 700mm × 700mm. EinScan-S weighs 3.5kg in total, so it is light and portable. The operation process of data collection using desktop scanner EinScan-S is shown in Figure 1-3.

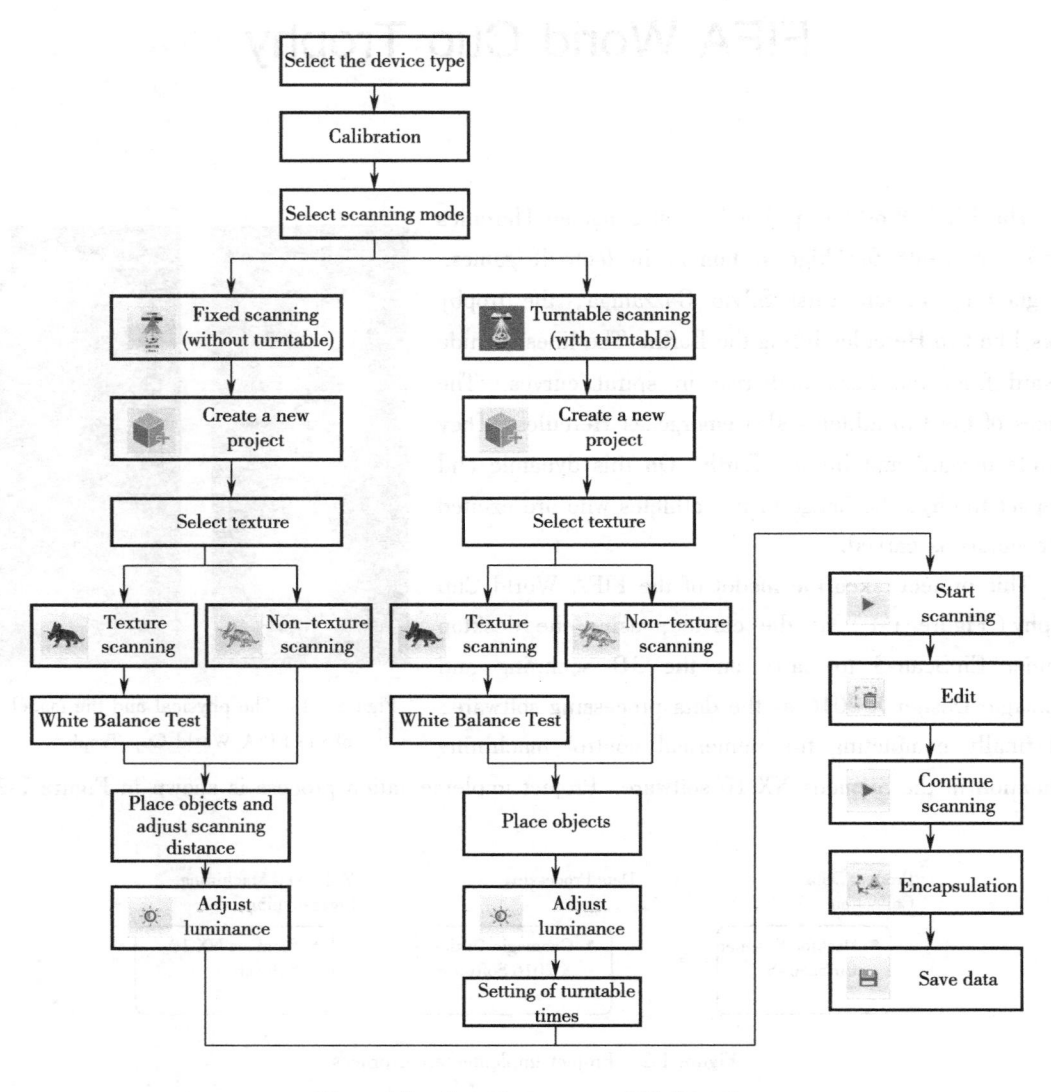

Figure 1-3   Operation process of EinScan-S

Transparent or reflective objects cannot be scanned directly, and contrast enhancers need to be sprayed. There is a layer of electroplating on the outer surface of the model of the FIFA World Cup Trophy, which is glittering and reflective. If no contrast enhancer is sprayed, only 420 points and 267 meshes are obtained for one scanning. The scanning effect is not good, as shown in Figure 1-4a. After spraying contrast enhancer, 334207 points and 322738 meshes can be obtained for one scanning, as shown in Figure 1-4b, the scanning effect is much better.

▶ 2

Project 1　Model of the FIFA World Cup Trophy

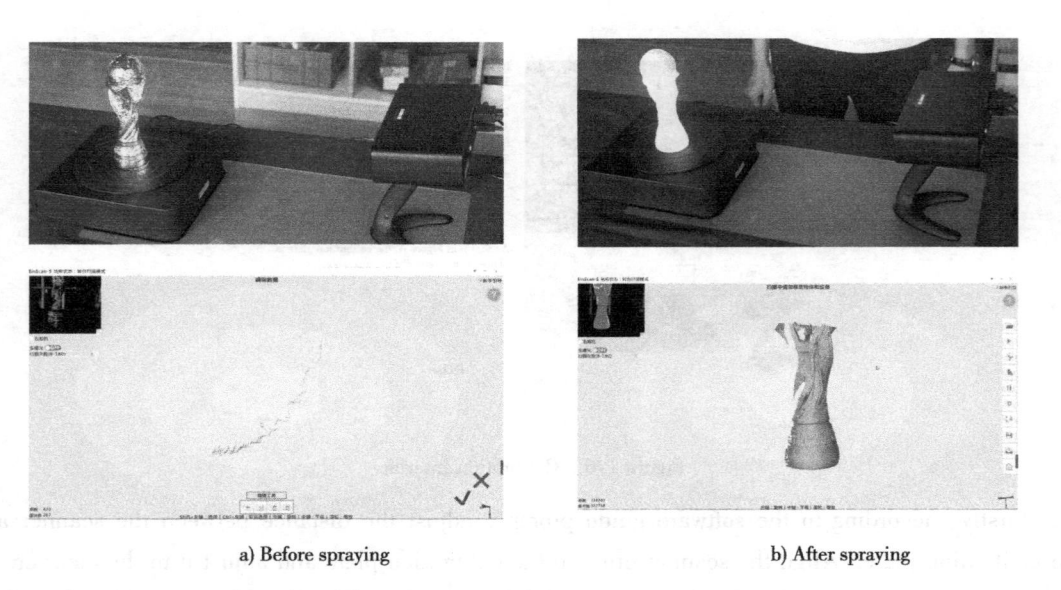

a) Before spraying　　　　　　　　　　　b) After spraying

Figure 1-4　Comparison of scanning effect before and after spraying contrast enhancer

1.1.1　Scanner Calibration

After installing the scanning software, calibration should be carried out before the first scan. Without calibration, the scanning mode cannot be entered. If there is no calibration, the software will prompt "if there is no calibration data, please calibrate first". Before calibration, attention should be paid to the alignment of edge of the calibration plate with line at the fingered place shown in Figure 1-5 as required, and also with the edge of light.

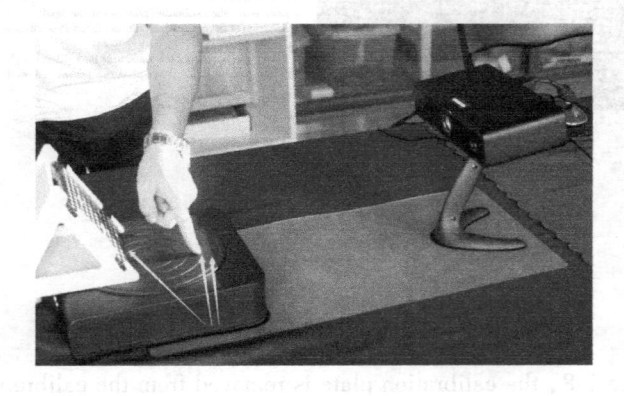

Figure 1-5　Placement of calibration plate

When using EinScan-S software for the first time, it is prompted to select the device type. Select 【EinScan-S】, then click 【Next】, and the calibration interface is displayed, as shown in Figure 1-6.

EinScan-S needs to calibrate three directions, so the calibration plate needs to be placed three times, and placement is carried out according to the software wizard prompts.

3 ◄

3D Digital Design and Manufacturing

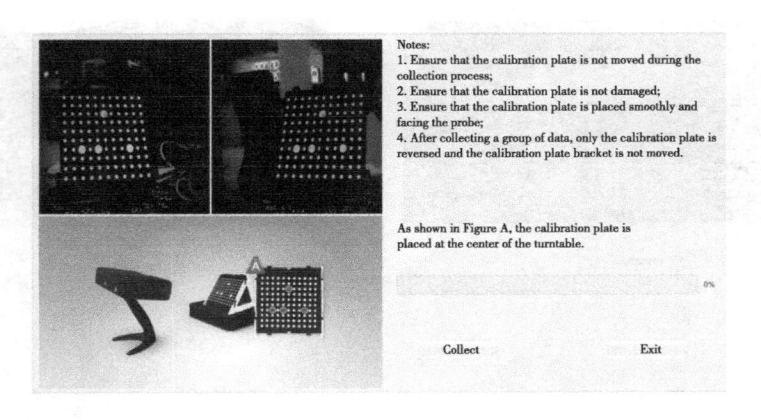

Figure 1-6　Calibration interface

Firstly, according to the software guide prompt, adjust the distance between the scanner and the calibration plate. Align the scanner cross to the calibration plate and adjust it to the clear cross. Place the calibration plate bracket in the center of the turntable. The first placement direction of the calibration plate is shown in Figure 1-6. Ensure that the calibration plate is placed smoothly and facing the probe, then click 【Collect】. The turntable rotates automatically for one circle to collect data. Do not move the calibration plate in the process of collection.

After the collection, the turntable stops and the software interface prompts for Calibration B, as shown in Figure 1-7.

Figure 1-7　Software interface shows Calibration B

As shown in Figure 1-8, the calibration plate is removed from the calibration plate bracket, and the calibration plate is rotated counterclockwise for 90° and embedded in the calibration plate bracket groove. Note: Only the calibration plate is reversed and the calibration plate bracket is not moved.

After Calibration B is completed, the calibration plate is removed from the calibration plate bracket, and the calibration plate is rotated 90° counterclockwise and embedded in the calibration plate bracket groove, as shown in Figure 1-9.

Project 1   Model of the FIFA World Cup Trophy

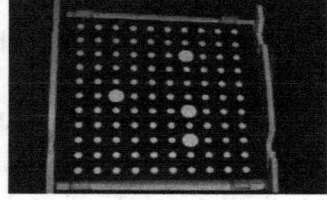
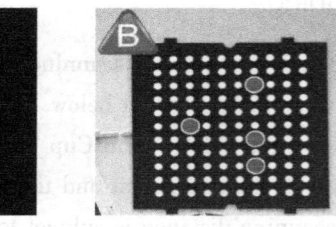

Figure 1-8   Placement position of calibration plate in Calibration B

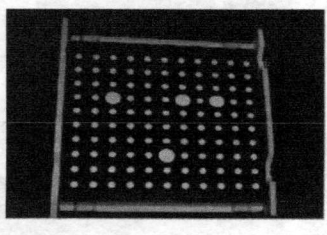
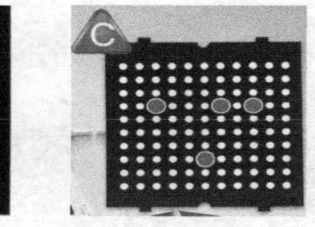

Figure 1-9   Placement position of calibration plate in Calibration C

The collection is completed, and the calibration calculation is carried out, as shown in Figure 1-10.

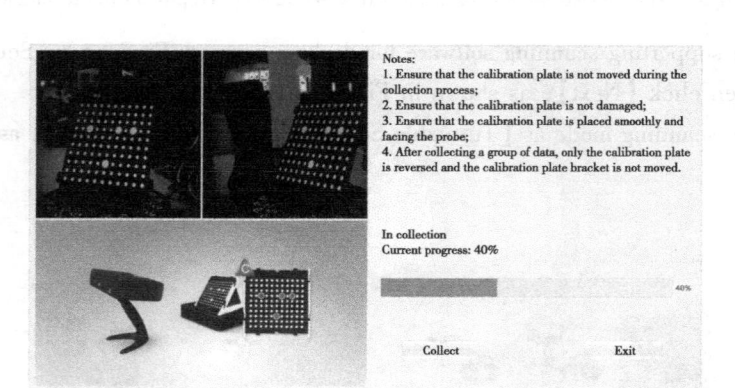

Figure 1-10   Calibration calculation

If the calibration is successful, it will automatically return to the scanning interface. If failure of the calibration is prompted, please follow the above steps and re-calibrate.

The initial calibration must be completed in accordance with the process. In the following situations, the scanner needs to be calibrated again:

① The scanner is used for the first time or used after long time in idle.

② The scanner vibrates seriously during transport.

③ In the process of scanning, there are frequent splicing errors and splicing failures, etc.

④ In the process of scanning, the scanning data is incomplete and the quality of the data is seriously degraded.

3D Digital Design and Manufacturing

1.1.2 Data Scanning

When the calibration is completed, scanning can be started. The turntable scanning mode is selected. The specific steps are described below.

1) Place the model of the FIFA World Cup Trophy on the turntable, as shown in Figure 1-11. Adjust the distance between the equipment and the object to the appropriate working distance (290 ~480mm). The best scanning distance is subject to the distance of projected cross on the scanned object.

Figure 1-11   Place the model of the FIFA World Cup Trophy on the turntable

2) Open the supporting scanning software for desktop scanner EinScan-S. Select the device as 【EinScan-S】, then click 【Next】, as shown in Figure 1-12a.

3) Select the scanning mode as 【Turntable Scanning】, then click 【Next】, as shown in Figure 1-12b.

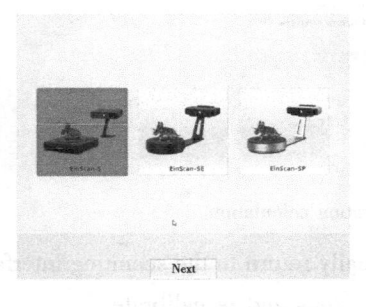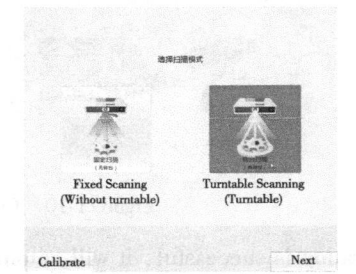

a) Select scanning equipment   b) Select scanning mode

Figure 1-12   Select scanning equipment and scanning mode

4) After clicking 【Create A New Project】, enter the project name, as shown in Figure 1-13.

5) After clicking 【Non-texture Scanning】, click 【Apply】, as shown in Figure 1-14.

6) According to the brightness of the object, select the appropriate brightness setting. You can view the current brightness in real time in the left preview window. It is preferable that the color is white or a little reddish. Set it to the brightest, as shown in Figure 1-15, then click 【Apply】.

▶ 6

Project 1　Model of the FIFA World Cup Trophy

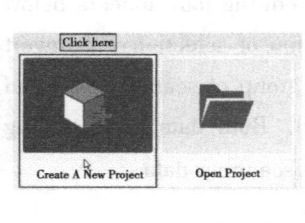

a) Create a new project

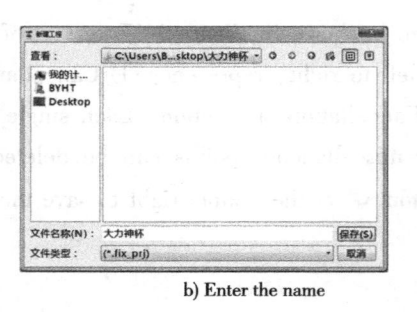

b) Enter the name

Figure 1-13　Create a new project and enter name

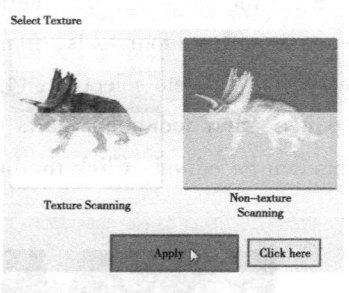

Figure 1-14　Select non-texture scanning

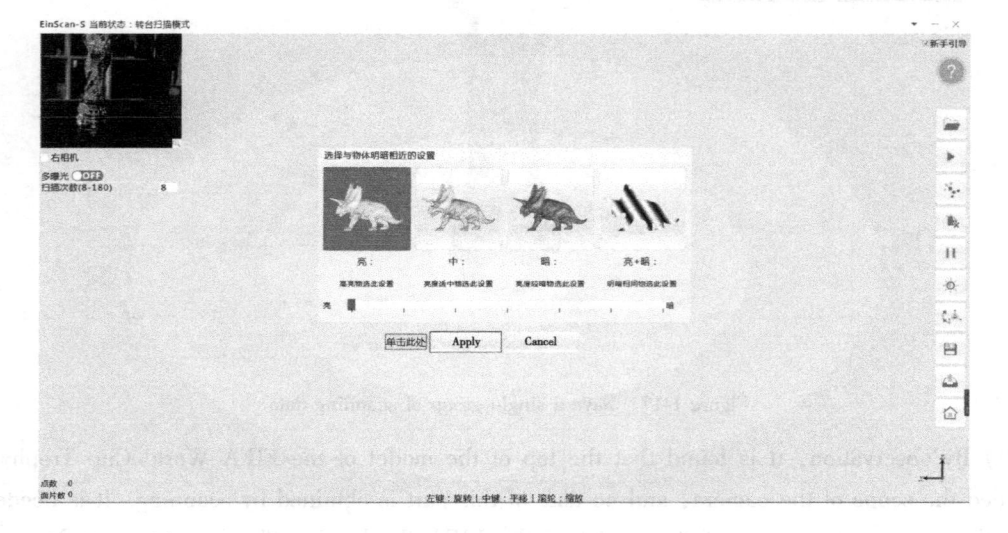

Figure 1-15　Adjust brightness

7) The number of scanning shall be the default value, i. e. 8. Click the "Scan" button to start scanning. Do not move objects and devices in the process of scanning, as shown in Figure 1-16.

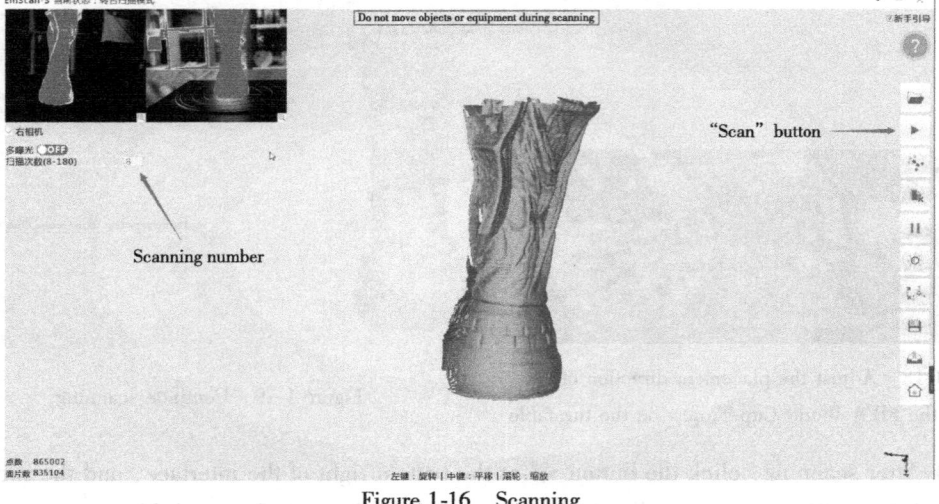

Figure 1-16　Scanning

7 ◀

3D Digital Design and Manufacturing

8) After scanning, the interface is shown in Figure 1-17. A set of editing tool appears below the screen. These four tools, from left to right, represent: ① Cancellation of selection; ② Invert selection; ③ Delete selection; ④ Cancellation of deletion. Each single group of scanned data can be edited, and redundant parts or miscellaneous points can be deleted. Both data and marking points can be edited. Click the button ✔ at the bottom right to save the scanning data.

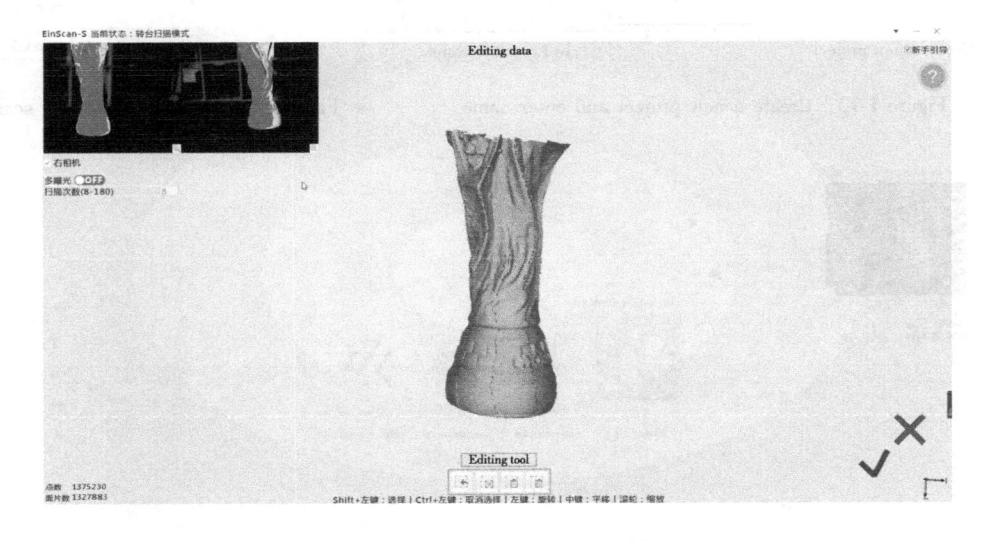

Figure 1-17   Save a single group of scanning data

9) By observation, it is found that the top of the model of the FIFA World Cup Trophy has exceeded the scope of the camera, and no data of this part is obtained by scanning. It is needed to adjust the placement position of the model of the FIFA World Cup Trophy. Place it flat on the turntable, as shown in Figure 1-18, and then click the "Scan" button ▶ to continue scanning, as shown in Figure 1-19.

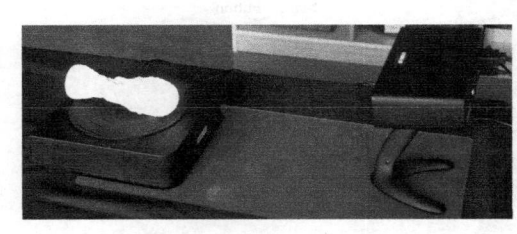

Figure 1-18   Adjust the placement direction of the model of the FIFA World Cup Trophy on the turntable

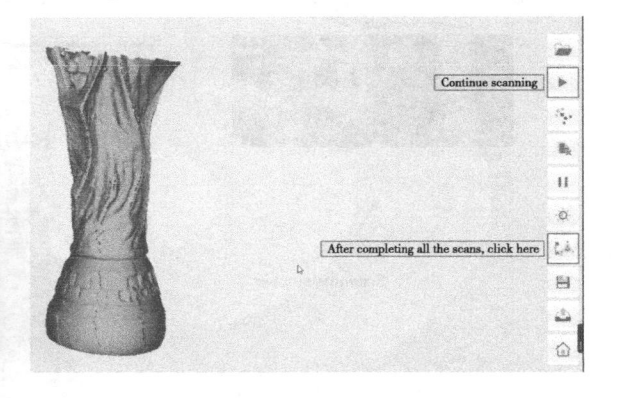

Figure 1-19   Continue scanning

10) After scanning, click the button ✔ at the bottom right of the interface, and the two groups of scanned data will be automatically stitched together, as shown in Figure 1-20.

▶ 8

Project 1　Model of the FIFA World Cup Trophy

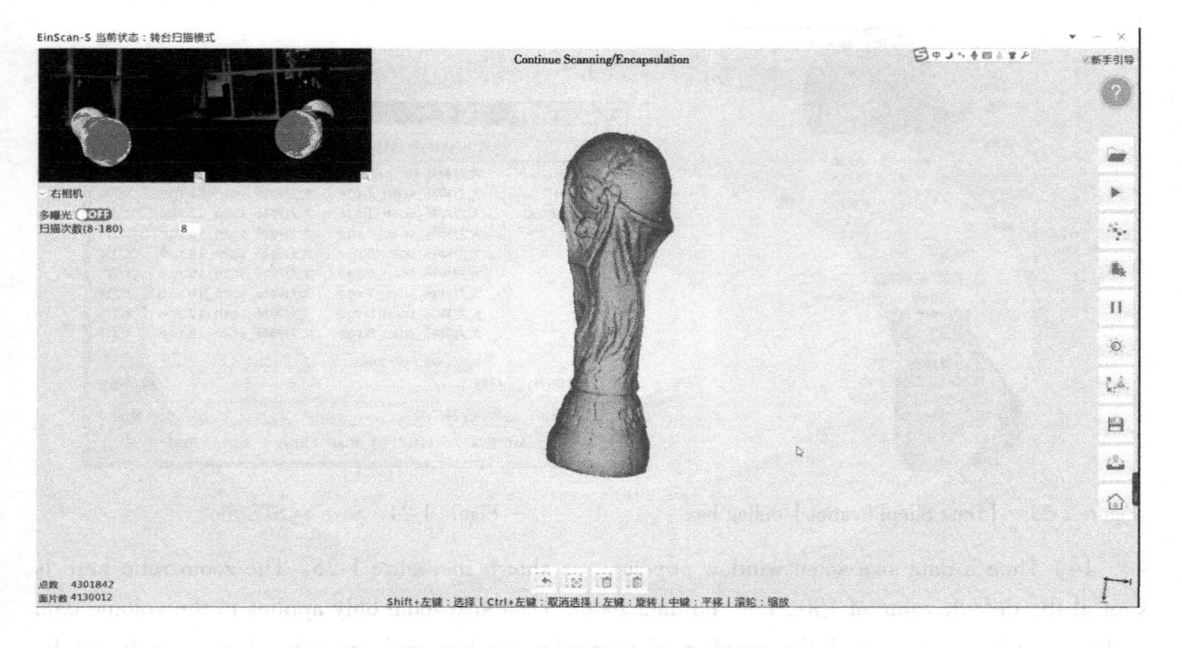

Figure 1-20　Automatic data stitching

11）After the data scanning is completed, click the "Generate Mesh" button ⚒ to encapsulate the data, including closed and non-closed models. Closed encapsulation can be used directly for 3D printing models. Here, select the 【Closed model】, as shown in Figure 1-21, then select 【High Details】, as shown in Figure 1-22. The progress bar of data encapsulation is displayed on the bottom of the software interface.

Figure 1-21　Select closed model　　　　　　　Figure 1-22　Select high details

12）After encapsulation, the 【Data Simplification】 dialog box appears, as shown in Figure 1-23. There is no need to simplify the data here. Just click 【Apply】. It is needed to wait for a certain time due to the amount of data is large, and the progress bar of data processing will also be displayed on the bottom of the software interface.

13）After data processing is completed, save the model data as a ". stl" file, as shown in Figure 1-24.

9 ◀

3D Digital Design and Manufacturing

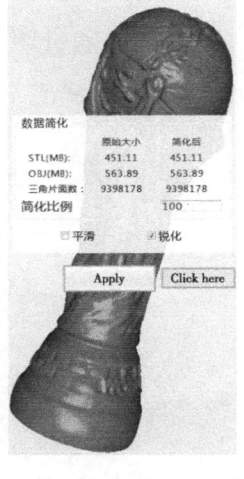

Figure 1-23　【Data Simplification】dialog box

Figure 1-24　Save as STL file

14) Then a data size zoom window appears, as shown in Figure 1-25. The zoom ratio here is kept at the default value of 100, i. e. no data zoom. Data size zoom only applies to the volume data of the scanning object, and the number of triangular meshes and the size of data shall not be reduced.

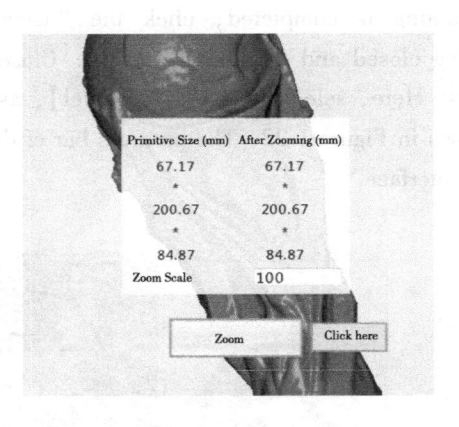

Figure 1-25　Data size zoom window

So far, the data collection of the model of the FIFA World Cup Trophy has been completed.

Task 2　Data Processing

Data processing refers to the optimization of scanned data, such as removal of the miscellaneous points brought on the model surface in the scanning process and noise caused by the external environment, filling the holes caused by the marking points and the data not obtained by the scanner, etc. Here Geomagic Design X is selected as data processing software, and data processing flow is shown in Figure 1-26. The specific steps are described below.

▶ 10

Project 1   Model of the FIFA World Cup Trophy

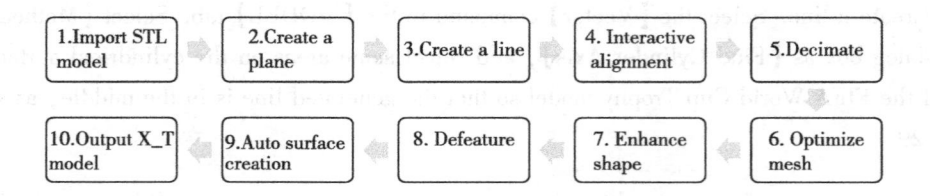

Figure 1-26   Data processing flow of FIFA World Cup Trophy

1) Import STL model. Open the Geomagic Design X 2016 software, and click the "Import" button ⊳ at the top left of the interface. Select the STL file of the FIFA World Cup Trophy digital model generated by scanning before, and click【Import Only】, as shown in Figure 1-27.

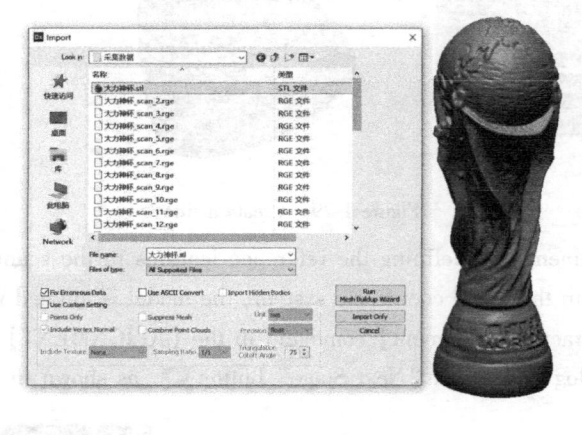

Figure 1-27   Import data

2) Create a plane. Select the【Plane】command in the【MODEL】tab, and set【Method】in the pop-up dialog box to【Pick Multiple Points】. Create a plane by selecting three points on the bottom of the FIFA World Cup Trophy model. Click the "OK" button ✓ to complete the creation of the plane, as shown in Figure 1-28.

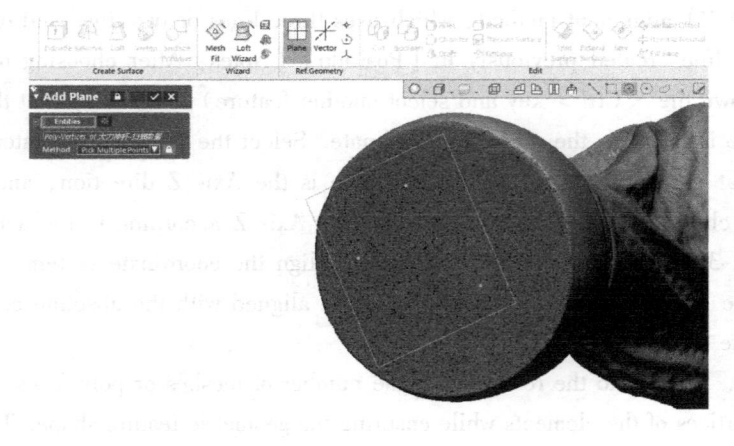

Figure 1-28   Create a plane

11 ◀

3D Digital Design and Manufacturing

3) Create a line. Select the 【Vector】 command in the 【MODEL】 tab. Select 【Method】 in the pop-up dialog box as 【Find Cylinder Axis】, and select some areas on the cylindrical surface at the bottom of the FIFA World Cup Trophy model so that the generated line is in the middle, as shown in Figure 1-29.

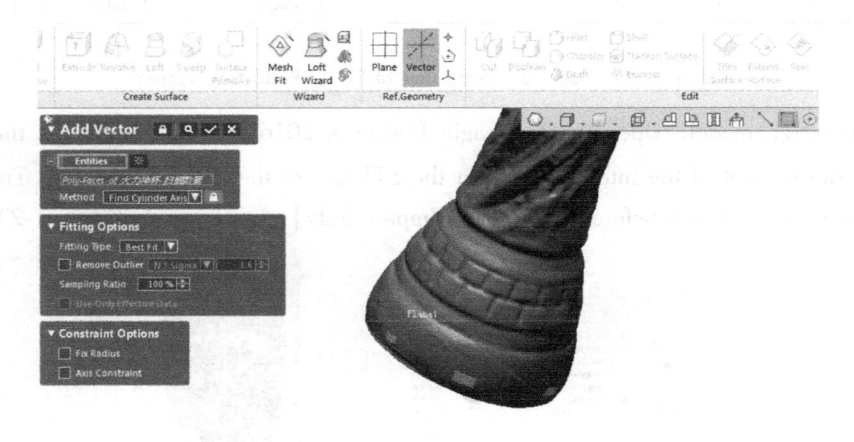

Figure 1-29　Create a line

4) Interactive alignment. By defining the reference features in the scanning model to match the coordinate axis or plane in the world coordinate system, the model is aligned with the world coordinate system. Select the 【Interactive Alignment】 command in the 【ALIGNMENT】 tab, as shown in Figure 1-30. In the pop-up dialog box, click "Next Stage" button ➡, as shown in Figure 1-31.

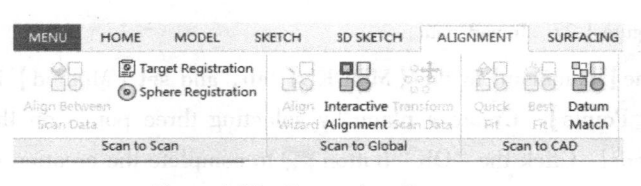

Figure 1-30　Interactive alignment　　　　　　Figure 1-31　Click "Next Stage" button

Choose 【X-Y-Z】 alignment method, which uses three lines or two lines and one origin. Select the plane and the line created previously in 【Position】 (Notes: After choosing one feature, it is needed to hold down the < Ctrl > key and select another feature), indicating that the intersection of the plane and line is taken as the origin of coordinate. Select the previously created plane in 【Axis Z】 to indicate that the normal direction of the plane is the Axis Z direction, and then determine whether or not to click the Axis Z arrow to reverse the Axis Z according to the actual situation, as shown in Figure 1-32. Click the "OK" button ✓ to align the coordinate system so that the bottom and the axis of the FIFA World Cup Trophy model are aligned with the absolute coordinate system, as shown in Figure 1-33.

5) Decimate. It refers to the reduction of the number of meshes or poly-faces of selected region by merging the vertices of the elements while ensuring the geometric feature shape. There are 4699088 poly-faces generated by scanning the model of the FIFA World Cup Trophy, and it is a large amount of data. Here, the 【Decimate】 command in the 【POLYGONS】 tab is used to reduce 50%, as shown in

▶ 12

Project 1   Model of the FIFA World Cup Trophy

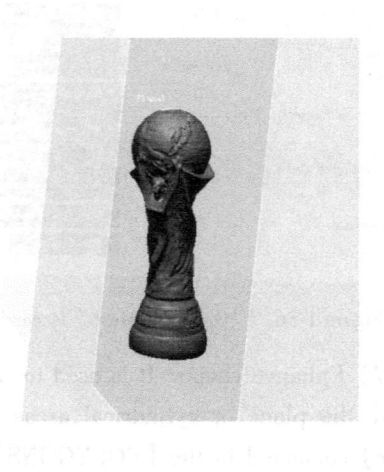

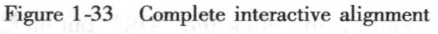

Figure 1-32   Specify the position
to be manually aligned with Axis Z

Figure 1-33   Complete interactive alignment

Figure 1-34. (The time needed for decimate varies according to the performance of the computer, and it usually takes a few minutes for ordinary computers.)

6) Optimize mesh. According to the characteristic shape of the mesh, the length and smoothness of the element edge are set to optimize the mesh. By pressing < F8 > in the default shortcut, enter the edge rendering mode and enlarge the thickness to observe meshes. It can be found that the triangular meshes are not uniform, as shown in Figure 1-35. Select the 【Optimize Mesh】 command in the 【POLYGONS】 tab; keep the parameters in the pop-up dialog box at the default value; and click the "OK" button ✓ to optimize the mesh, as shown in Figure 1-36. (The optimization process also takes a certain amount of time. Please wait with patience.)

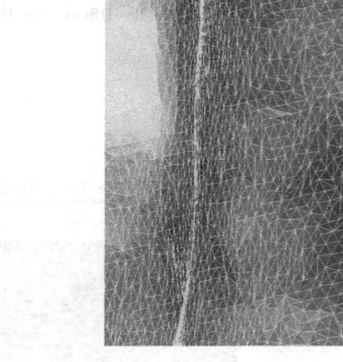

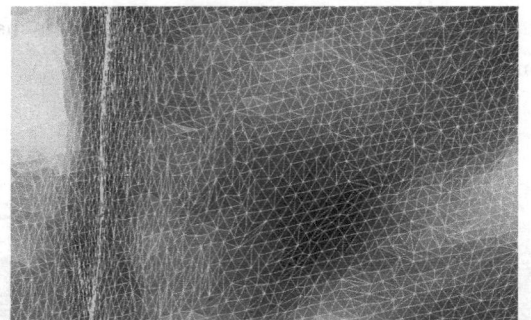

Figure 1-34   Reduce the mesh by 50%   Figure 1-35   Observe and discover that the meshes are not uniform

After the optimization of the meshes is completed, the triangular meshes are observed again and found to be more regular and symmetrical than before. Switch back to the rendering state by pressing < F7 > (rendering), as shown in Figure 1-37.

13

3D Digital Design and Manufacturing

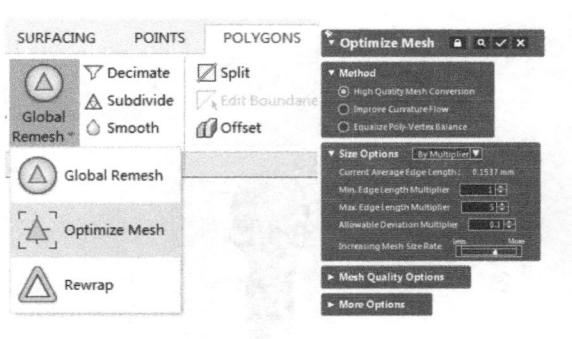
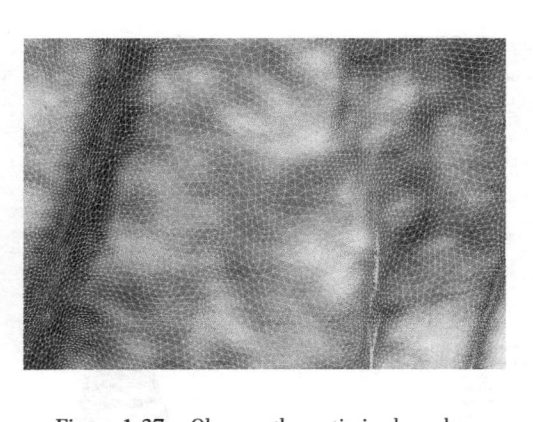

Figure 1-36　"Optimize Mesh" is carried out　　　　Figure 1-37　Observe the optimized meshes

7) Enhance shape. It is used to sharpen the sharp areas (edges and corners) on the mesh and smooth the plane or cylindrical areas to improve the quality of the mesh. Select the 【Enhance Shape】 command in the 【POLYGONS】 tab; keep the parameters in the pop-up dialog box at the default value, and click the "OK" button ✓, as shown in Figure 1-38.

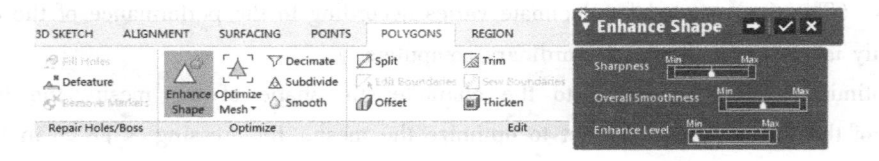

Figure 1-38　Sharpen the overall smoothness by 【Enhance Shape】

8) Defeature. It is used to delete feature shapes or irregular protrusions on the mesh and reconstruct the poly-faces. If it is observed and found that there are small bumps in the FIFA World Cup model, select the 【Defeature】 command in the 【POLYGONS】 tab and select the bump feature in the lasso selection mode. Click the "OK" button ✓ to delete the bump features, as shown in Figure 1-39, Figure 1-40. The same method is used to deal with the bump features at other locations.

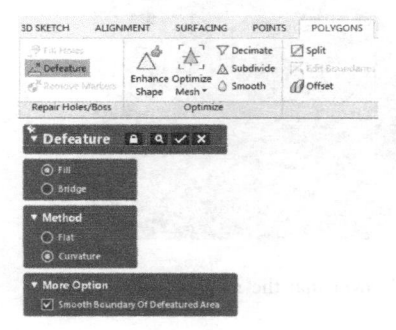
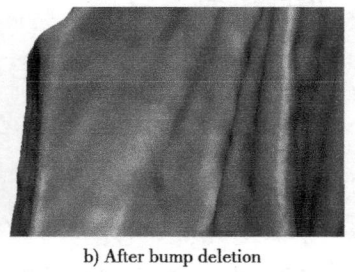

a) Before bump deletion　　　　　　　b) After bump deletion

Figure 1-39　【Defeature】 command　　　　　　Figure 1-40　Contrast before and after defeature

▶ 14

Project 1　Model of the FIFA World Cup Trophy

9) Auto surface creation. It is used to automatically generate NURBS surface. Select the【Auto Surface】command in the【SURFACING】tab; keep the parameters in the pop-up dialog box at the default value; and click "Next Stage" button ➡, as shown in Figure 1-41a. Observe whether the facet shown in Figure 1-41b meets the requirements. Click the "OK" button ☑ to complete the surface creation after confirming that they are correct.

When the auto surface creation is completed, it can be seen that a "Solid Bodies" will be generated in the model tree on the left, as shown in Figure 1-42.

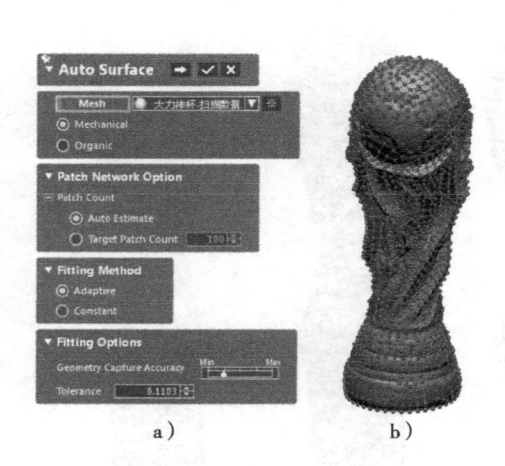

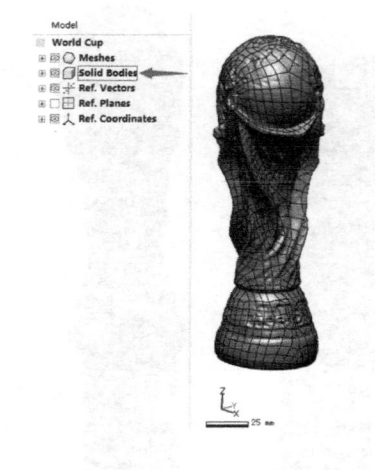

Figure 1-41　【Auto Surface】command

Figure 1-42　Generating corresponding "Solid Bodies" in the model tree

10) Export x_t Model. After selecting the "Solid Bodies" in the model tree, right-click it, then select【Export】from the pop-up context menu and select【Save as type】as "*.x_t", as shown in Figure 1-43.

Figure 1-43　Export x_t Model

15

3D Digital Design and Manufacturing

Task 3　Machining Programming

The shape of the FIFA World Cup Trophy is quite complex, and turning or three-axis milling cannot complete the machining of the artware, so multi-axis machining is adopted. There are four processes in multi-axis machining of the artware, which are roughing, second roughing, semi-finishing and finishing, as shown in Figure 1-44.

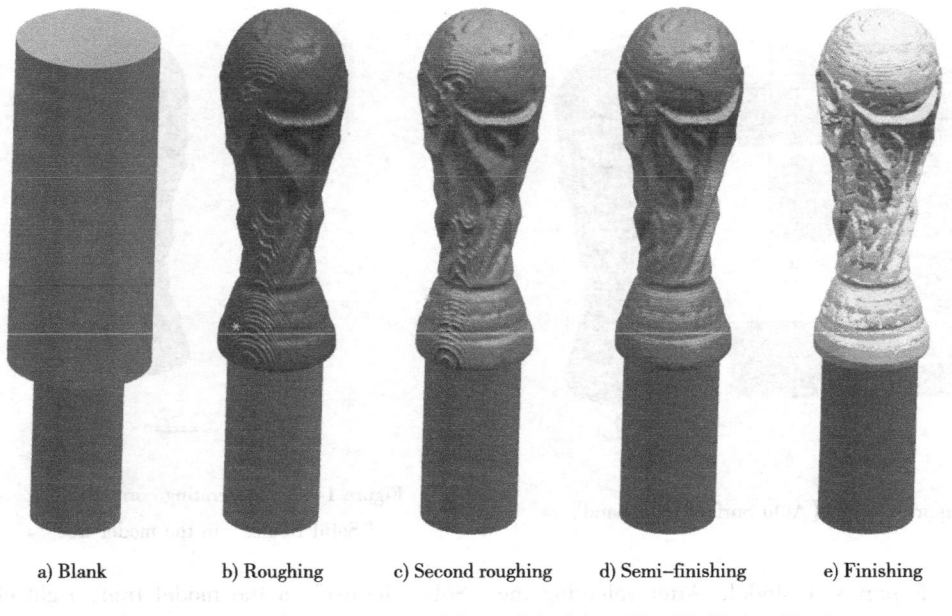

a) Blank　　b) Roughing　　c) Second roughing　　d) Semi−finishing　　e) Finishing

Figure 1-44　Machining Process of the model of the FIFA World Cup Trophy

1.3.1　Initial Setting

Import the model. Import the data model of the FIFA World Cup Trophy in x_ t format by the instruction of 【File】/【Import】/【Parasolid】.

Initial settings are required before creating a multi-axis machining program, as shown in Figure 1-45. The specific steps are described below.

Create datum ➡ Create blanks ➡ Create the clamping section ➡ Create the drive surface ➡ Enter the machining module ➡ Set workpiece ➡ Create MCS

Figure 1-45　Main steps of initial setting

1) Create datum. Enter the 【Modeling】 application module. Select the 【Insert】/【Datum/Point】/【Datum CSYS】 command, and the 【Datum CSYS】 dialog box is popped up. Click 【OK】, and create the benchmark directly from the origin, as shown in Figure 1-46.

2) Create blanks. Use the 【Extrude】 command, and click the "Sketch Section" button 🔀 in the 【Section】 group of 【Extrude】 dialog box to enter sketch mode. Draw a circle on the XC-YC

▶ 16

Project 1   Model of the FIFA World Cup Trophy

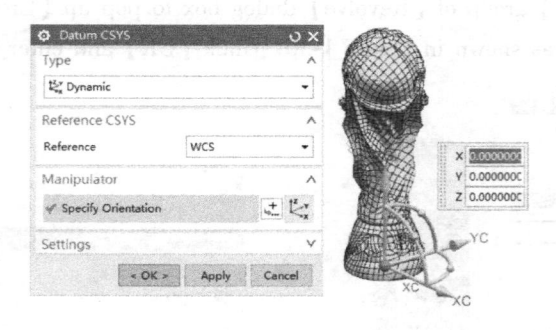

Figure 1-46   Create datum

plane. The circle needs to cover the whole part. The diameter of the circle is integer 80 mm. Then, exit the sketch mode. The extruding height needs to cover the whole workpiece. Input 200mm, click 【OK】, and then extrude to a 80mm × 200mm cylindrical blank, as shown in Figure 1-47.

3) Create the clamping section. Parts need to be mounted on machine tools, so a clamping section needs to be created. Similarly, 【Extrude】 method is used to create the clamping section. Draw a circle with a diameter of 50 mm on the XC-YC plane, and extrude 100 mm downward. 【Boolean】 is selected as 【None】, as shown in Figure 1-48.

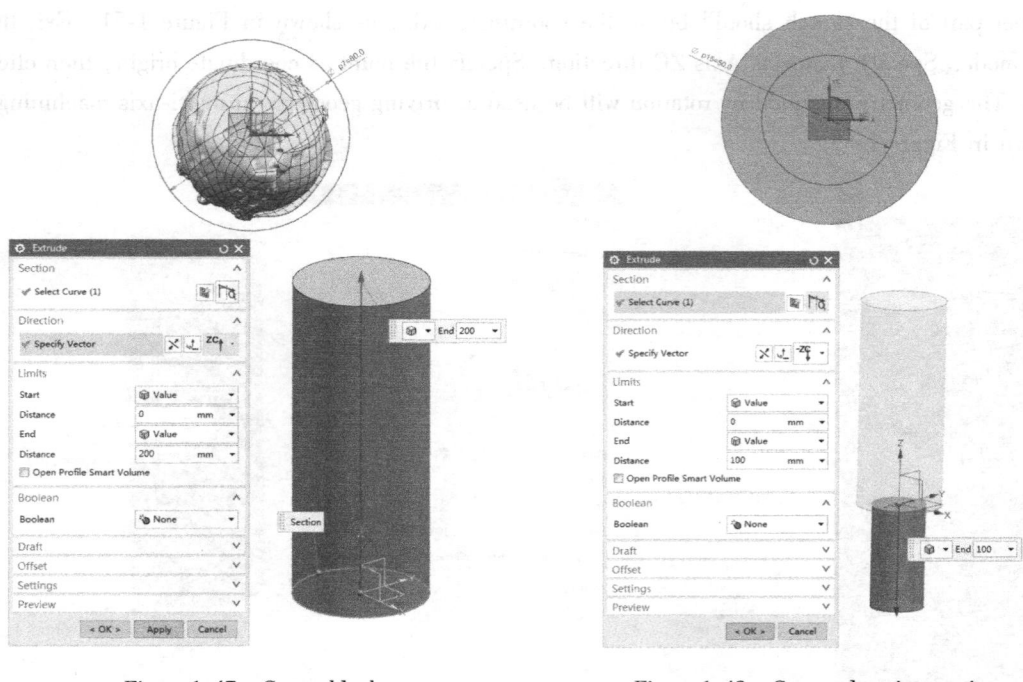

Figure 1-47   Create blanks   Figure 1-48   Create clamping section

Use the 【Edit Object Display】 command or press the shortcut key and < Ctrl + J > to change the color and transparency of the blank and clamping parts, as shown in Figure 1-49.

4) Create the drive surface. Select the 【Revolve】 command, and click the "Sketch Section"

17

3D Digital Design and Manufacturing

button 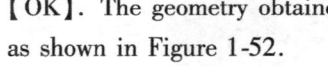 in the 【Section】 group of 【Revolve】 dialog box to pop up 【Create Sketch】 dialog box. Select the XC-ZC plane, as shown in Figure 1-50. Click 【OK】 and enter the sketch mode.

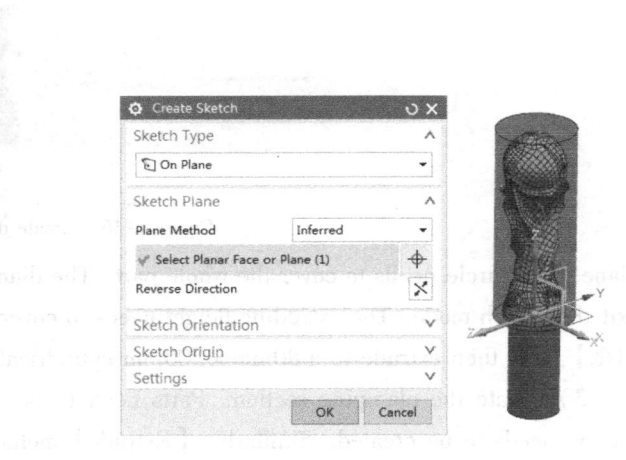

Figure 1-49 Change color and transparency   Figure 1-50 Select sketch plane

When drawing a line and an arc, it should be noted that the end point and center of the arc on the upper part of the sketch should be on the coordinate axis, as shown in Figure 1-51. Exit the sketch mode. Specify vector as Axis ZC direction. Specify the point as coordinate origin, then click 【OK】. The geometry obtained by rotation will be used as driving geometry in multi-axis machining, as shown in Figure 1-52.

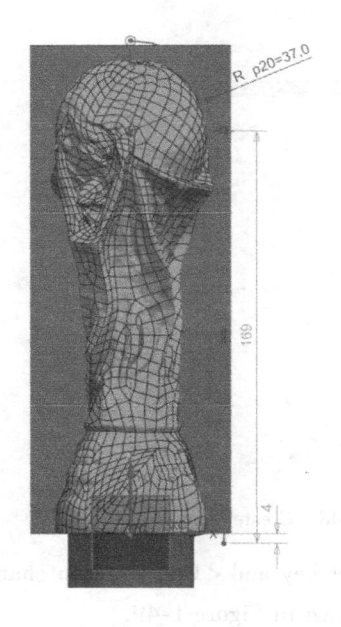

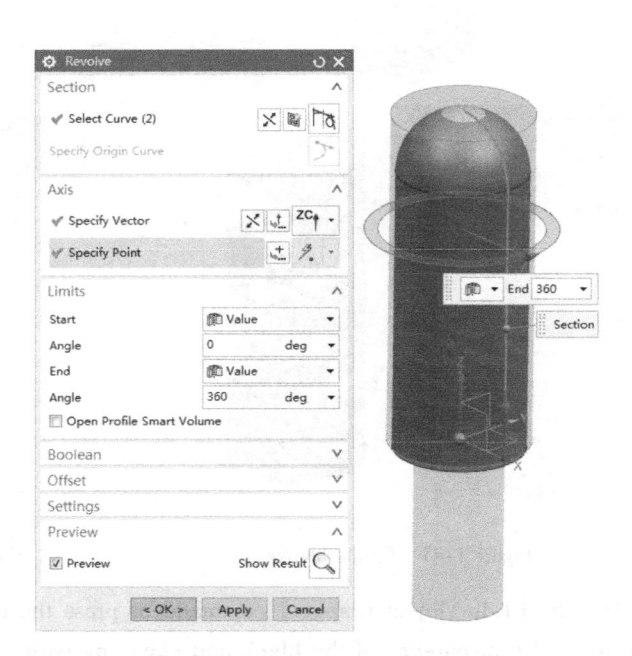

Figure 1-51 Create sketch   Figure 1-52 Rotate sketch to generate driving geometry

▶ 18

Project 1   Model of the FIFA World Cup Trophy

5) Enter the machining module. Select 【CAM Session Configuration】 in the pop-up 【Machining Environment】 dialog box as 【cam_general】. Select 【CAM Setup to Create】 as 【mill_contour】, and click 【OK】, as shown in Figure 1-53.

6) Set workpiece. Click on the "Geometry" icon in 【Operation Navigator】, then click the "+" in front of 【MCS_MILL】 to expand as shown in Figure 1-54. Double-click 【WORKPIECE】 to pop up the 【Workpiece】 dialog box. Specify part, blank and check in turn, as shown in Figure 1-55.

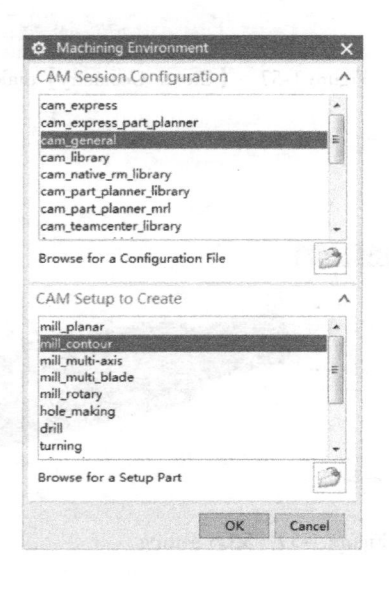

Figure 1-53   【Machining Environment】 dialog box

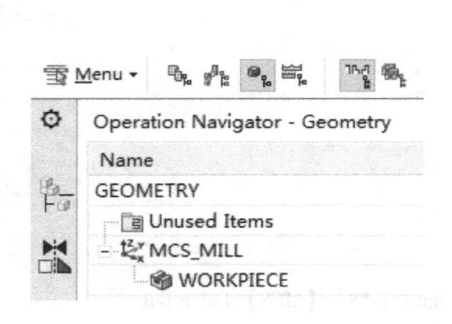

Figure 1-54   Operation Navigator-Geometry

7) Create MCS (Machine Coordinate System). In roughing, the part is machined on one side. After rotated 180°it is machined again, as shown in Figure 1-56. Two MCS need to be created.

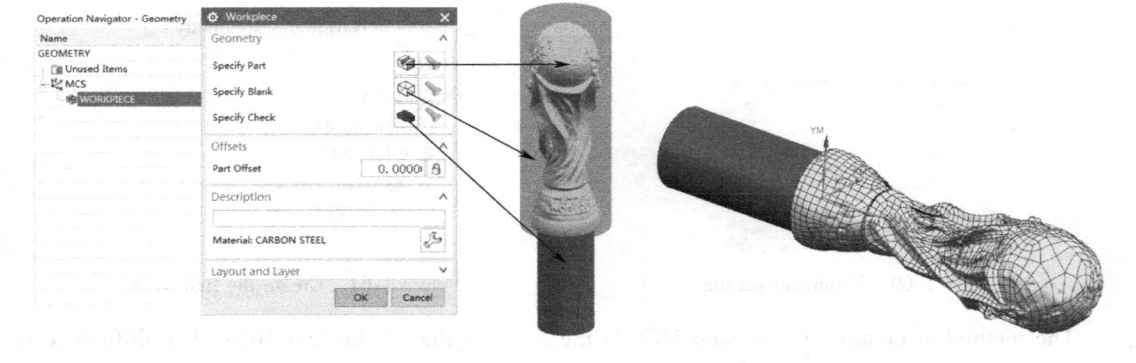

Figure 1-55   Set workpiece

Figure 1-56   Analyze workpiece

Click the 【Create Geometry】 command, and select "MCS" in the pop-up dialog box as 【Geometry Subtype】. Select 【WORKPIECE】 in 【Geometry】, and enter 【Name】 as "MCS-1", as

19

3D Digital Design and Manufacturing

shown in Figure 1-57. Click【OK】and then pop up the 【MCS】dialog box, as shown in Figure 1-58. Click "CSYS Dialog box" button ⬚ in 【Machine Coordinate System】group to pop up the 【CSYS】dialog box. Rotate the coordinate system to make Axis ZM upward, Axis XM towards the top of the artware, as shown in Figure 1-59. Select【Use Main MCS】in【Special Output】and input 【Safe Clearance Distance】to "100", as shown in Figure 1-60. At this point, the creation of the first MCS is completed, as shown in Figure 1-61.

Figure 1-57 【Create Geometry】dialog box

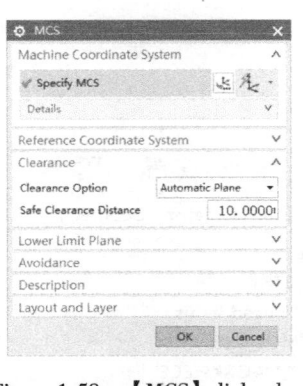

Figure 1-58 【MCS】dialog box

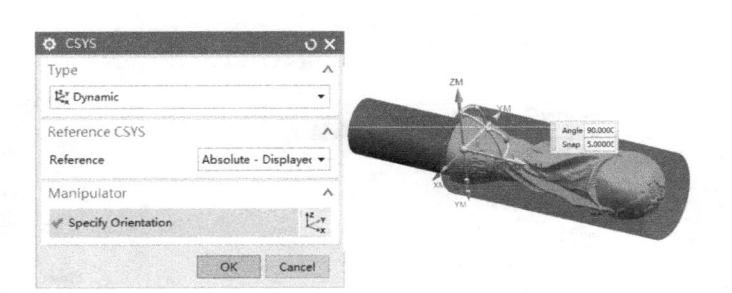

Figure 1-59 MCS setting

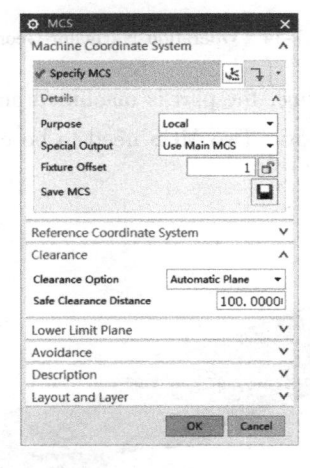

Figure 1-60 Parameter setting

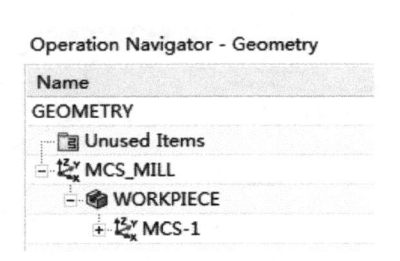

Figure 1-61 Create the first MCS

The method of creating the second MCS is the same as that of the first MCS. The difference is that the coordinate system direction needs to be adjusted to the Axis ZM downward, and the Axis XM faces to the top, so as to ensure that the Axis ZM's direction of the second MCS is opposite to the Axis ZM's direction of the first machine coordinate system, as shown in Figure 1-62.

▶ 20

Project 1　Model of the FIFA World Cup Trophy

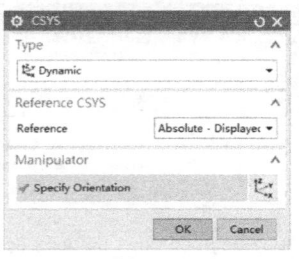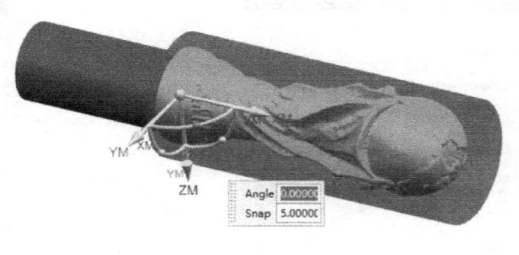

Figure 1-62　Rotating coordinate system

As shown in Figure 1-63, double-click 【MCS Mill】 in 【Operation Navigator-Geometry】. In the pop-up 【MCS_Mill】 dialog box, select 【Main】 in 【Purpose】and input "200" in 【Safe Clearance Distance】, as shown in Figure 1-64.

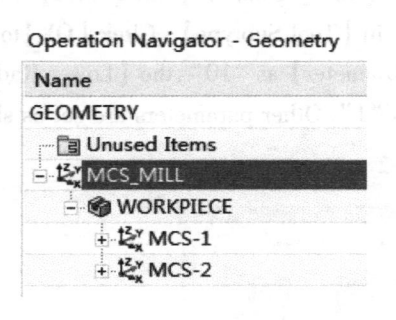

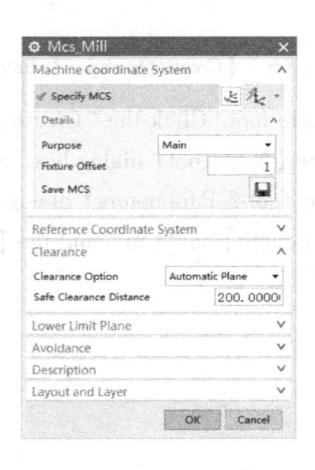

Figure 1-63　Completion of MCS setting

Figure 1-64　MCS_Mill setting

1.3.2　Roughing

Roughing uses cavity milling, i. e. fixed axis machining. This machining method has the highest material removal efficiency and relatively simple programming. The basic idea is: compiling the roughing program on one side in MCS-1, copying it to MCS-2, and getting the program on the other side of workpiece so as to complete the roughing of the whole workpiece. The specific steps are described below.

1) Create operation. Click the 【 Create Operation 】 command as shown in Figure 1-65 to pop up the dialog box as shown in Figure 1-66. Select 【mill_contour】 in 【Type】; select "cavity mill" in 【Operation Subtype】; and select 【MCS-1】 in 【Geometry】. Click 【OK】, and the【Cavity Mill】 dialog box is popped up, as shown in Figure 1-67.

Figure 1-65　【Create Operation】
command

21 ◀

3D Digital Design and Manufacturing

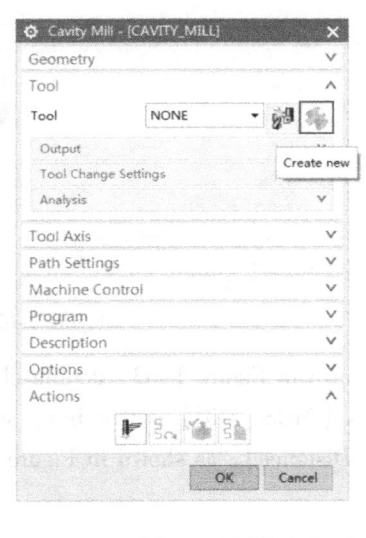

Figure 1-66 【Create Operation】dialog box   Figure 1-67 【Cavity_Mill】dialog box

2) Create tool. Click the "Create new" button in the 【Tool】group of 【Cavity Mill】dialog box to pop up the 【New Tool】dialog box. Select the "MILL" in 【Tool Subtype】. Click 【OK】to pop up the 【Milling Tool-5 Parameters】dialog box. Input the 【Diameter】as "10", the 【Lower Radius】as "1", the 【Length】as "150" and the 【Tool Number】as "1". Other parameters are set as shown in Figure 1-68.

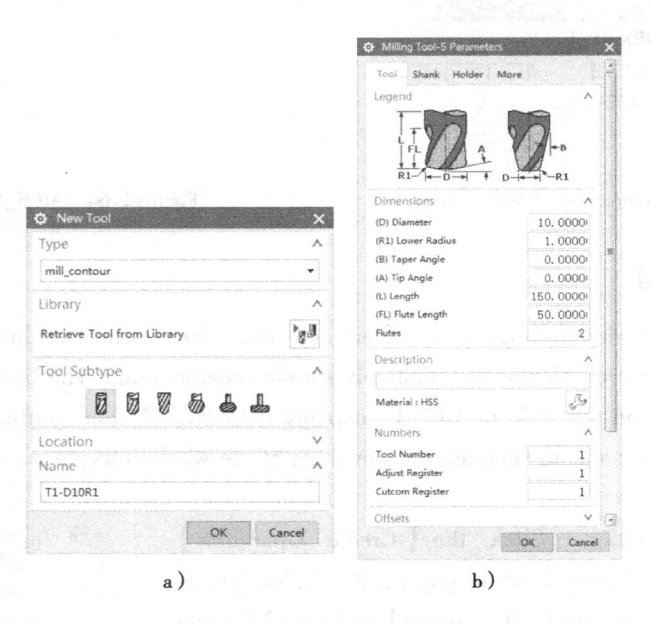

a)   b)

Figure 1-68   Create D10R1 tool

3) Create a tool holder. Click the 【Holder】tab in the 【Milling Tool-5 Parameters】dialog box, and click the "Retrieve Holder from Library" button in the 【Library】group. The steps to create a tool holder is shown in Figure 1-69.

▶ 22

Project 1   Model of the FIFA World Cup Trophy

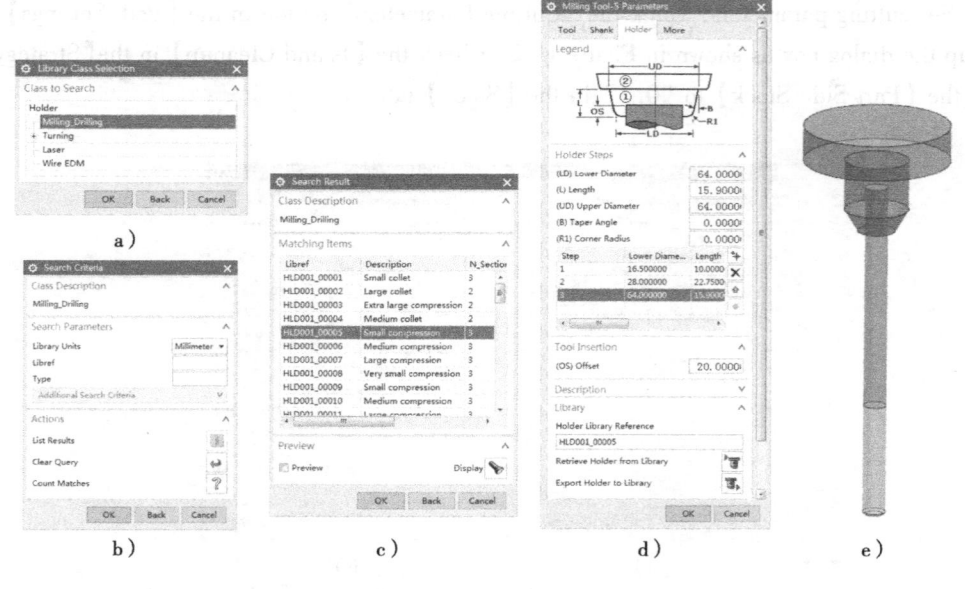

a)

b)

c)

d)

e)

Figure 1-69   Add a tool holder

4) Set the tool path. In the 【Path Settings】 group of 【Cavity Mill】 dialog box, the 【Cut Pattern】 is set to 【Follow Periphery】, and the 【Maximum Distance】 of cutting depth per tool is set to "1mm", as shown in Figure 1-70.

5) Set the cut levels. Click the "Cut Levels" button in the 【Path Settings】 group to pop up the dialog box as shown in Figure 1-71. The diameter of the blank is 80 mm, so the initial depth is 80 mm. As second roughing is carried out in two sides, the depth in the range is appropriately more than 40mm, so the depth is set to 42 mm in order to ensure that there is no residue in the middle, as shown in Figure 1-71.

Figure 1-70   Path settings

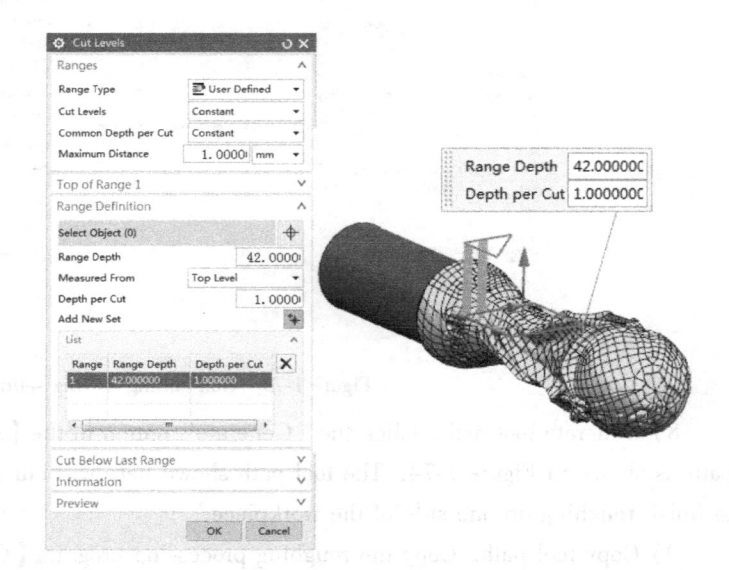

Figure 1-71   Cut levels setting

23 ◀

3D Digital Design and Manufacturing

6) Set cutting parameters. Click the "Cutting Parameters" button in the 【Path Settings】 group to pop up the dialog box as shown in Figure 1-72. Check the 【Island Cleanup】 in the 【Strategy】 tab and set the 【Part Side Stock】 to "0. 5" in the 【Stock】 tab.

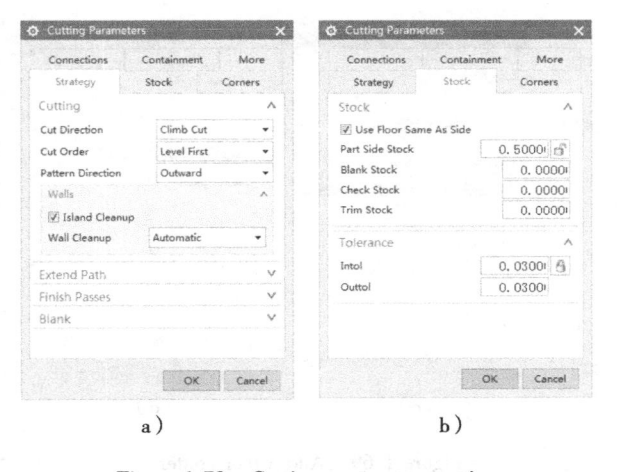

a)   b)

Figure 1-72   Cutting parameters setting

7) Set feeds and speeds. Click the "Feeds and Speeds" button in the 【Path Settings】 group to pop up the dialog box as shown in Figure 1-73a. Set the spindle speed to "6000" and the cutting feed rate to "1500". Click the "Calculate Feeds and Speeds based on this value" button behind the 【Spindle Speed】 input box. The results are shown in Figure 1-73b.

a)   b)

Figure 1-73   Non-cutting moving setting

8) Generate tool path. Click the "Generate" button in the 【Actions】 group to generate the tool path as shown in Figure 1-74. The tool path shown here will cut off the upper part of the workpiece to finish roughing on one side of the workpiece.

9) Copy tool path. Copy the roughing processing program 【CAVITY_MILL】 created in 【MCS-1】, and 【Paste Inside】 to 【MCS-2】, as shown in Figure 1-75. Double-click 【CAVITY_MILL_

▶ 24

Project 1　Model of the FIFA World Cup Trophy

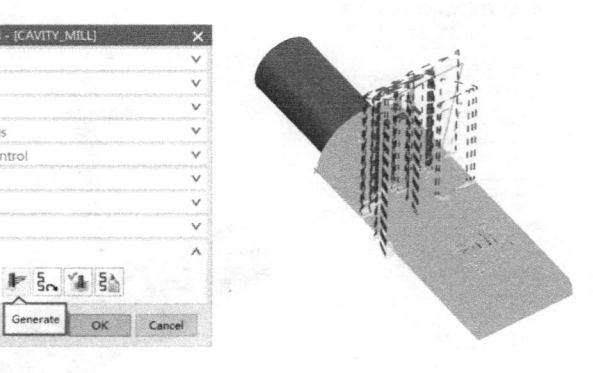

Figure 1-74　Generate tool path

COPY】. Change the cutting depth to 42 mm, and the remaining parameters remain unchanged, as shown in Figure 1-76. Click the "Generate" button in the 【Actions】 group to generate the tool path on the other side.

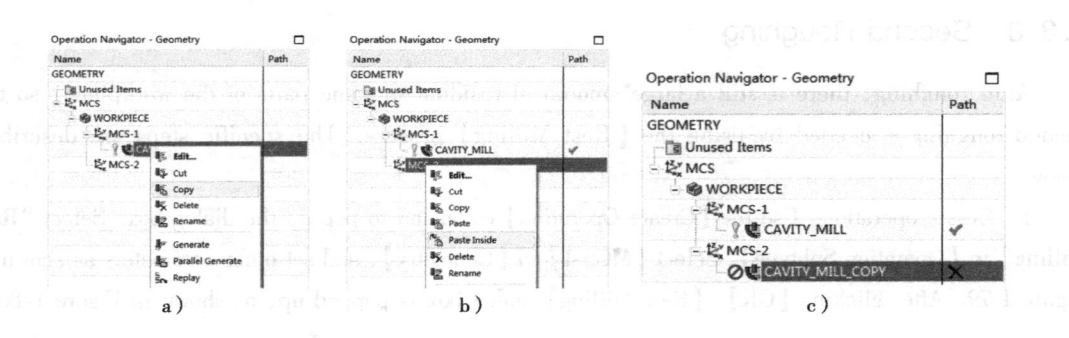

a)　　　　　　　　　　　b)　　　　　　　　　　　c)

Figure 1-75　Copy tool path

10) Verify tool path. Select 【MCS_MILL】 in the geometry view of the operation navigator. Click the 【Verify Tool Path】 command to pop up the 【Tool Path Visualization】 dialog box, and click the "Play" button in the 【3D Dynamic】 tab to see the cutting effect, as shown in Figure 1-77 and Figure 1-78.

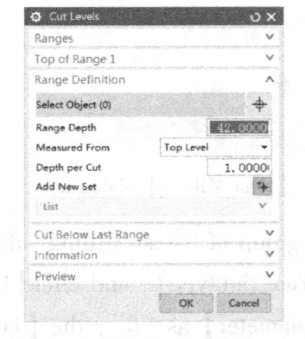

Figure 1-76　Cut levels setting

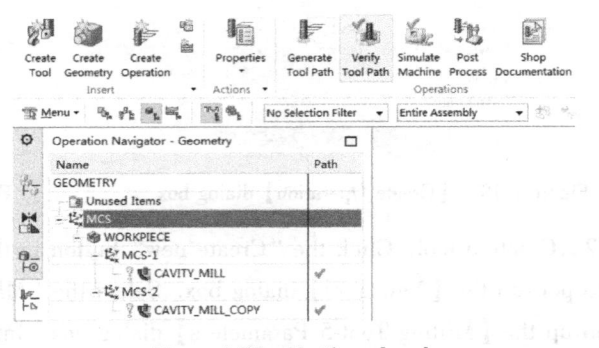

Figure 1-77　Verify tool path

25

3D Digital Design and Manufacturing

Figure 1-78　3D dynamic simulation

1.3.3　Second Roughing

After roughing, there is still a large amount of residual in some parts of the workpiece, so the second roughing is needed by using the【Rest Milling】process. The specific steps are described below.

1) Create operation. Use the【Create Operation】command to pop up the dialog box. Select "Rest Milling" in【Operation Subtype】. Select【MCS-1】in【Geometry】, and set other parameters as shown in Figure 1-79. After clicking【OK】,【Rest Milling】dialog box is popped up, as shown in Figure 1-80.

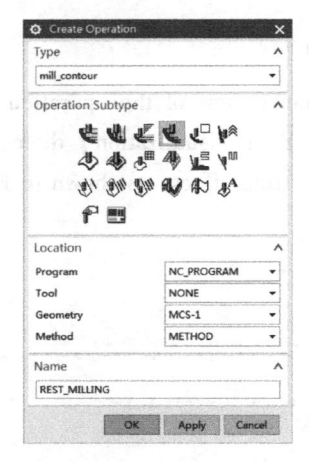

Figure 1-79　【Create Operation】dialog box

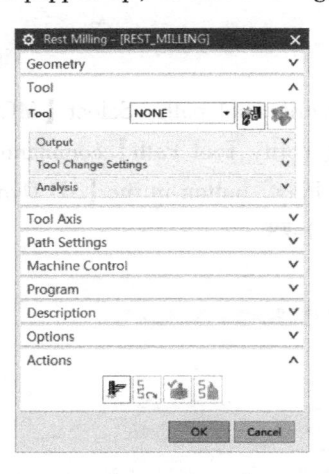

Figure 1-80　【Rest Milling】dialog box

2) Create a tool. Click the "Create new" button in the【Tool】group of "Rest Milling" dialog box to pop up the【New Tool】dialog box. Select the "MILL" in【Tool Subtype】, and click【OK】to pop up the【Milling Tool-5 Parameters】dialog box. Input the【Diameter】as "6", the【Lower Radius】as "3", the【Length】as "125", and the【Tool Number】as "2". Other parameters are

26

Project 1   Model of the FIFA World Cup Trophy

set as shown in Figure 1-81.

3) Create a tool holder. Select the tool holder from the 【Library】in the 【Holder】tab, as shown in Figure 1-82.

a)   b)

Figure 1-81   Create a ball tool with a diameter of 6mm   Figure 1-82   Add a tool holer

4) Set the tool path. Select 【Follow Periphery】in 【Cut Pattern】 and input 【Maximum Distance】as "1", as shown in Figure 1-83.

5) Set the cut levels. Click the "Cut Levels" button in the 【Path Settings】group and enter 【Range Depth】as "44", slightly greater than the depth at roughing, as shown in Figure 1-84.

6) Set cutting parameters. Click the "Cutting Parameters" button in the 【Path Settings】group and enter 【Part Side Stock】as "0. 3"in the【Stock】tab, as shown in Figure 1-85.

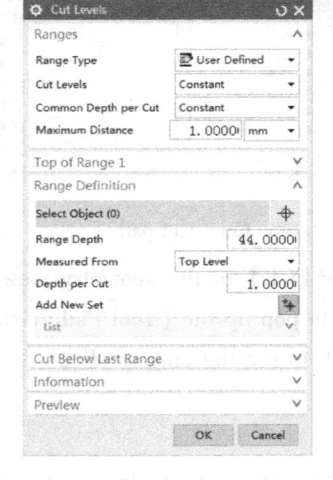

Figure 1-83   Set path   Figure 1-84   Set cut levels   Figure 1-85   Set cutting parameters

27

3D Digital Design and Manufacturing

7) Set feeds and speeds. Click the "Feeds and Speeds" button in the 【Path Settings】 group. Set the spindle speed to "6000" and the cutting feed rate to "1000", and click the "Calculate Feeds and Speeds based on this value" button behind the 【Spindle Speed】 input box.

8) Generate tool path. Click the "Generate" button in the 【Actions】 group to generate the tool path as shown in Figure 1-86. After the second roughing, the machining stock will be more uniform in all parts of the workpiece.

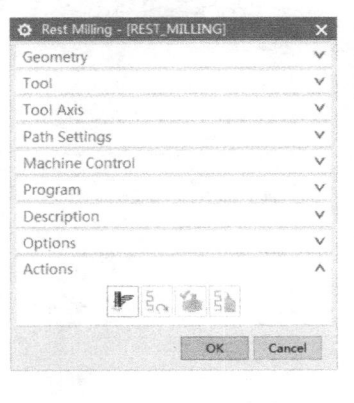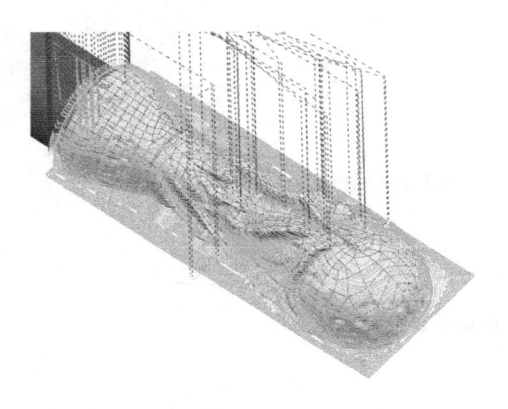

Figure 1-86　Generate tool path

9) Copy tool path. Copy the rest milling program 【REST_MILLING】 created in 【MCS-1】, and 【Paste Inside】 to 【MCS-2】, as shown in Figure 1- 87. Double-click 【CAVITY _ MILL _ COPY】; change the cutting depth to "44"; and the other parameters remain unchanged. Click the "Generate" button in the 【Actions】 group to generate the tool path on the other side.

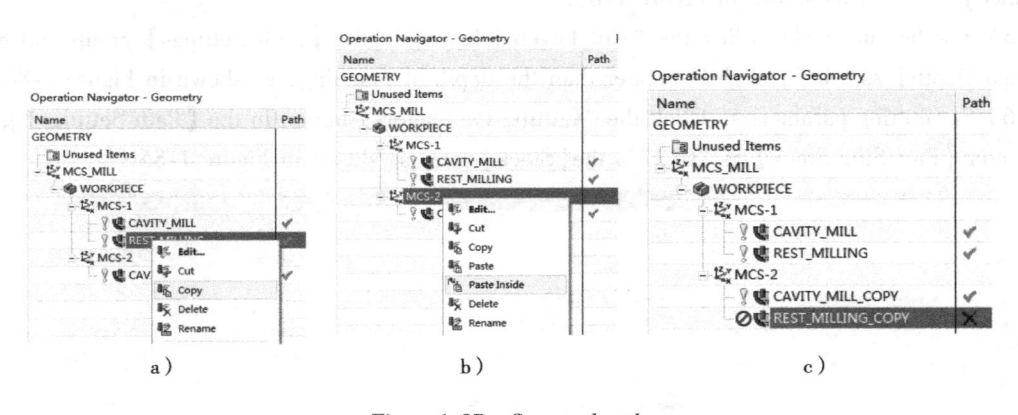

Figure 1-87　Copy tool path

10) Verify tool path. Select 【WORKPIECE】 in the geometry view of the operation navigator, then click the 【Verify Tool Path】command to pop up the 【Tool Path Visualization】 dialog box. Click the "Play" button in the 【3D Dynamic】 tab to see the cutting effect, as shown in Figure 1-88.

1.3.4　Semi-finishing

Because there are uneven and large amount of residuals in some areas of the workpiece after

28

Project 1   Model of the FIFA World Cup Trophy

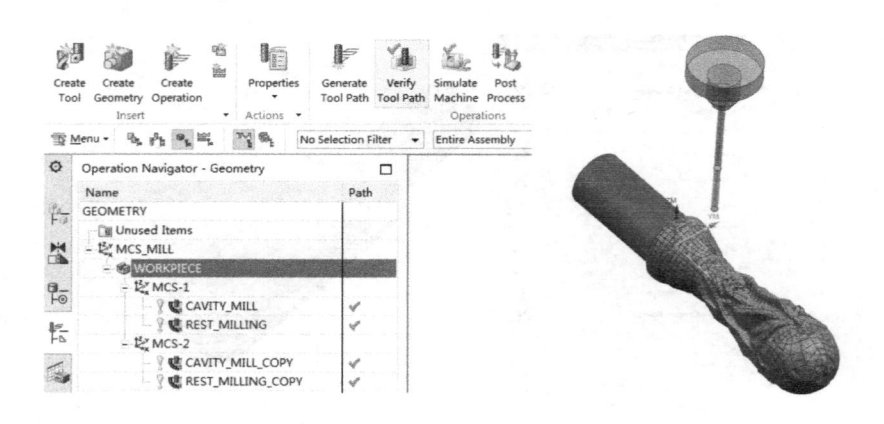

Figure 1-88   Verify tool pathing

roughing, semi-finishing is required by multi-axis machining. The specific steps are as follows.

1) Create operation. Use the 【Create Operation】 command to pop up the 【Create Operation】 dialog box. Select 【mill＿multi-axis】 in 【Type】. Select "Variable Contour" in 【Operation Subtype】. Select 【T2-B6】 in 【Tool】. Select 【WORKPIECE】 in 【Geometry】. Set other parameters as shown in Figure 1-89. Click 【OK】 to pop up the【Variable Contour】dialog box.

2) Set drive method. Select 【Method】 to 【Surface Area】 in the 【Drive Method】 group of 【Variable_Contour】 dialog box as shown in Figure 1-90. Specify the drive geometry as shown in Figure 1-91.

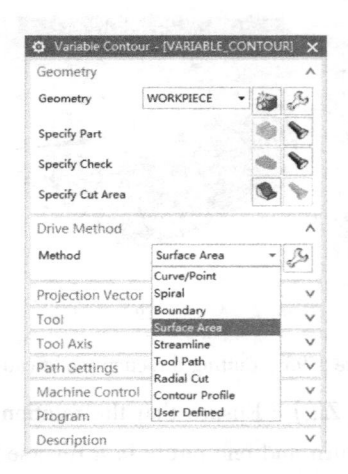

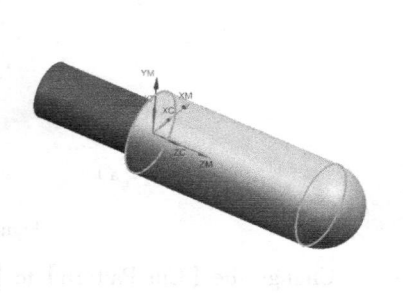

Figure 1-89   Create variable contour mill operation

Figure 1-90   Select surface area for drive method

Figure 1-91   Specify drive geometry

In order to observe the cutting direction conveniently, select 【Cut Pattern】 as 【Zig Zag】. Select 【Stepover】 as 【Number】, and input 【Number of Stepovers】 as "10". Click the "Display" button, and observe that the cutting direction is vertical, as shown in Figure 1-92. It needs to be changed into a horizontal way of winding up circle by circle.

Click the "Cut Direction" button and select the driving direction as shown in Figure 1-93b.

29

3D Digital Design and Manufacturing

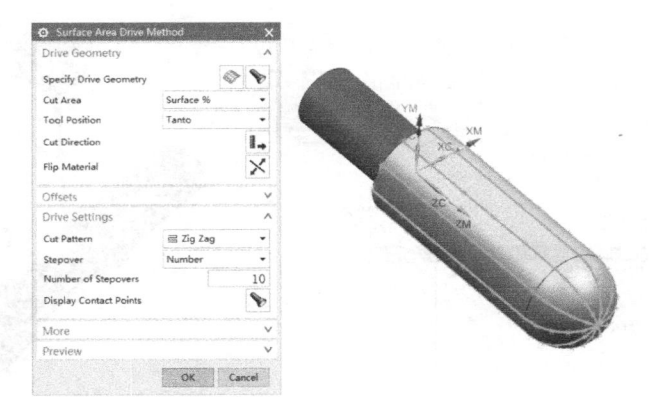

Figure 1-92　Observe cut direction

Click the "Display" button again for observation, as shown in Figure 1-93c. When the width of two sides differs a lot in the way of 【Number】, there will be fewer rows on one side and more rows on the other side.

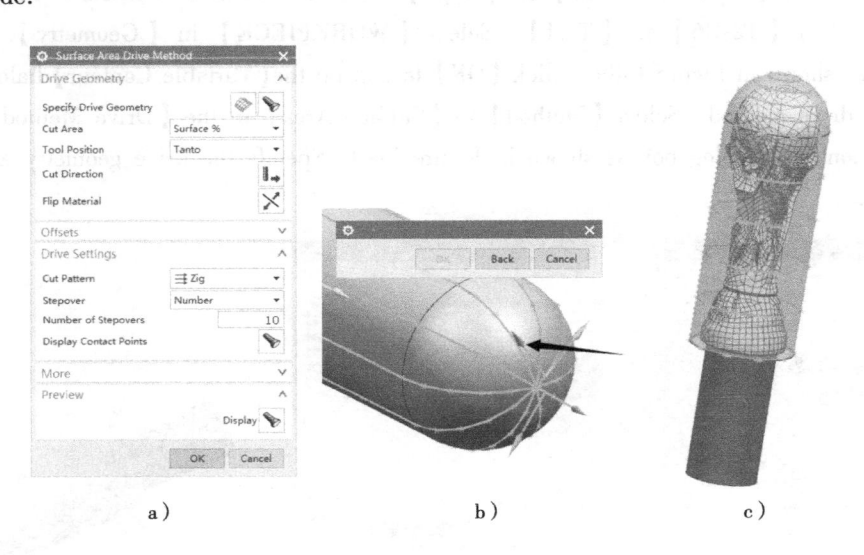

a)　　　　　　　　　　　　　b)　　　　　　　　　　　　　c)

Figure 1-93　Cutting direction modification

Change the 【Cut Pattern】 to 【Zig】(Ensure that the rotation is in the same direction, and no situation of one circle in a positive turn and one circle in a reverse turn). Change the 【Stepover】 to 【Scallop】, and set the 【Maximum Scallop Height】 to "0. 1". Click the "Display" button in the preview again, as shown in Figure 1-94 (The rotator can be hidden for easy observation). Click 【OK】 to complete the drive method setting.

3) Set tool axis. Choose the 【Axis】 as 【Normal to Driver】 in the 【Tool Axis】 group to indicate that the tool axis will be perpendicular to the drive surface, as shown in Figure 1-95.

4) Set cutting parameters. Click the "Cutting Parameters" button in the 【Path Settings】 group and enter 【Part Stock】 in the 【Stock】 tab of 【Cutting Parameters】 dialog box as "0. 1", as shown in Figure 1-96 and Figure 1-97.

▶ 30

Project 1   Model of the FIFA World Cup Trophy

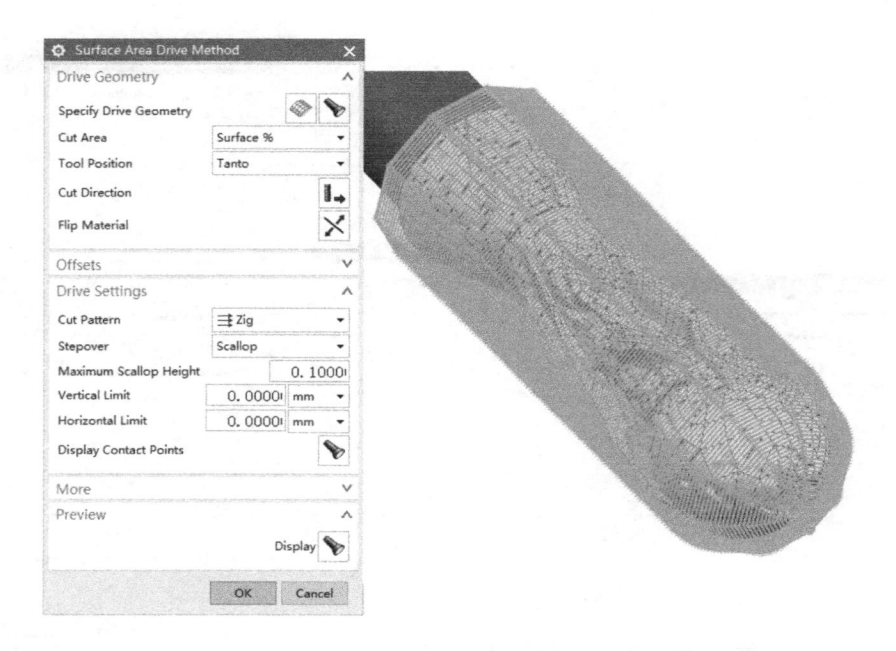

Figure 1-94   Drive method setting

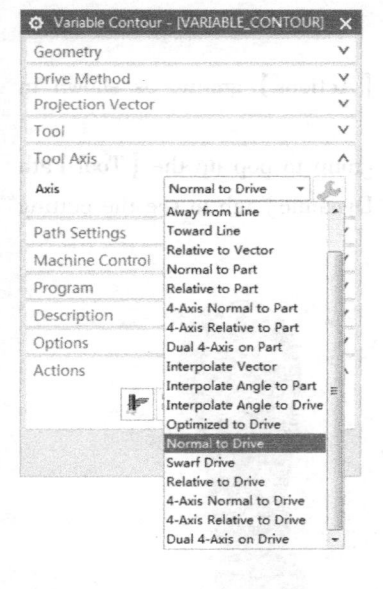

Figure 1-95   Tool axis setting

Figure 1-96   Set cutting parameters

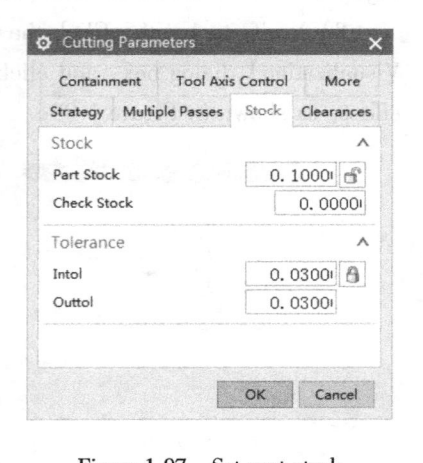

Figure 1-97   Set part stock

5) Set non cutting moves. Click the "Non Cutting Moves" button in the 【Path Settings】 group. Set 【Engage Type】 to 【None】; set 【Retract Type】 to 【Same as Engage】; and set 【Traverse Type】 within regions to 【Direct】, as shown in Figure 1-98.

6) Set feeds and speeds. Click the "Feeds and Speeds" button in the 【Path Settings】 group. Set the spindle speed to "8000" and the cutting feed rate to "1200". Click the "Calculate Feeds and Speeds based on this value" button behind the 【Spindle Speed】 input box.

31

3D Digital Design and Manufacturing

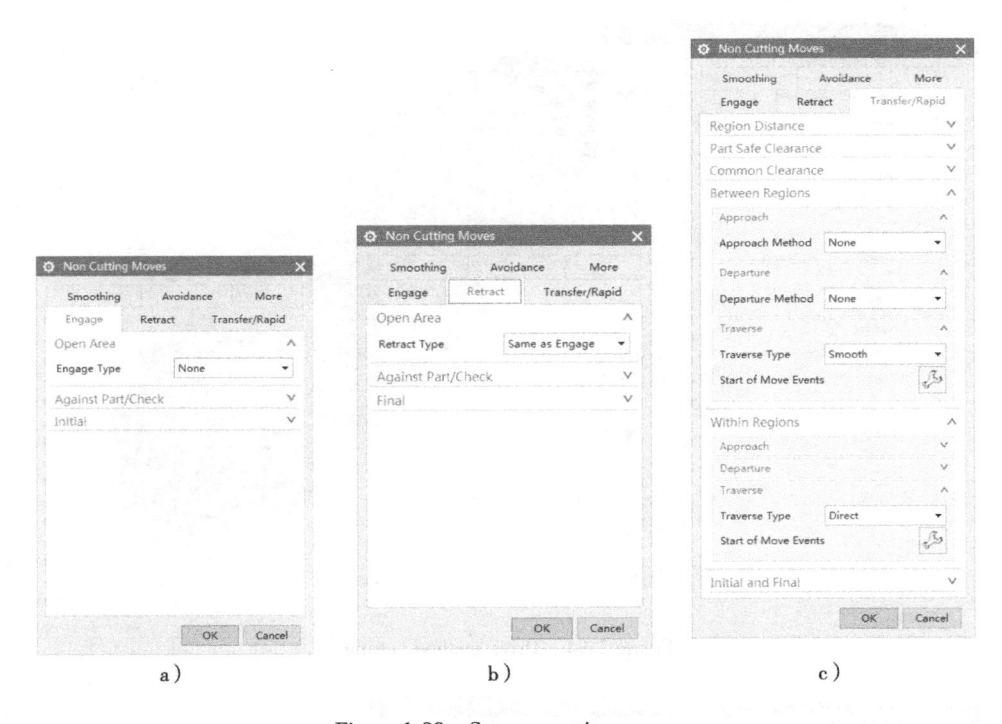

Figure 1-98   Set non cutting moves

7) Generate tool path. Click the "Generate" button in the 【Actions】 group, as shown in Figure 1-99a. Generate the tool path as shown in Figure 1-99b.

8) Verify tool path. Click the "OK" button in the 【Actions】 group to pop up the 【Tool Path Visualization】 dialog box, and click the "Play" button in the 【3D Dynamic】 tab to see the cutting effect, as shown in Figure 1-99c.

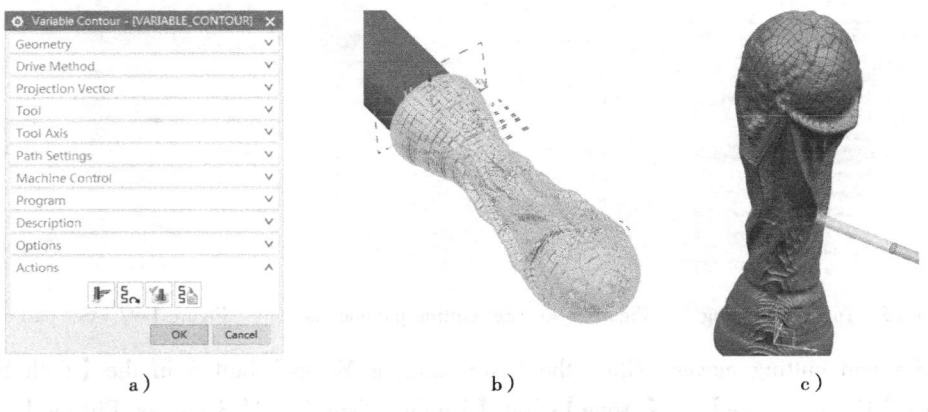

Figure 1-99   Generate and verify tool path

1.3.5 Finishing

After semi-finishing, the basic shape of the model can be observed, but there still is some stock, especially in some pits, so more fine finishing is needed. The specific steps are described below.

▶ 32

Project 1　Model of the FIFA World Cup Trophy

1) Create operation. Click the 【Create Operation】 command to pop up the dialog box. Select 【Tool】 as 【NONE】 to indicate that no tool is selected, then the tool is recreated, as shown in Figure 1-100.

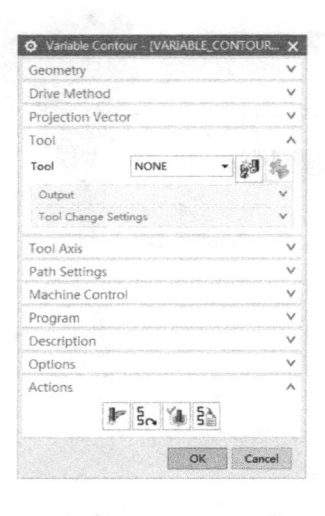

Figure 1-100　【Create Operation】dialog box　　　Figure 1-101　【Variable Contour】dialog box

2) Create a tool. Click the "Create new" button in the 【Tool】 group of 【Variable Contour】 dialog box to pop up the 【New Tool】 dialog box, and select "BALL MILL" in 【Tool Subtype】, as shown in Figure 1-102a. Click 【OK】, and the 【Milling Tool-5 Parameters】 dialog box is popped up. Input the 【Ball Diameter】 as "2", the【Length】 as "25", the【Tool Number】 as"3". The rest of the parameters are shown in Figure 1-102b. Click the 【Shank】 tab, and the parameters of the tool handle are shown in Figure 1-102c.

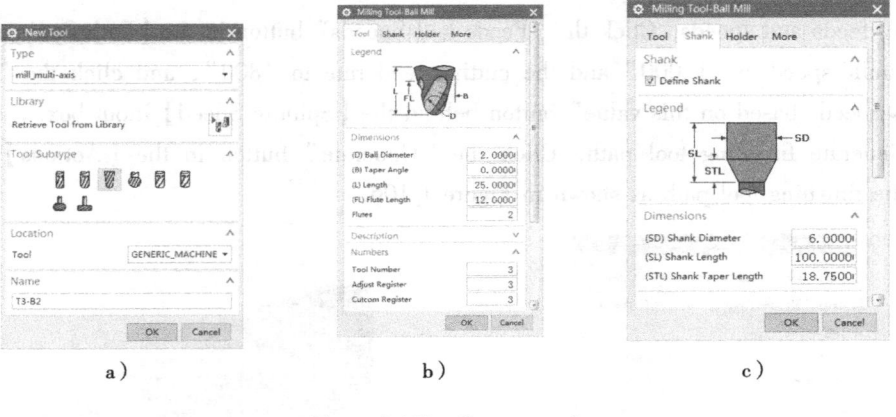

Figure 1-102　Create a tool

3) Create a tool holder. Select tool holder in 【Library】 as shown in Figure 1-103.

4) Set drive method. Select the drive method of 【Surface Area】, as shown in Figure 1-104a. The drive body and cutting direction are specified as the same as in semi-finishing. Change the 【Cut Pattern】 to 【Helical】 (It is spiral descent without interlayer switching). Set the 【Stepover】 to 【Scallop】 and the

33

3D Digital Design and Manufacturing

【Maximum Scallop Height】 to "0.01". The other parameters are set as shown in Figure 1-104b.

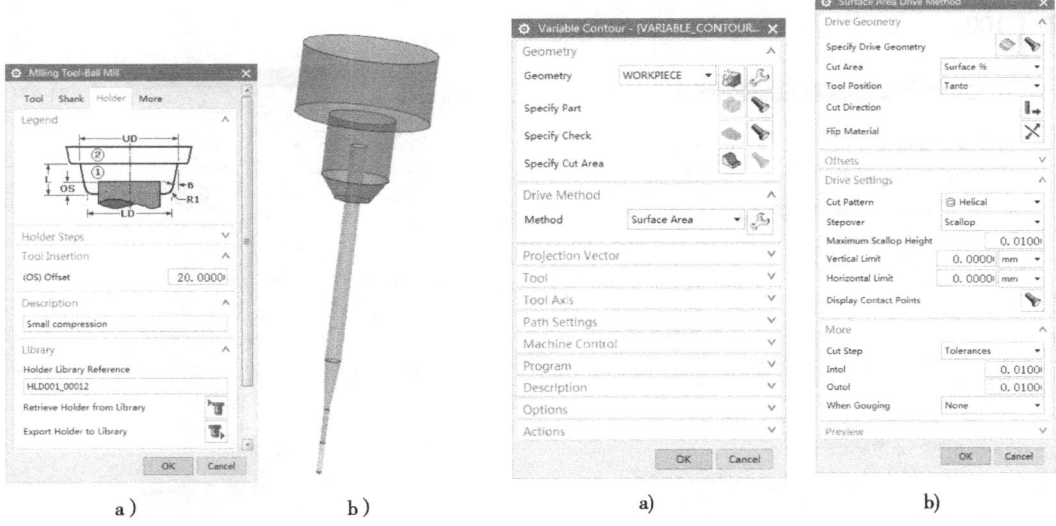

a)	b)

Figure 1-103　Create a tool holder

a)	b)

Figure 1-104　Drive method setting

5) Set tool axis. Choose the 【Axis】 as 【Normal to Driver】 in the 【Tool Axis】 group to indicate that the tool axis will be perpendicular to the drive surface.

6) Set cutting parameters. Click the "Cutting Parameters" button in the 【Path Settings】 group and enter 【Part Stock】 in the 【Stock】 tab of 【Cutting Parameters】 dialog box as "0", as shown in Figure 1-105.

7) Set non cutting moves. Because it is in helical mode, only one Engage and Retract are needed, and the rest are continuous, so it can be set to the default value.

8) Set feeds and speeds. Click the "Feeds and Speeds" button in the 【Path Settings】 group. Set the spindle speed to "12000" and the cutting feed rate to "800", and click the "Calculate Feeds and Speeds based on this value" button behind the 【Spindle Speed】 input box.

9) Generate finishing tool path. Click the "Generate" button in the 【Actions】 group to generate the finishing tool path as shown in Figure 1-106.

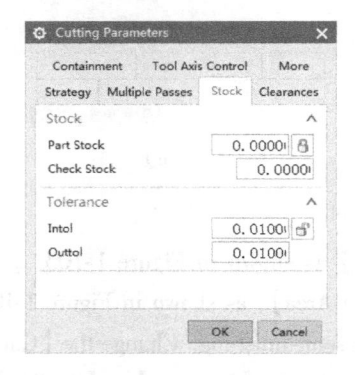

Figure 1-105　Cutting parameters setting

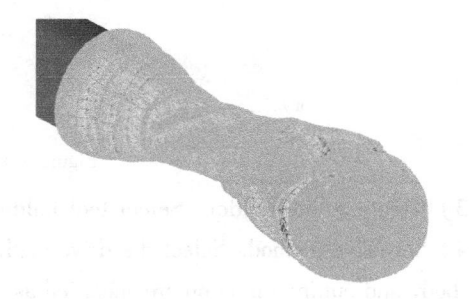

Figure 1-106　Generate finishing tool path

Project 1   Model of the FIFA World Cup Trophy

1.3.6   Machining Simulation

NX 10 is used to build simulation model of the machine tool, the assemble tool, and the workpiece and fixture. The software can read NC code of the machine tool and interpret and execute it. 3D motion simulation of the machining process is carried out, which mainly solves the following problems:

① Verifying the correctness of numerical control program, reducing the machining error of the first part and increasing the credibility of the program.

② Simulating the actual movement of NC machine tools, checking potential collisions, and reducing the risk of machine tool collision.

③ Optimizing the program, improving the machining efficiency, and extending the tool life.

The specific steps of the machining simulation are as follows.

1) Switch to the machine tool view of the operation navigator, then select 【GENERIC_MACHINE】 and right-click it. Select 【Edit】 to pop up the 【Generic Machine】 dialog box, as shown in Figure 1-107 and Figure 1-108.

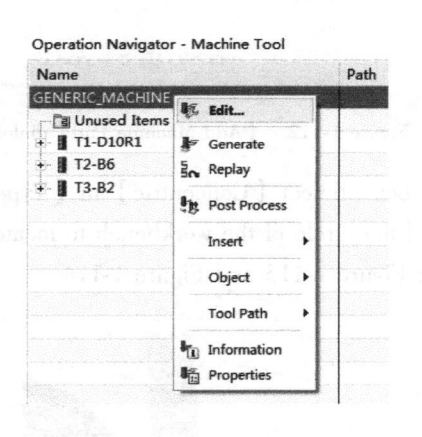

Figure 1-107   【Operation Navigator-Machine Tool】view   Figure 1-108   【Generic Machine】 dialog box

2) Click the "Retrieve Machine from Library" button to pop up the 【Library Class Selection】 dialog box. Select 【MILL】 and click 【OK】, as shown in Figure 1-109.

3) Pop up the 【Search Result】 dialog box. Select 【sim06_mill_5ax_sinumerik_mm】 and click 【OK】, as shown in Figure 1-110.

4) Pop up the 【Part Mounting】 dialog box. Select 【Use Assembly Positioning】 in 【Positioning】. Select parts, blanks, clamping parts and drive surfaces, and click 【OK】, as shown in Figure 1-111.

5) Pop up the 【Add Machine Part】 dialog box. Select 【By Constraints】 in 【Positioning】, as shown in Figure 1-112.

35

3D Digital Design and Manufacturing

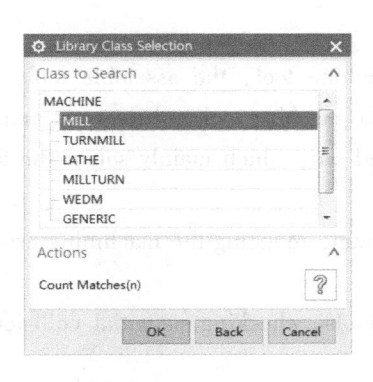

Figure 1-109 【Library Class Selection】 dialog box

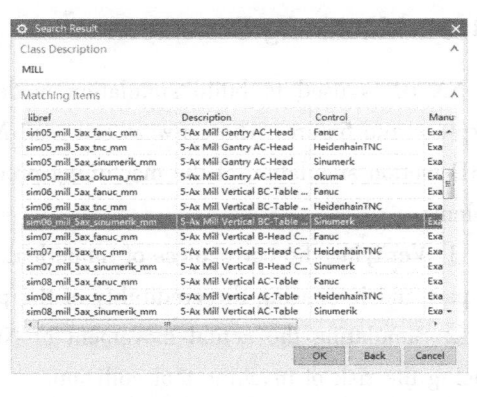

Figure 1-110 【Search Result】 dialog box

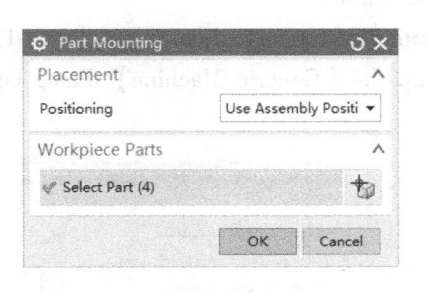

Figure 1-111 【Part Mounting】 dialog box

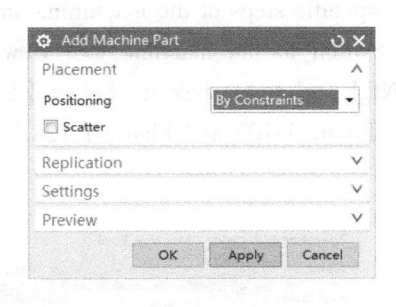

Figure 1-112 【Add Machine Part】 dialog box

6) Pop up the 【Assembly Constraints】 dialog box. Select 【Concentric】 in 【Type】, then select the bottom circle of the blank clamping part and the circle of the workbench to locate through the centers of the two, and click 【OK】, as shown in Figure 1-113 and Figure 1-114.

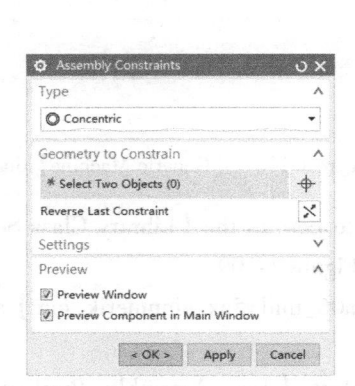

Figure 1-113 【Assembly Constraints】dialog box

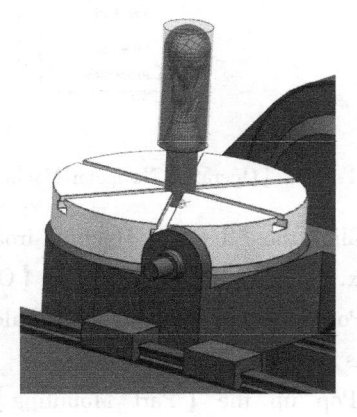

Figure 1-114 Concentric locating

7) The 【Operation Navigator-machine Tool】view at this time is shown in Figure 1-115. Click the 【Simulate Machine】 command to pop up the 【Simulation Control Panel】 dialog box. Check 【Show 3D Material Removal】, and click the "Play" button, as shown in Figure 1-116.

8) Machining simulation process is as shown in Figure 1-117.

▶ 36

Project 1　Model of the FIFA World Cup Trophy

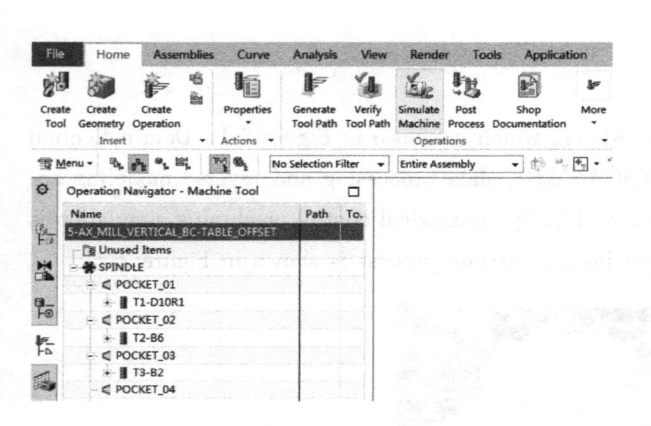

Figure 1-115　【Operation Navigator-Machine Tool】view　　Figure 1-116　【Simulation Control Panel】dialog box

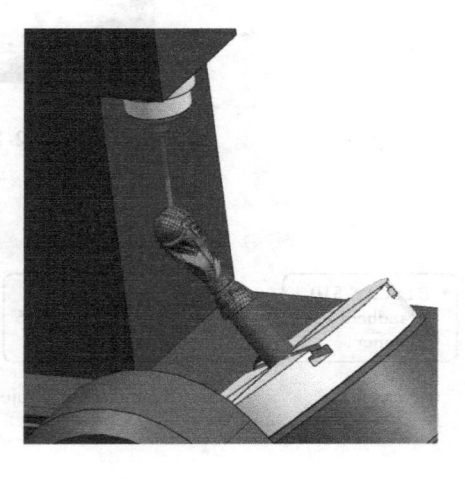

Figure 1-117　Machining simulation process

37

Project 2　Turbine Wheel

This project is based on physical turbine wheel, which is shown in Figure 2-1. Data collection is carried out by hand-held laser scanner BYSCAN 510, data processing and reverse modeling are carried out by Geomagic Design X 2016 software. Finally, numerical control machining simulation is completed in Siemens NX 10 software. Project implementation process is shown in Figure 2-2.

Figure 2-1　Turbine wheel

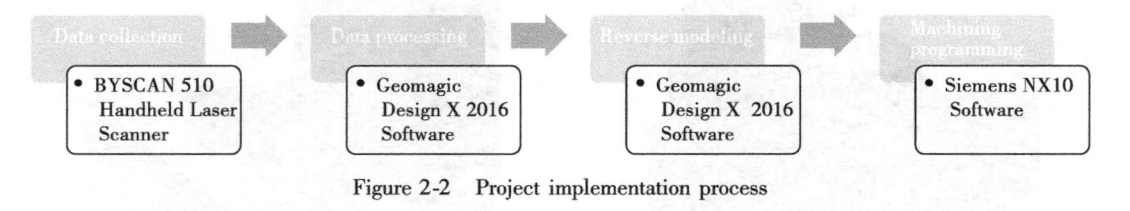

Figure 2-2　Project implementation process

Task 1　Data Collection

Data collection of this project is carried out using hand-held laser scanner BYSCAN 510. The supporting scanning software is ScanViewer, as shown in Figure 2-3.

2.1.1　Scanner Calibration

Quick calibration is required when the scanner has not been used for a long time or has experienced severe shock in transportation. Generally speaking, the scanner can be used for a long time after one calibration. The calibration steps are as follows.

Project 2　Turbine Wheel

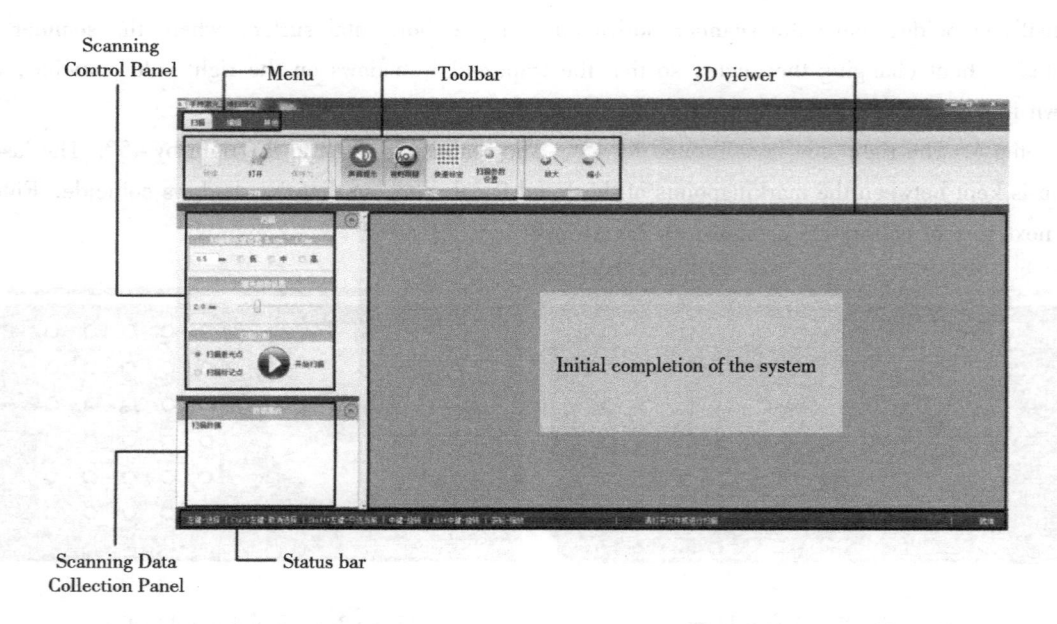

Figure 2-3　Introduction of ScanViewer software interface

Step 1. Click the【Quick Calibration】command, as shown in Figure 2-4. A quick calibration interface appears after clicking, as shown in Figure 2-5.

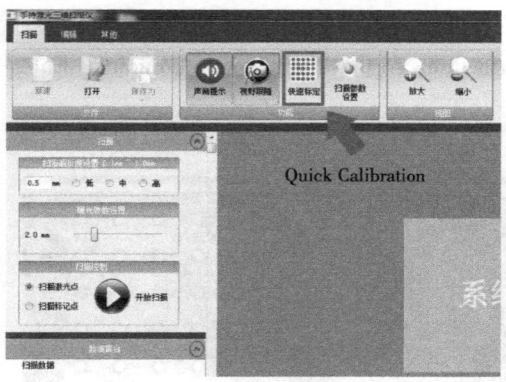

Figure 2-4　Click【Quick Calibration】

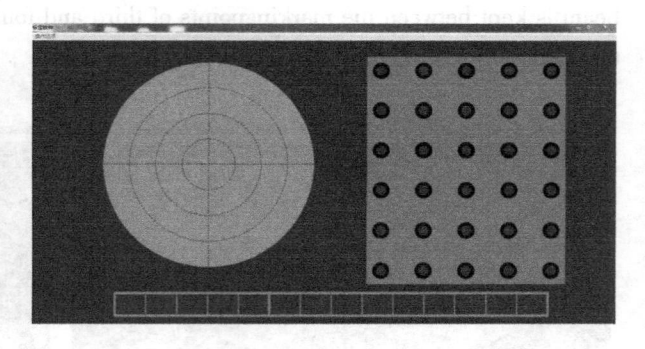

Figure 2-5　Quick calibration interface

Step 2. Place the calibration plate on a stable surface. The scanner is facing the calibration plate at a distance of 30 cm. Press the third key (hereinafter referred to as the scan key) from the top to the bottom of the scanner, and emit a laser beam (taking three parallel lasers as an example), as shown in Figure 2-6.

Step 3. Control the angle of the scanner, and adjust the distance between the scanner and the calibration plate to make the left shadow circles coincide. Ensuring that the left shadow circles

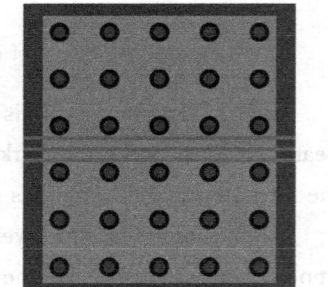

Figure 2-6　Press the scan key

39 ◀

3D Digital Design and Manufacturing

basically coincide, move the scanner horizontally on the horizontal surface where the scanner is located without changing the angle, so that the trapezoidal shadows on the right side coincide, as shown in Figure 2-7. Enter the next step of calibration.

Step 4. The right side is calibrated by 45°. The scanner is tilted to the right by 45°. The laser beam is kept between the markingpoints of third and fourth line, so that the shadows coincide. Enter the next step of calibration as shown in Figure 2-8.

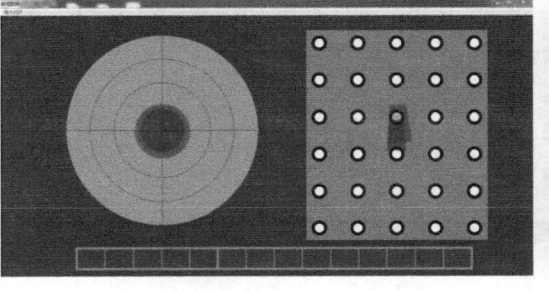 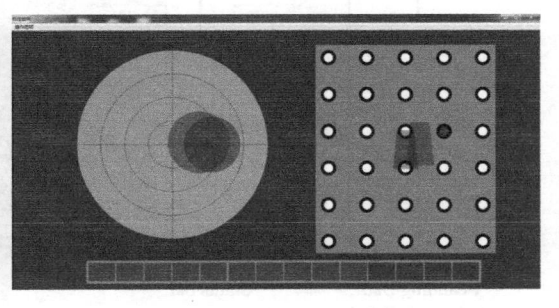

Figure 2-7　Step 3 of calibration　　　　　　　　Figure 2-8　Step 4 of calibration

Step 5. The left side is calibrated by 45°. The scanner is tilted to the left by 45°. The laser beam is kept between the markingpoints of third and fourth line, so that the shadows coincide. Enter the next step of calibration as shown in Figure 2-9.

Step 6. The upper side is calibrated by 45°. The scanner is tilted upwards by 45°. The laser beam is kept between the markingpoints of third and fourth line, so that the shadows coincide. Enter the next step of calibration as shown in Figure 2-10.

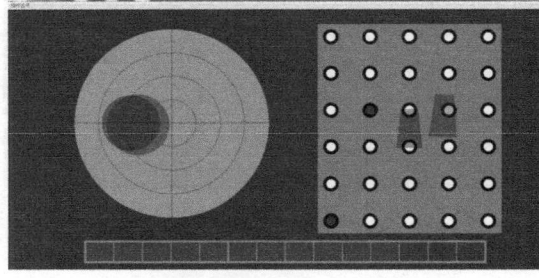 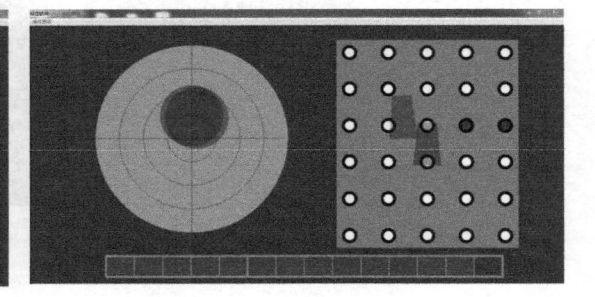

Figure 2-9　Step 5 of calibration　　　　　　　　Figure 2-10　Step 6 of calibration

Step 7. The lower side is calibrated by 45°. The scanner is tilted downwards by 45°. The laser beam is kept between the markingpoints of third and fourth line, so that the shadows coincide. Enter the next step of calibration as shown in Figure 2-11.

After completing the seventh step, the interface prompts are shown in Figure 2-12. Click the upper right corner to close the calibration window. The calibration is completed here.

▶ 40

Project 2   Turbine Wheel

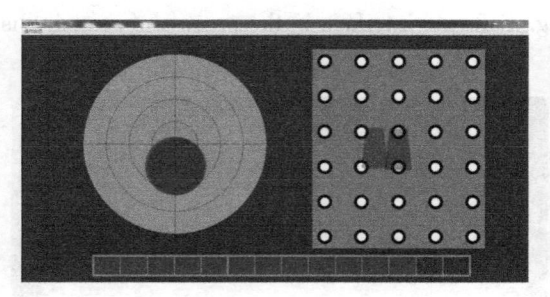
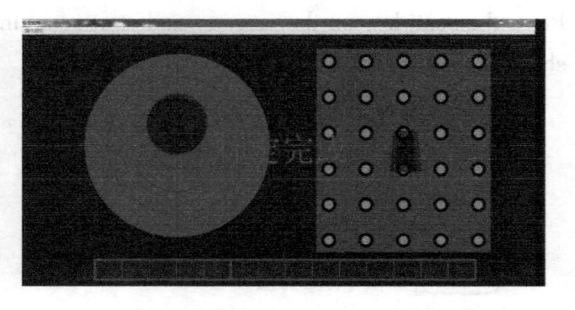

Figure 2-11   The seventh step of calibration   Figure 2-12   Calibration is completed

2.1.2   Stick Mark Points

Stick mark points on the front and back sides of the turbine wheel before the scanning starts, as shown in Figure 2-13. Five to six mark points are stuck to the front and back transition areas. When scanning the back after the front scanning is completed, attention should be paid not to touch or move the mark points in the transition area, otherwise the accuracy of stitching may be affected.

a) Front mark points   a) Front mark points

Figure 2-13   Stick mark points

2.1.3   Scan Mark Points

When using hand-held laser scanner BYSCAN 510 to scan objects, the laser points can be scanned directly, or the mark points can be scanned before scanning the laser points. The latter has higher scanning accuracy and convenient transition in the scanning process. The front and back transition area of turbine wheel is a narrow circumferential area perpendicular to the bottom. Practical operation shows that direct scanning of laser points can easily lead to splicing failure, so the method of scanning mark points first is adopted here. The auxiliary plate is used when scanning mark points.

1) Turn on ScanViewer, the scanning software of BYSCAN 510 hand-held laser scanner.

2) In the scanning control panel, set the 【Scanning Resolution Settings】 to "1mm", and the 【Exposure Parameter Settings】 to "1ms". Select the 【Mark Point】 option, and then click the 【Start】 button, as shown in Figure 2-14.

3) Place the turbine wheel as shown in Figure 2-15. The scanner is facing the turbine wheel.

41 ◄

3D Digital Design and Manufacturing

Press the scan key on the scanner and start scanning. Software interface in the process of scanning is shown in Figure 2-16.

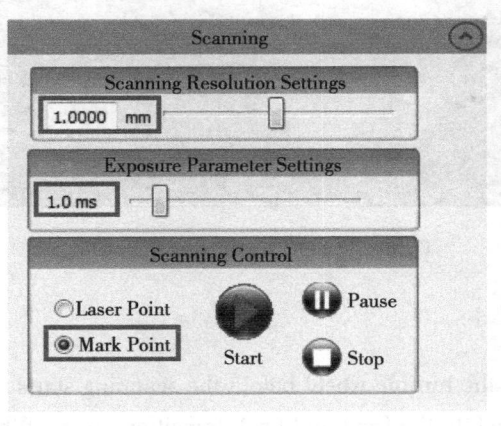

Figure 2-14　Parameter setting of mark points scanning　　　Figure 2-15　Placement of the turbine wheel

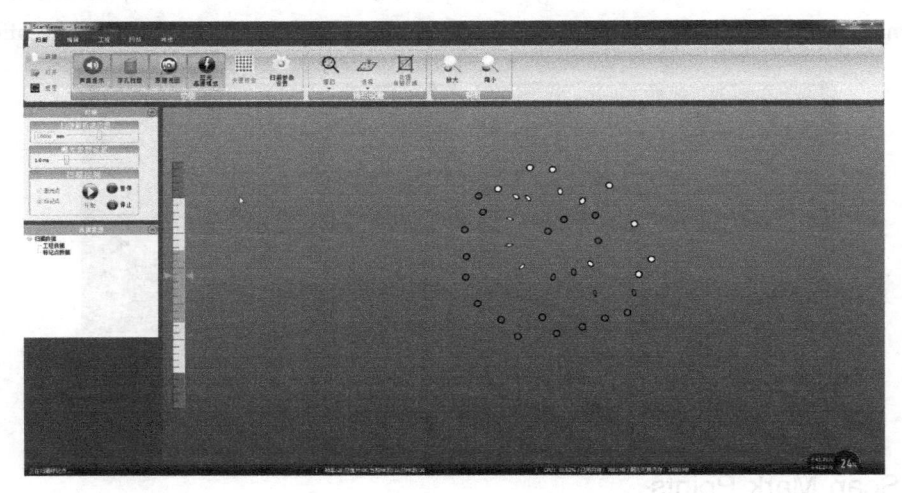

Figure 2-16　Software interface when scanning mark points

4) After scanning the mark points on the front of the turbine wheel, place the auxiliary plate on the workbench and adjust the placement position of the turbine wheel, so that the mark points on the front and back sides can be scanned, as shown in Figure 2-17. Because the transition between the front and the back side of the turbine wheel is 90°, it is not easy to determine the relative position of the mark points on the front and the back side, so the auxiliary plate is used here. Firstly, the front side of the turbine wheel and the auxiliary plate are scanned to determine the relative position relationship between the mark points on the front side of the turbine wheel and the mark points on the auxiliary plate. Then, the back side of the turbine wheel and the auxiliary plate are scanned to determine the relative position relationship between the mark points on the back side of the turbine wheel and the mark points on the auxiliary plate. In this way, the relative position relationship between the mark points on the front side and on the back side of the turbine wheel is determined by using the mark points on the auxiliary plate as "bridge".

▶ 42

Project 2　Turbine Wheel

5) Finally, delete the mark points on the scanned auxiliary plate, as shown in Figure 2-18. The mark point scanning is completed.

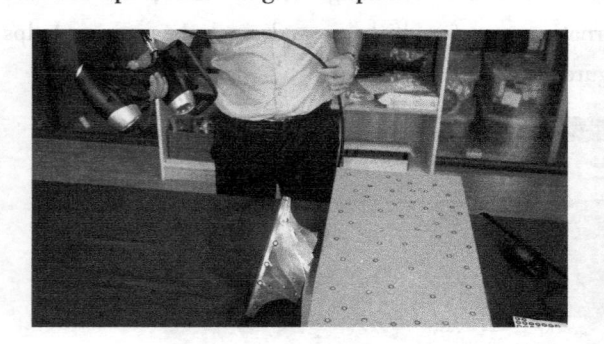

Figure 2-17　Use auxiliary plate

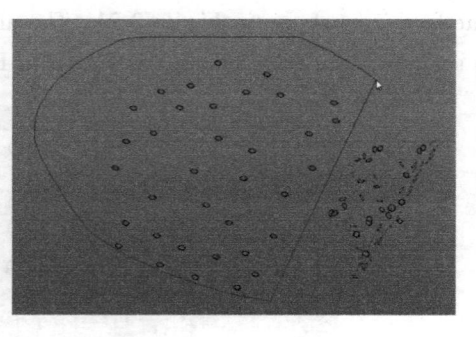

Figure 2-18　Select mark points on the auxiliary plate and delete them

2.1.4　Scan Laser Points

Laser point scanning is performed after mark point scanning is completed. It is difficult to scan the area between turbine blades. In order to scan easily, a single beam scanning method can be used.

1) In the scanning control panel, set 【Scanning Resolution Settings】 to "0.5mm", 【Exposure Parameter Settings】 to "1ms". Select the 【Laser Point】 option, and then click the "Start" button. Select【Red Light】 in the pop-up drop-down list, as shown in Figure 2-19.

2) Place the turbine wheel to make it face up, and make the scanner face the turbine blade. Press the scan key on the scanner to start scanning. Software interface in the process of scanning is shown in Figure 2-20.

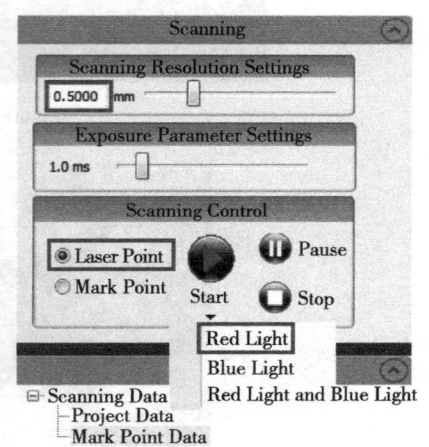

Figure 2-19　Parameter setting of laser point scanning

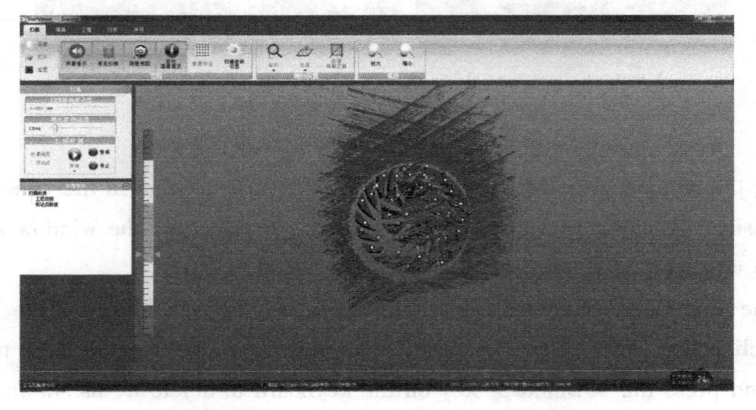

Figure 2-20　Software interface when scanning laser points

3D Digital Design and Manufacturing

It is difficult to scan the area between blades. After double-clicking the scan key on the scanner, it can be switch to the mode of single-beam laser scanning. At this time, the software interface is as shown in Figure 2-21. The alternating use of multi-beam and single-beam laser helps to scan areas between blades, as shown in Figure 2-22.

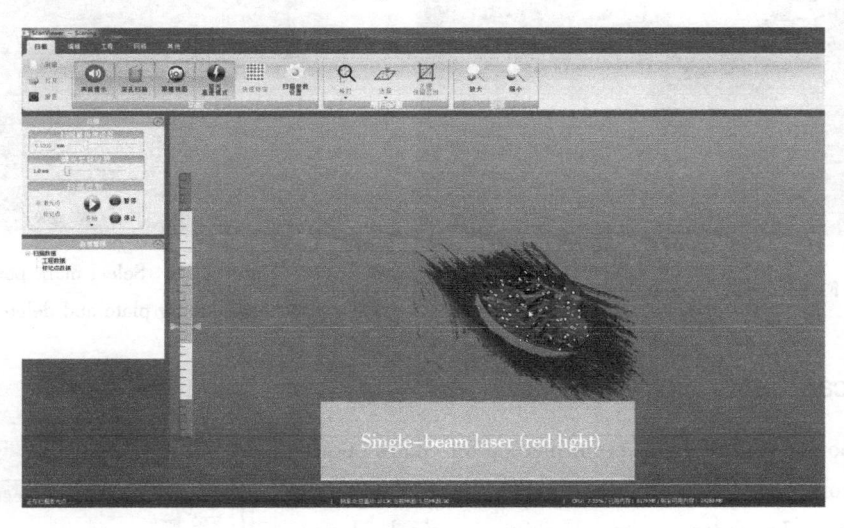

Figure 2-21　Single-beam laser scanning

a) Multi-beam scanning　　　　　b) Single-beam scanning

Figure 2-22　Multi-beam and Single-beam Scanning

By pressing the window zoom-in key on the scanner, you can zoom in the view and observe the scanning area easily, as shown in Figure 2-23. Similarly, by pressing the window zoom-out key on the scanner, the view can be zoomed out.

3) After the front side scanning is completed, press the scan key on the scanner to stop scanning, then click the "Pause" button on the scanning software. Use the lasso tool to select the irrelevant data and press the < Delete > key on the keyboard to delete it, as shown in Figure 2-24.

4) Adjust the placement position of the turbine wheel to make the back side face up. Click the

▶ 44

Project 2　Turbine Wheel

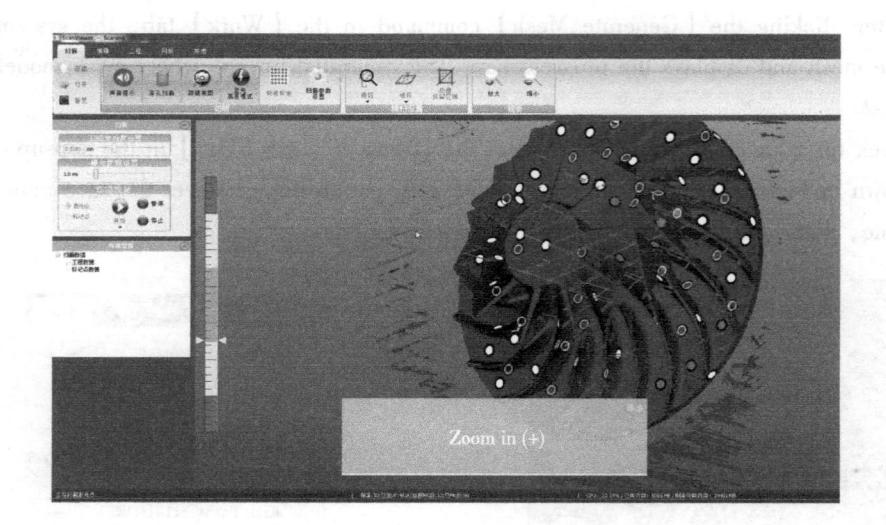

Figure 2-23　Zoom in the view

Figure 2-24　Select and delete the irrelevant data

"Start" button on the software interface, and select 【Red Light】 in the pop-up drop-down list, as shown in Figure 2-19. Then press the scan key on the scanner to start scanning, as shown in Figure 2-25.

5) After the back side scanning is completed, press the scan key on the scanner to stop scanning, then click the "Pause" button on the scanning software. Use the lasso tool to select the irrelevant data and press the < Delete > key on the keyboard to delete it, as shown in Figure 2-26.

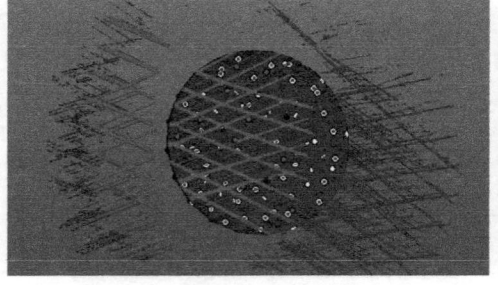

Figure 2-25　Scan the back side of the turbine wheel

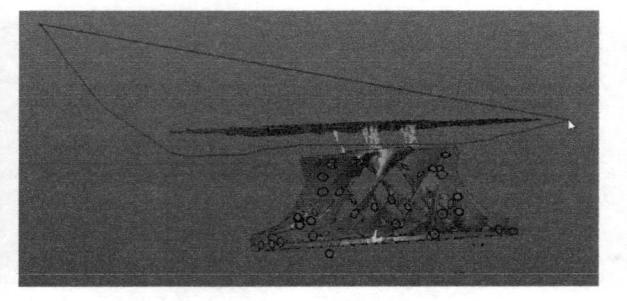

Figure 2-26　Delete the irrelevant scanning data

45

3D Digital Design and Manufacturing

6) After clicking the 【Generate Mesh】 command in the 【Work】 tab, the system starts to generate the mesh and displays the progress bar. The generated turbine wheel mesh model is shown in Figure 2-27.

7) Click the 【Save】 command and select the 【Mesh File (*STL)】 in the pop-up drop-down list, as shown in Figure 2-28. In the pop-up 【Save As】 dialog box, select the save path and enter the file name, then click the 【Save】 button.

Figure 2-27　Generated turbine wheel mesh model

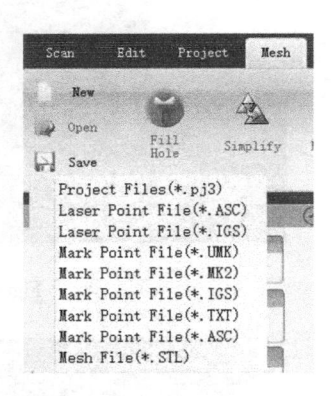

Figure 2-28　Save as STL file

Task 2　Data Processing

1) Open the software. Click the Geomagic Design X 2016 on the "Start" menu, or double-click the icon 🗗 on the desktop, to start the Geomagic Design X 2016 application software.

2) Import the model. Click the "Import" button 📁 at the top left of the interface, as shown in Figure 2-29. Select the generated model data of scanning, and click 【Import Only】, as shown in Figure 2-30. Turbine wheel model is successfully imported into Geomagic Design X 2016 software, as shown in Figure 2-31.

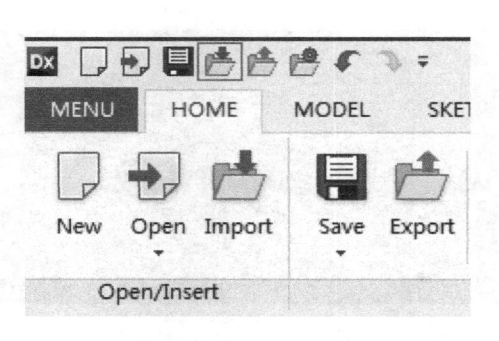

Figure 2-29　Click the "Import" button

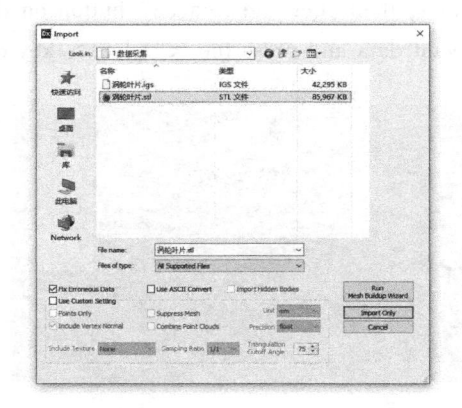

Figure 2-30　【Import】 dialog box

Project 2　Turbine Wheel

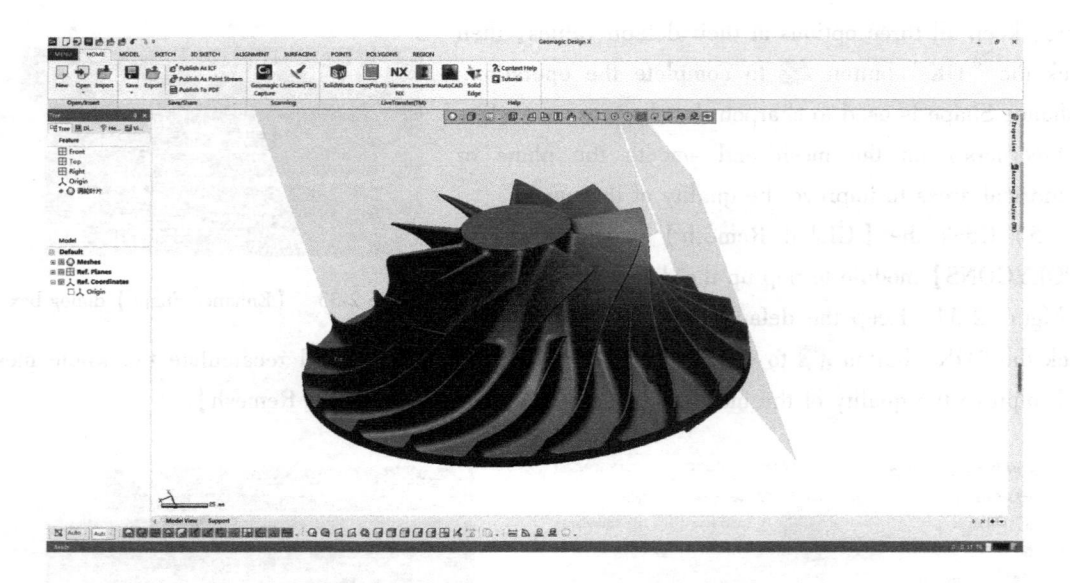

Figure 2-31　Import turbine wheel model

3）Click the【Healing Wizard】command in the【POLYGONS】module to pop up the【Healing Wizard】dialog box. The software automatically retrieves various defects in the mesh model, such as non-manifold vertices, overlapping poly-faces, suspended poly-faces, crossing poly-faces, etc., as shown in Figure 2-32. After clicking the "OK" button ✔, the software automatically fixes the defect retrieved.

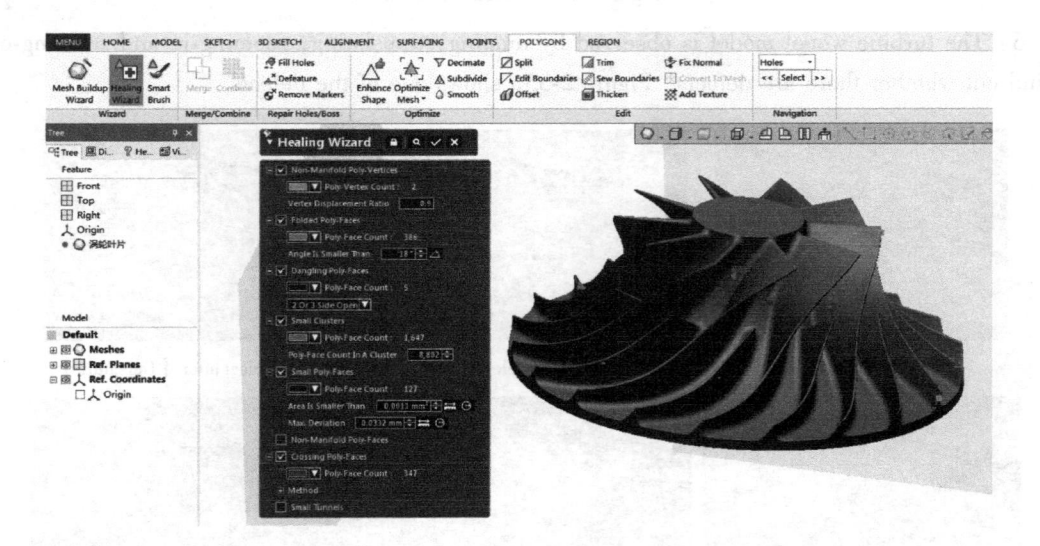

Figure 2-32　Repair mesh defects automatically with【Healing Wizard】

4）Click the【Enhance Shape】command in the【POLYGONS】module to pop up the dialog box as shown in Figure 2-33.【Sharpness】indicates setting the range of sharp areas to perform sharpening.【Overall Smoothness】indicates setting the range of fillet area to perform smoothing.【Enhance Level】indicates the number of iterations that are set to perform the operation.

47

3D Digital Design and Manufacturing

Here, keep all three options at their default values, then click the "OK" button ☑ to complete the operation. Enhance Shape is used to sharpen the sharp areas (edges and corners) on the mesh and smooth the plane or cylindrical areas to improve the quality of the mesh.

5) Click the 【Global Remesh】 command in the 【POLYGONS】 module to pop up the dialog box as shown in Figure 2-34. Keep the default parameter settings and click the "OK" button ☑ to complete the operation. The system will recalculate the whole mesh and improve the quality of the mesh by using the command of 【Global Remesh】.

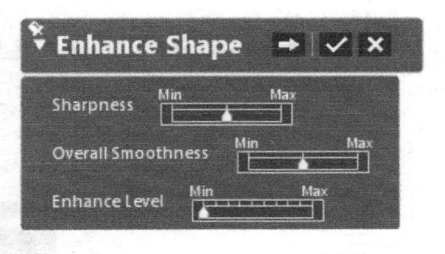

Figure 2-33　【Enhance Shape】 dialog box

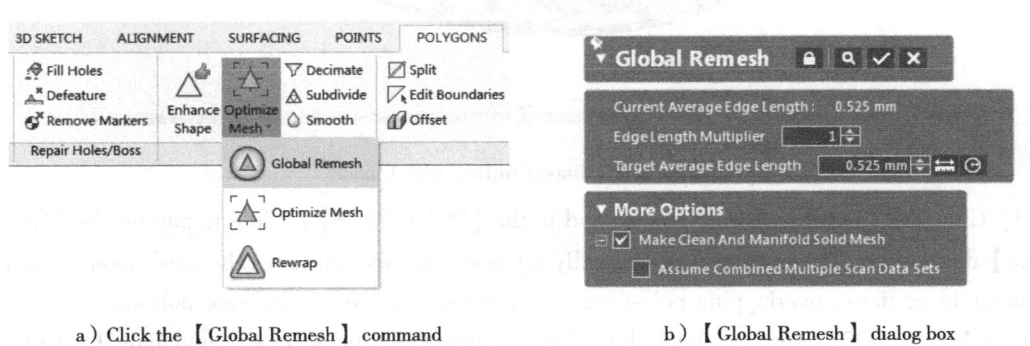

a）Click the 【Global Remesh】 command　　　　b）【Global Remesh】 dialog box

Figure 2-34　Global Remesh

6) The turbine wheel model is observed by rotating, translating, zooming-in and zooming-out, to find out whether there are defects. Figure 2-35a shows one of the defects.

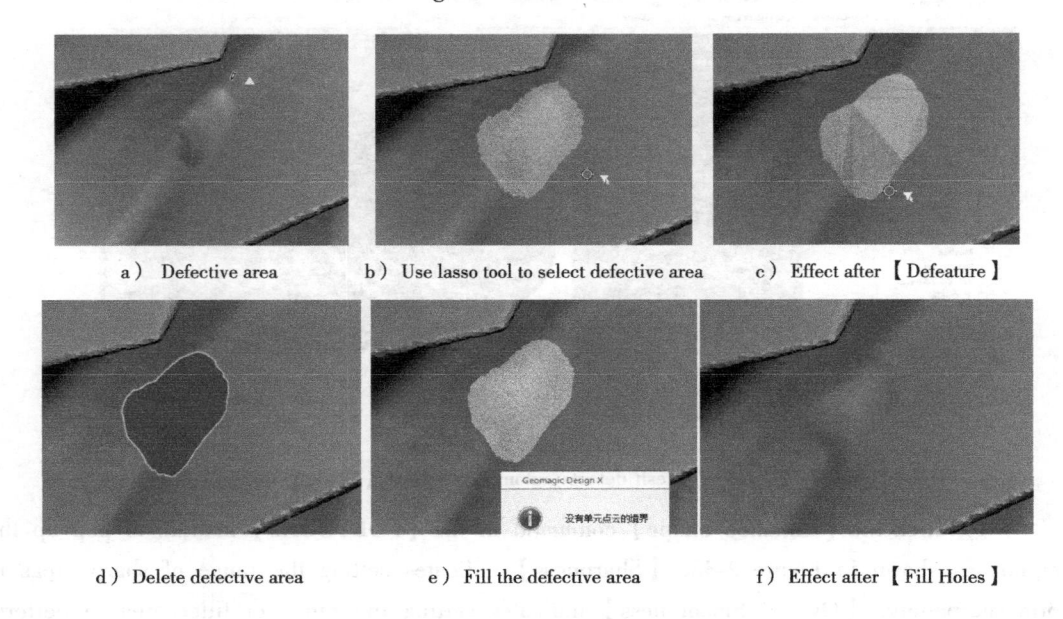

a）Defective area　　　b）Use lasso tool to select defective area　　　c）Effect after 【Defeature】

d）Delete defective area　　　e）Fill the defective area　　　f）Effect after 【Fill Holes】

Figure 2-35　Repair defective area

▶ 48

Project 2　Turbine Wheel

First, try to repair the defect with the 【Defeature】 command, which may remove the selected poly-faces and repair the area through intelligent filling. The lasso tool shown in Figure 2-36 is used to select the defect area. Click the 【Defeature】 command in the 【POLYGONS】 module to pop up the dialog box shown in Figure 2-37. The parameters are set by default, then click the "OK" button ☑ to complete the operation. The effect is shown in Figure 2-35c. Because the repair effect is not good, press < Ctrl + Z > on the keyboard to cancel the operation of Defeature.

Figure 2-36　Select lasso tool and select visible only

Next, try to repair the defect with the 【Fill Holes】 command. Use the lasso tool to select the defective area and press the Delete key on the keyboard to delete it. Click the 【Fill Holes】 command in the 【POLYGONS】 module to pop up the 【Fill Holes】 dialog box as shown in Figure 2-38. Select the hole shown in Figure 2-35d. The parameters are set by default, then click the "OK" button ☑ to complete the operation. The effect is shown in Figure 2-35f. The repair method of re-filling the hole after deleting the defect area has a better effect.

Continue to find out if there are any defects in the turbine wheel model, and if any, continue to repair them with the method of filling holes after deleting.

Figure 2-37　【Defeature】 dialog box

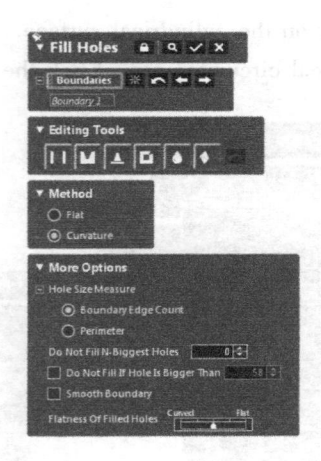

Figure 2-38　【Fill Holes】 dialog box

7) Click the 【Plane】command in the 【MODEL】 module to pop up the 【Add Plane】 dialog box. As shown in Figure 2-39, click drop-down arrow and select 【Pick Multiple Points】 in the pop-up drop-down list. As shown in Figure 2-40, select four points at the bottom of the turbine wheel model and click the "OK" button ☑. A plane can be created with three points, and a plane created with four points has higher fitting accuracy. As shown in Figure 2-41, the newly created "Plane 1" is selected by default. Press < Esc > key on the keyboard to cancel the selection.

49 ◀

3D Digital Design and Manufacturing

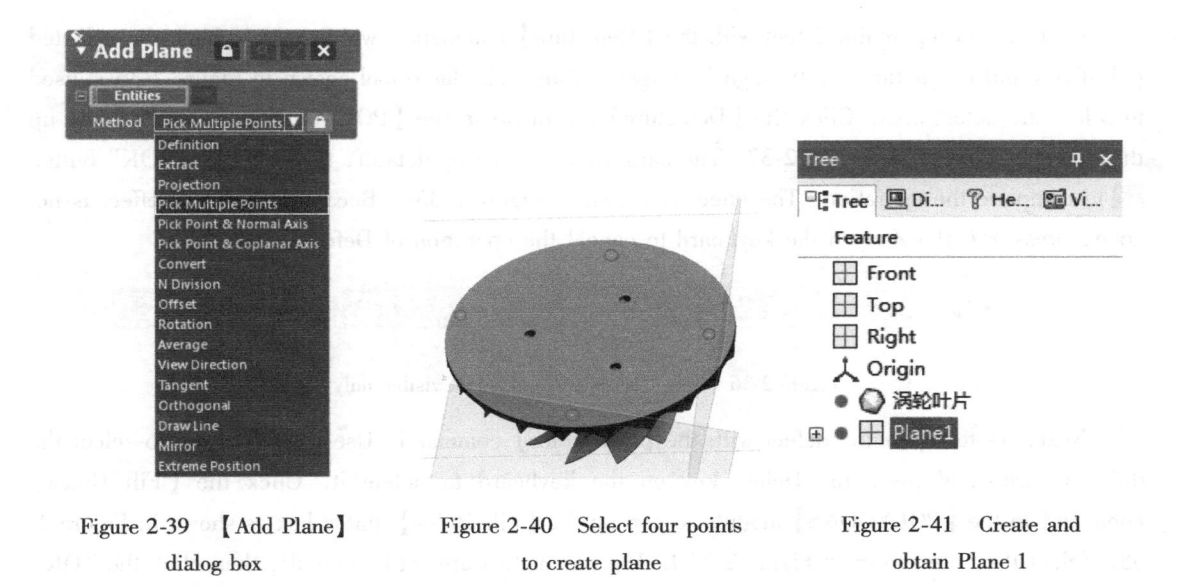

Figure 2-39　【Add Plane】
dialog box

Figure 2-40　Select four points
to create plane

Figure 2-41　Create and
obtain Plane 1

8) Click the 【Vector】 command in the 【MODEL】 module to pop up the 【Add Vector】 dialog box. As shown in Figure 2-42, click the drop-down arrow and select 【Find Cylinder Axis】 in the pop-up drop-down list. As shown in Figure 2-43, use the lasso tool to select the area on the cylindrical surface at the bottom of the turbine wheel model. Be careful not to select the area with poor quality on the cylindrical surface. Meanwhile, the selected area is roughly distributed around the cylindrical circumference. Click the "OK" button ☑ to create "Vector 1".

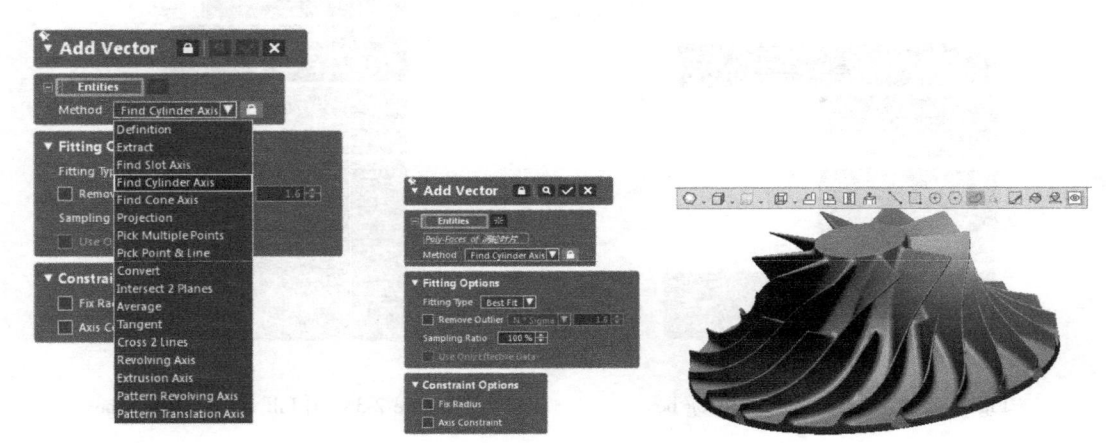

Figure 2-42　【Add Vector】 dialog box

Figure 2-43　Select the area on the
bottom cylindrical surface of the turbine whell

9) Click the 【Interactive Alignment】 command in the 【ALIGNMENT】 module to pop up the dialog box, as shown in Figure 2-44a. After clicking the "Next" button ➡, the dialog box is as shown in Figure 2-44b. Select the 【X-Y-Z】 option, then click 【Position】 in the dialog box and select "Vector 1" and "Plane 1", which can be selected either in the graphics window or in the feature tree. The location is the intersection of "Vector 1" and "Plane 1". Click the 【Axis Z】 in

▶ 50

Project 2   Turbine Wheel

the dialog box, then select "Plane 1". Axis Z is the normal direction of "Plane 1". Check whether the Axis Z of the coordinate system is pointed to the head from the bottom of the turbine wheel. If not, double-click the Axis Z arrow to reverse it. After clicking the "OK" button ✓, a coordinate system is created as shown in Figure 2-45.

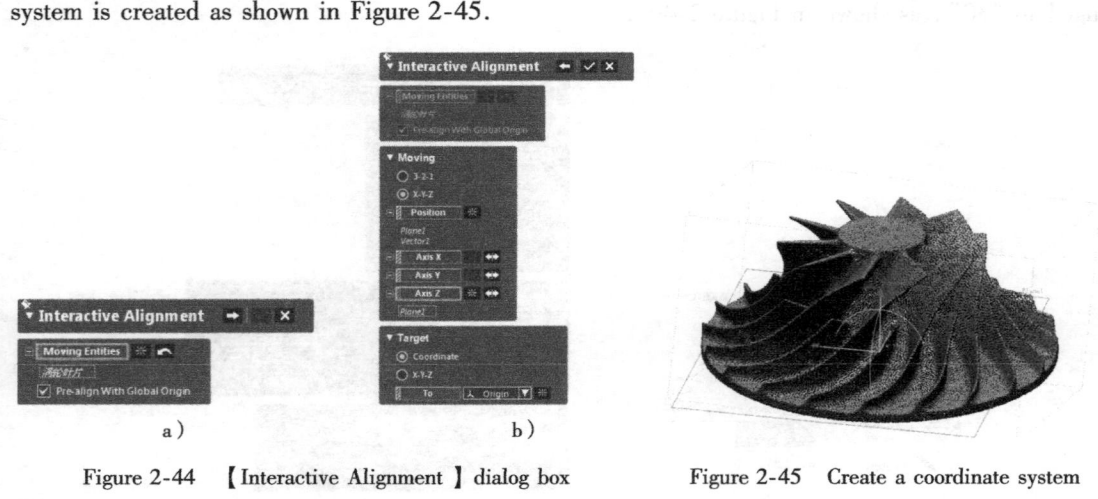

Figure 2-44   【Interactive Alignment】 dialog box   Figure 2-45   Create a coordinate system

Task 3   Reverse Modeling

2.3.1   Main Body Modeling

1) Select "Plane 1" and "Vector 1" in the feature tree and click 【Delete】, then click 【Yes】 in the pop-up dialog box, as shown in Figure 2-46.

2) Click the 【Vector】 command in the 【MODEL】 module to pop up the 【Add Vector】 dialog box. As shown in Figure 2-47, click drop-down arrow and select 【Intersect 2 Planes】 in the pop-up drop-down list. Select the 【Top】 plane and 【Right】 plane, then click the "OK" button ✓ to create the intersection of 【Top】 plane and 【Right】 plane, which is "Vector 1".

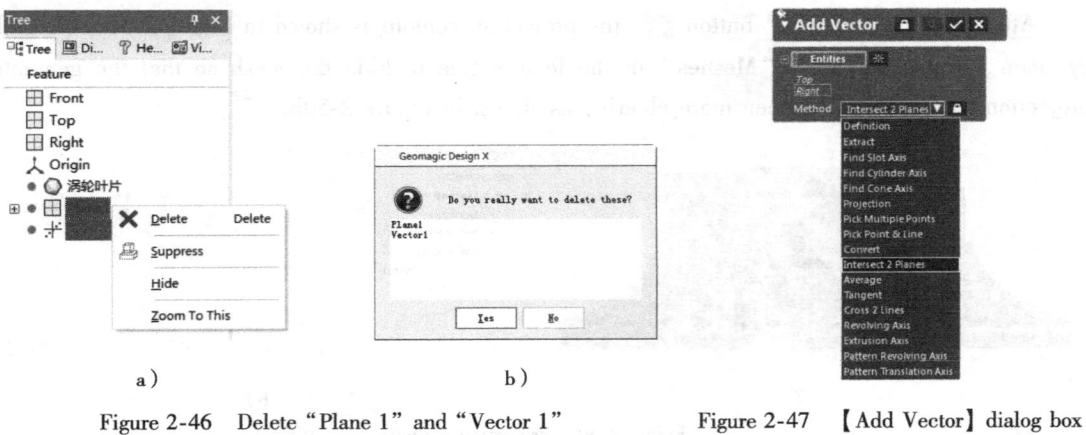

Figure 2-46   Delete "Plane 1" and "Vector 1"   Figure 2-47   【Add Vector】 dialog box

51 ◀

3D Digital Design and Manufacturing

3) Click the 【Mesh Sketch】 command in the 【SKETCH】 module to pop up the 【Mesh Sketch Setup】 dialog box, as shown in Figure 2-48. Select 【Rotational Method】. Click the 【Central Axis】 and then select "Vector 1". Click 【Base Plane】 and then select "Right plane". Set 【Silhouette Range】 to "30", as shown in Figure 2-49b.

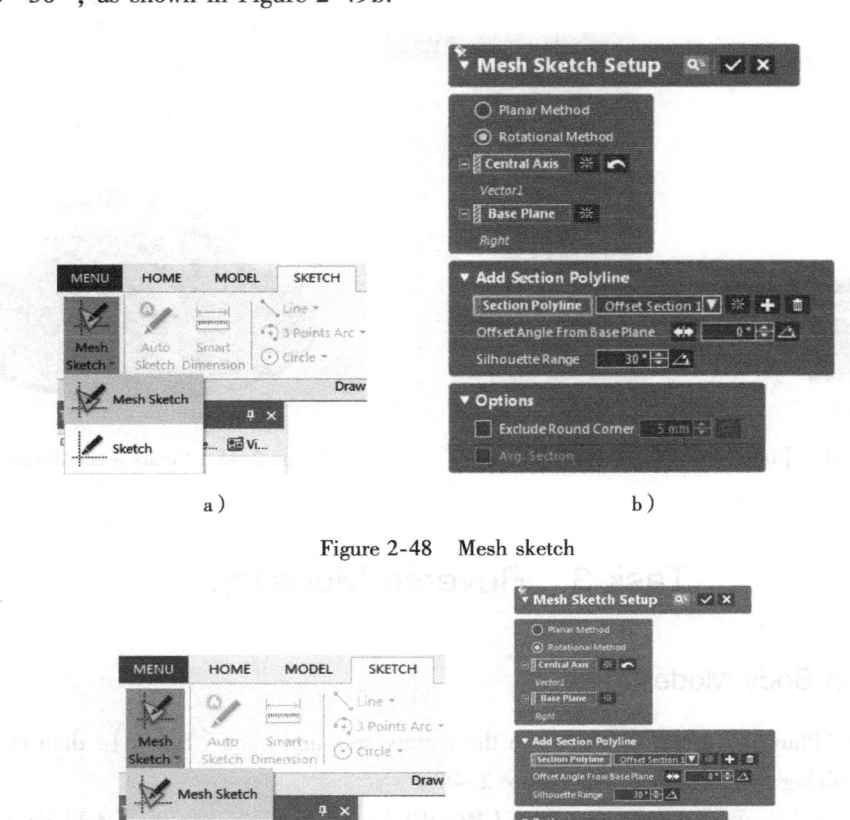

a)   b)

Figure 2-48   Mesh sketch

a) Silhouette range: 0   b) Silhouette range: 30

Figure 2-49   The effect of different silhouette range

After clicking the "OK" button ✓, the projection contour is shown in Figure 2-50. Close the eye icon ◉ in front of the "Meshes" in the feature tree to hide the mesh so that the generated projection contours can be seen more clearly, as shown in Figure 2-50b.

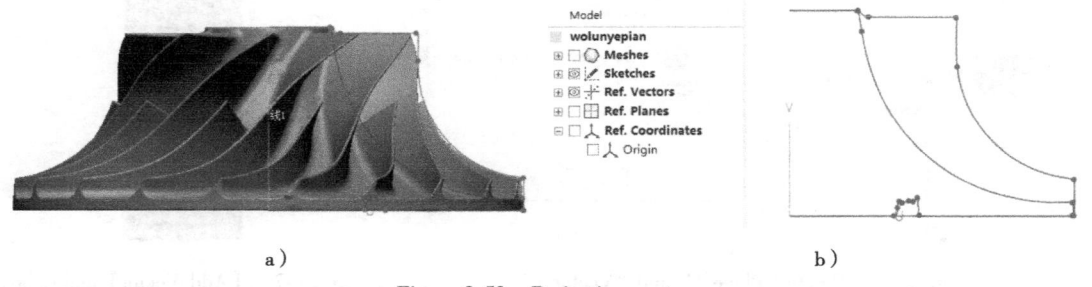

a)   b)

Figure 2-50   Projection contour

▶ 52

Project 2　Turbine Wheel

4) Click the 【Line】 command in the 【SKETCH】 module to pop up the 【Line】 dialog box. Select the origin of the coordinate system as the starting point of the line, and drag the line to the right. The system automatically captures the level constraint —. Then click and select the end of the line in the position beyond the projection contour, as shown in Figure 2-51. Press < Esc > key on the keyboard during creating a line to cancel the creation of a continuous line. By creating the second line in the same way, the line can be dragged left and right to make it pass through the end of the projection contour, as shown in Figure 2-52. Continue to create the remaining lines in the same way, as shown in Figure 2-53.

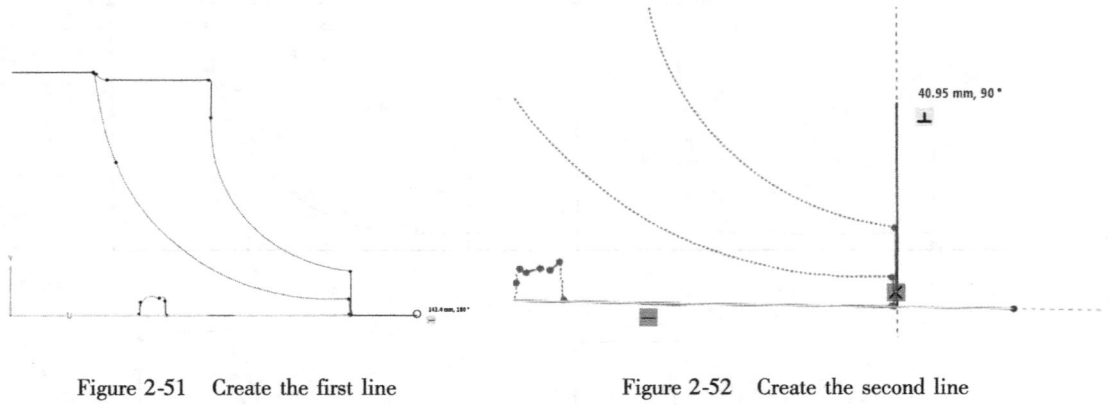

Figure 2-51　Create the first line　　　　　　Figure 2-52　Create the second line

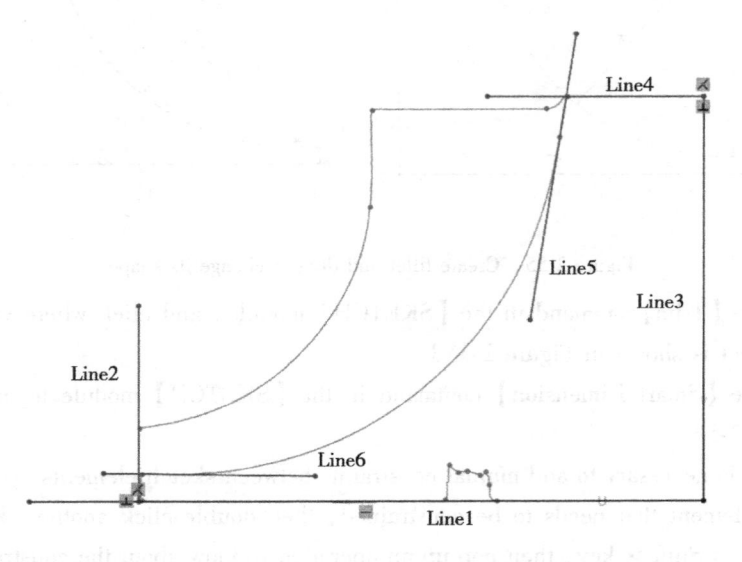

Figure 2-53　Continue to create line

5) Click the "Fillet" command in the 【SKETCH】 module, as shown in Figure 2-54. Select Line 6 and Line 5 as shown in Figure 2-53 in turn. Continue to hold down the left mouse button and drag the fillet to change its size so that they coincide with the contour, as shown in Figure 2-55a, b and c. After releasing the left mouse button, the radial dimension of the fillet will be automatically created. Double-click the radial dimension and enter "80. 8".

53

3D Digital Design and Manufacturing

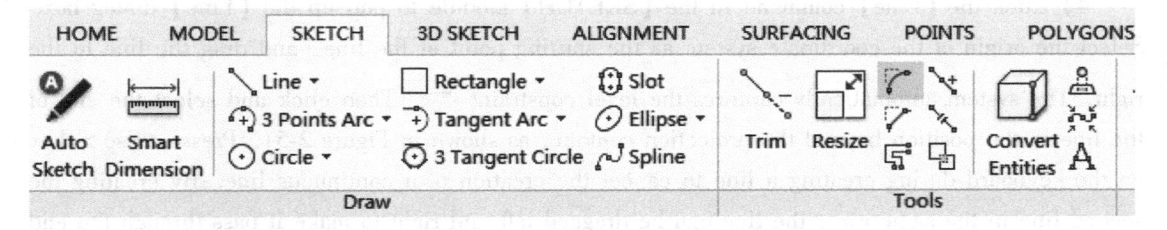

Figure 2-54　Click "Fillet" command

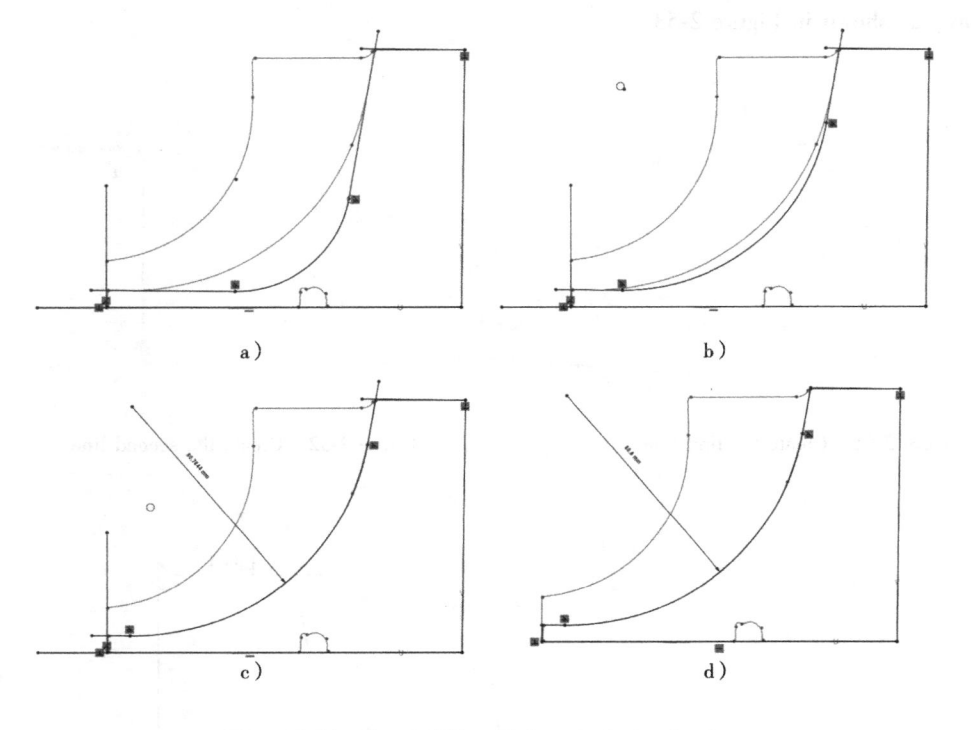

Figure 2-55　Create fillet and drag to change its shape

6) Click the 【Trim】command in the 【SKETCH】 module, and click where you need to trim. The trimmed effect is shown in Figure 2-55d.

7) Click the 【Smart Dimension】 command in the 【SKETCH】 module to mark the size, as shown in Figure 2-56.

8) When it is necessary to add mutual constraints between sketch elements, you can first select the first sketch element that needs to be constrained, then double-click another sketch element by holding down the <Shift> key, then pop up an operation window about the constraints between the two elements, as shown in Figure 2-57. The mutual constraints between the two elements are shown in the 【Common Constraint】, and the respective constraints of the two elements are shown in the 【Independent Constraint】.

In Figure 2-57, there are over-constraints, and the over-constraints are expressed in red. One of the three constraints: the vertical constraints of Line 3 and Line 4, the 90°angle size constraints of Line 1 and Line 3, and the parallel constraints between Line 1 and Line 4, can be removed to solve

▶ 54

Project 2   Turbine Wheel

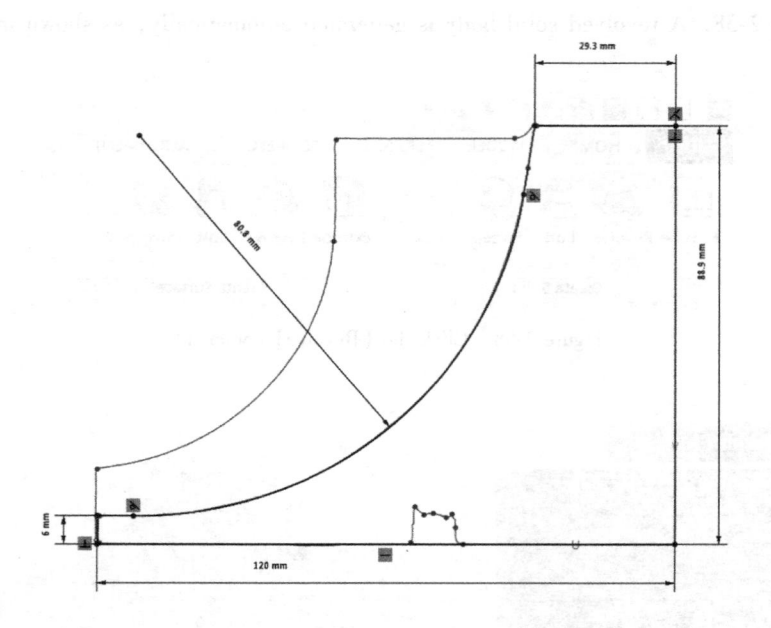

Figure 2-56   Mark size and set constraints

the over-constraints. You can select the constraints you want to delete in the 【Constraint】 dialog box
and then click the【delete constraint】button below.

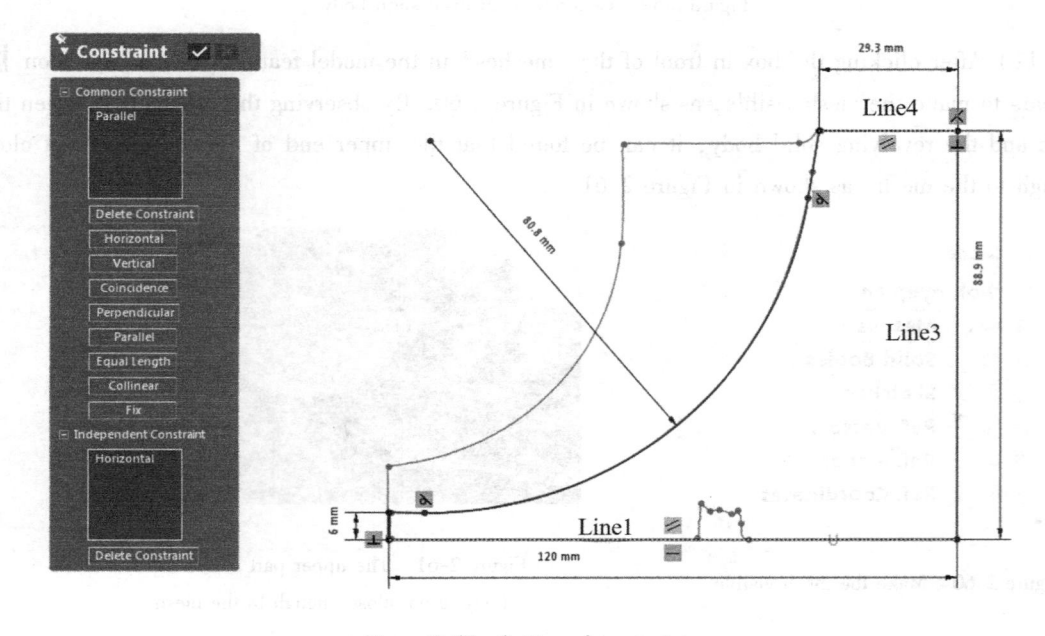

Figure 2-57   Setting of constraints

9) Click the "exit" button ⬅ in the upper left corner of the interface, or the "exit" button ⬅
in the lower right corner of the interface to exit the sketch.

10) Click the 【Revolve】 command in the 【Create Solid】 group in the 【MODEL】 module, as

55 ◀

3D Digital Design and Manufacturing

shown in Figure 2-58. A revolved solid body is generated automatically, as shown in Figure 2-59.

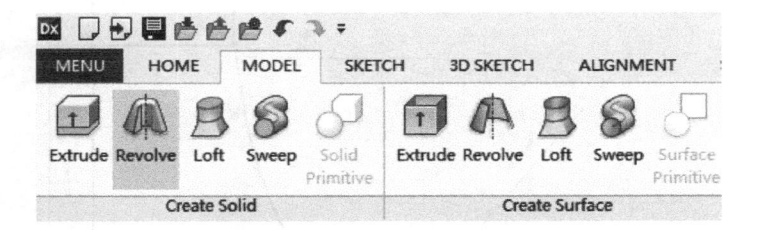

Figure 2-58 Click the 【Revolve】 command

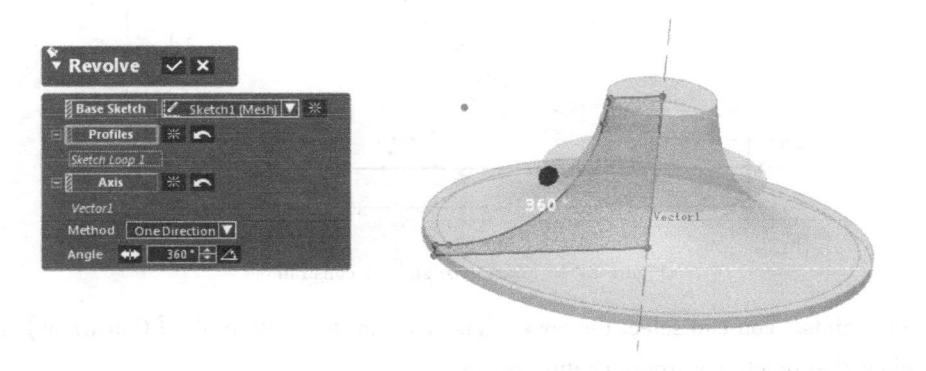

Figure 2-59 Generate a revolved solid body

11) After clicking the box in front of the "meshes" in the model feature tree, an eye icon ⊚ appears to make the mesh visible, as shown in Figure 2-60. By observing the proximity between the mesh and the revolving solid body, it can be found that the upper end of the turbine is not close enough to the mesh, as shown in Figure 2-61.

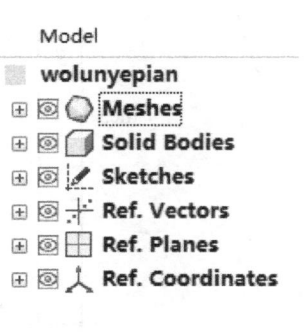

Figure 2-60 Make the mesh visible

Figure 2-61 The upper part of the solid body is not close enough to the mesh

12) Double-click on "Sketch 1 (Mesh)" in the feature tree, as shown in Figure 2-62, to enter the sketch module. Double-click the size shown in Figure 2-63, and change it from "29.3" to "29.5", then exit the sketch. You can see that the modified revolved solid body is close to the mesh, as shown in Figure 2-64.

▶ 56

Project 2　Turbine Wheel

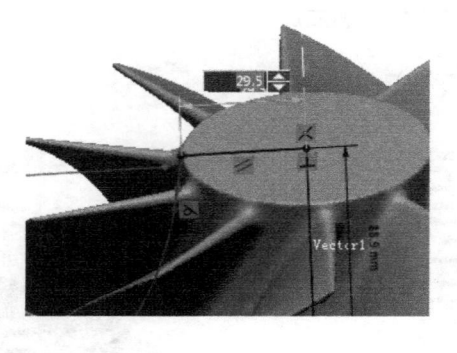

Figure 2-62　Double-click "Sketch 1 (Mesh)"　　　Figure 2-63　Modify dimension value to 29.5

13) The main part of the turbine wheel is modeled, as shown in Figure 2-65.

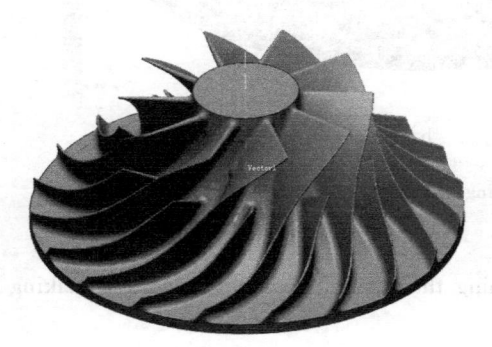

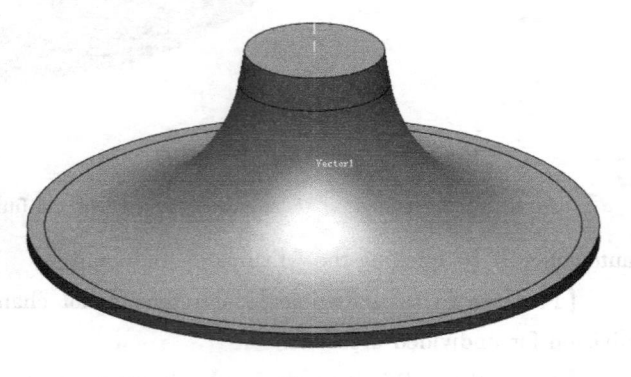

Figure 2-64　The modified revolved
solid body is close to the mesh

Figure 2-65　Main body of the turbine wheel

2.3.2　Blade Modeling

1) Check the fitting accuracy of turbine blade body. Click the【Accuracy Analyzer (TM)】on the right side of the interface, as shown in Figure 2-66a, and select【Deviation for Body】in the pop-up dialog box. The deviation diagram is shown in Figure 2-66b. From the deviation diagram, it can be seen that the fitting accuracy meets the requirements.

2) Click the【Auto Segment】command in the【REGION】module to pop up the dialog box as shown in Figure 2-67. Keep the default parameter settings and clicks the "OK" button ✔. The result is shown in Figure 2-68. As shown, different colors represent different regions.

The main options in the【Auto Segment】dialog box are described as follows.

【Sensitivity】: It refers to curvature sensitivity. The lower the sensitivity value is, the less regions are divided. Conversely, higher sensitivity value means more regions are divided. The range of the option is 0-100.

【Mesh Roughness】: It refers to the roughness of the current polygon model. When it is used to calculate curvature, the influence of roughness on region division is neglected. From smooth to rough, the roughness can be divided into four grades. Generally, the roughness can be calculated

57 ◀

3D Digital Design and Manufacturing

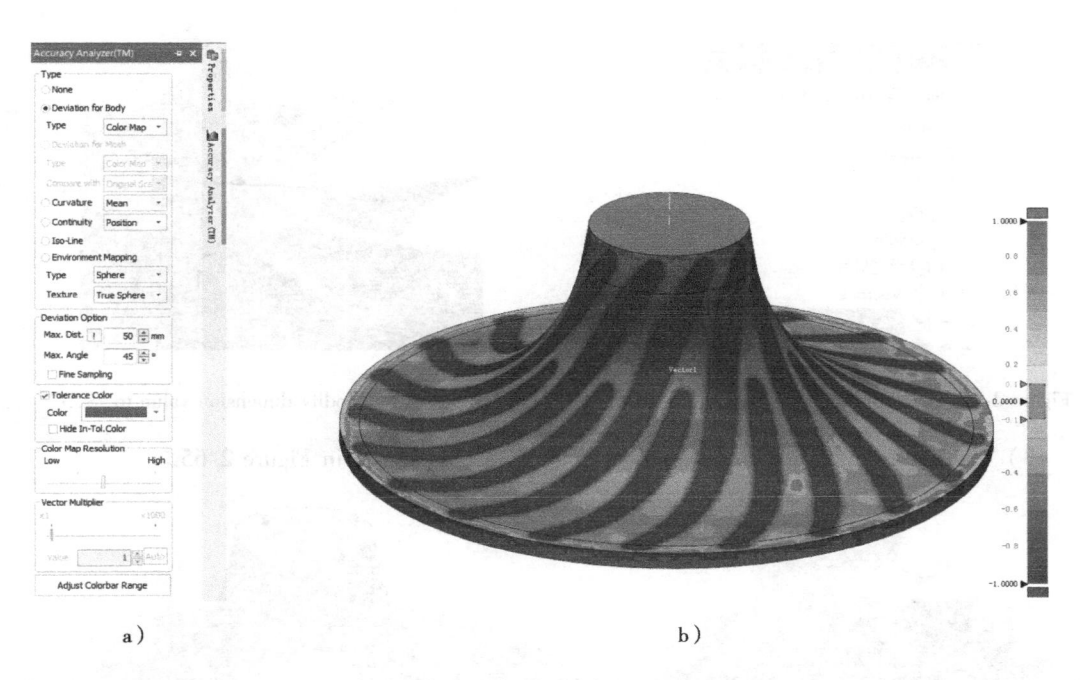

a)   b)

Figure 2-66   Check the fitting accuracy

automatically by clicking the "Estimate" button 🔆.

【Preserve Existing Regions】: It refers to not chaning the divided regions, and only making division for undivided areas.

【Merge Same Primitive Shapes】: It refers to the merging of regions of the same curvature change but not interconnected into the same region.

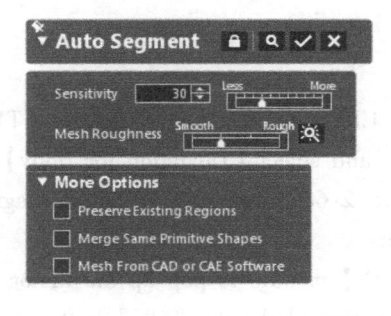

Figure 2-67   【Auto Segment】dialog box

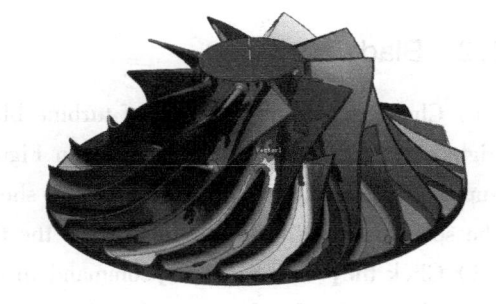

Figure 2-68   Fields after auto segmenting

3) Find a large blade and a small blade with good quality and adjacent to each other. Select the field on one side of the large blade, and click the 【Merge】command in the 【REGION】module to merge the three regions, as shown in Figure 2-69. Merge the regions on the other side of the large blade in the same way. The regions on both sides of the small blade should also be merged separately.

4) Click the 【Loft Wizard】command in the 【MODEL】module to pop up the 【Loft Wizard】dialog box. The parameter settings are shown in Figure 2-70. Select the regions as shown in Figure

▶ 58

Project 2 Turbine Wheel

2-71a. Click the "Next Stage" button ⇥ and observe the effect, then click the "OK" button ✓. The result is shown in Figure 2-71b. 【Loft Wizard】 is operated for the regions on the other side of the large blade in the same way.

Figure 2-69 The regions on one side of a large blade

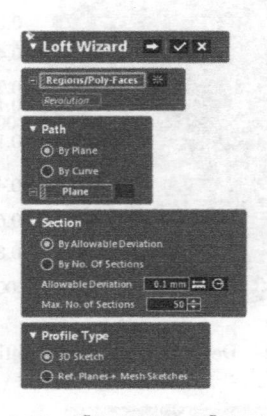

Figure 2-70 【Loft Wizard】 dialog box

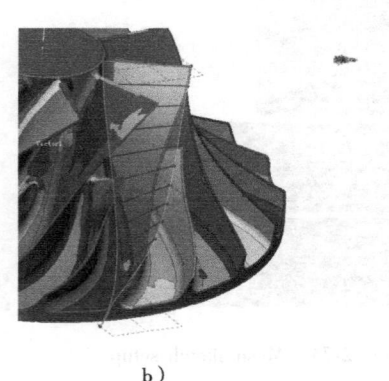

a)   b)

Figure 2-71 Extract lofting objects from regions

5) Click 【Accuracy Analyzer (TM)】 on the right side of the interface and select 【Deviation for Body】 in the pop-up dialog box. The deviation diagram is shown in Figure 2-72. From the deviation diagram, it can be seen that most of the lofting surface is green. One part of the lofting surface is red, indicating that the deviation is too large, but this part will be cut off later.

6) In the same way, 【Loft Wizard】 operations are carried out in the regions on both sides of the small blade, and the results are shown in: Figure 2-73.

7) Click the 【Mesh Sketch】 command in the 【SKETCH】 module to pop up the 【Mesh Sketch Setup】 dialog box, as shown in Figure 2-74. Select 【Rotational Method】. Click 【Central Axis】and then select "Upper Plane" and "Right Plane", taking the intersection of the two planes as the central axis. Click 【Base Plane】and then select "Right Plane". Set the 【Silhouette Range】 to "30".

After clicking the "OK" button ✓, close the eye icon ◎ in front of the "Meshes" in the feature tree to hide the mesh, so that the generated projection contours can be seen more clearly, as shown in Figure 2-75.

59 ◀

3D Digital Design and Manufacturing

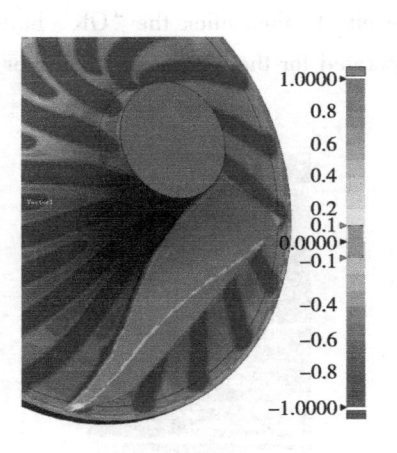

Figure 2-72　Deviation analysis of lofting surface

Figure 2-73　Four lofting surfaces created

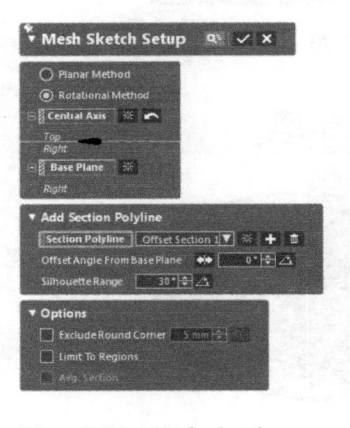

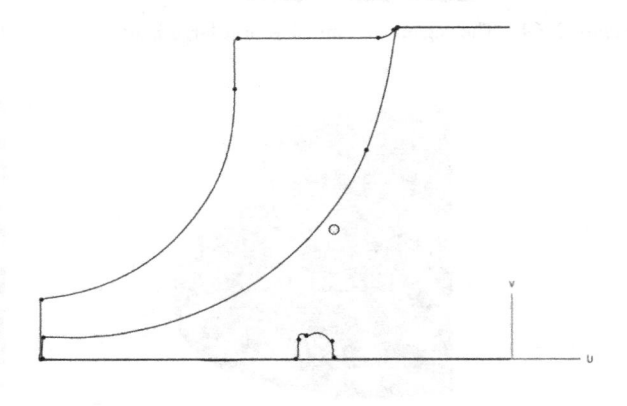

Figure 2-74　Mesh sketch setup

Figure 2-75　Projection contours

8) Click the 【Line】 command in the 【SKETCH】 module to pop up the 【Line】 dialog box, as shown in Figure 2-76. After selecting a line in the projection contour, click the button  (Accept Fitting) in the dialog box, and a line is created automatically. As shown in Figure 2-77, Line 1,

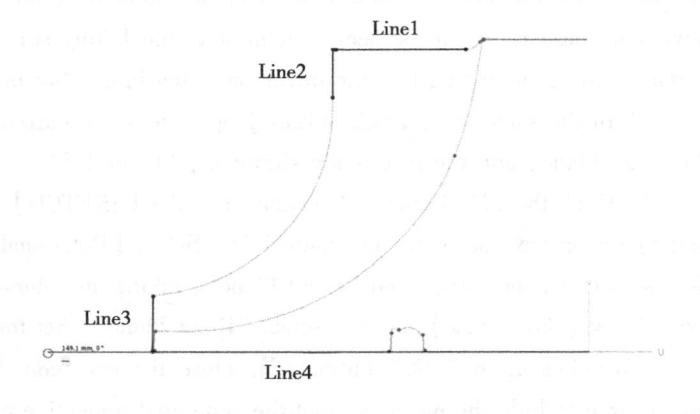

Figure 2-76　【Line】 dialog box

Figure 2-77　Create lines

Project 2 Turbine Wheel

Line 2 and Line 3 are created by this method. Line 4 is created by selecting two endpoints, and one of them is passing through the origin of the coordinate system.

9) Drag the end of Line 3 to extend it. Select Line 3 and press Shift key to select Line 4, then pop up the 【Constraint】 dialog box. Click 【Vertical】 in the【Common Constraint】, and click the "OK" button ✅ to add vertical constraints for Line 3 and Line 4, as shown in Figure 2-78.

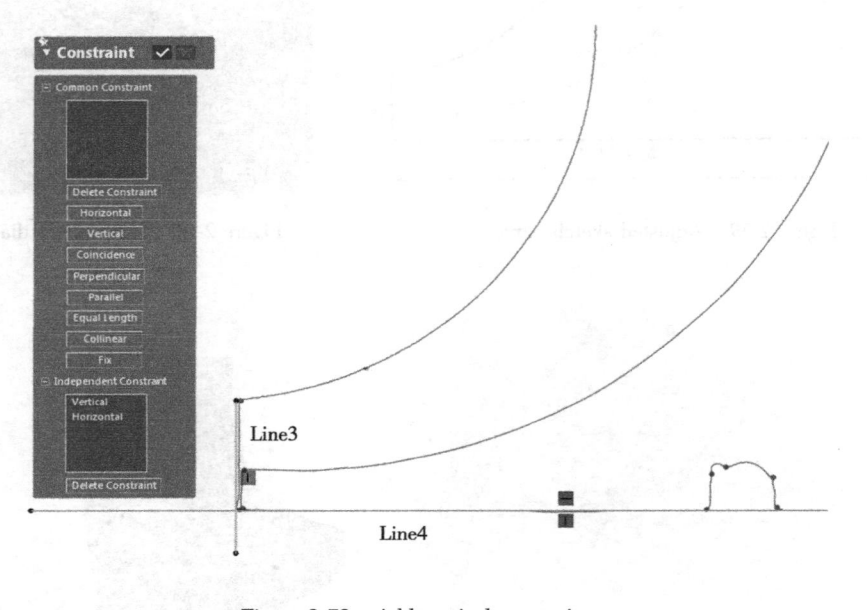

Figure 2-78　Add vertical constraints

10) Click the 【3 Points Arc】 command in the 【SKETCH】 module, then select the arc between Line 2 and Line 3 in Figure 2-77. After clicking the "OK" button ✅, the arc is created automatically.

11) Drag the endpoint of the line to lengthen it, and then trim it with the 【Trim】 command in the 【SKETCH】 module.

12) Click the 【Smart Dimension】 command in the 【SKETCH】 module to mark the size shown in Figure 2-79. The dimension value is 120 mm at the bottom and 85.8 mm at the right, and tangent and other constraints are also set. Click the "exit" button in the upper left corner of the interface, or the "exit" button in the lower right corner of the interface to exit the sketch.

13) Click the 【Revolve】 command in the 【Create Surface】 group in the 【MODEL】 module to pop up the【Revolve】 dialog box shown in Figure 2-80. Set 【Method】 to 【Both Directions】, then drag the two arrows shown in Figure 2-81a to enable the created rotation surface to intersect with the four lofting surfaces previously created. Click the "OK" button ✅ in the 【Revolve】 dialog box, and the result is shown in Figure 2-81b.

61 ◀

3D Digital Design and Manufacturing

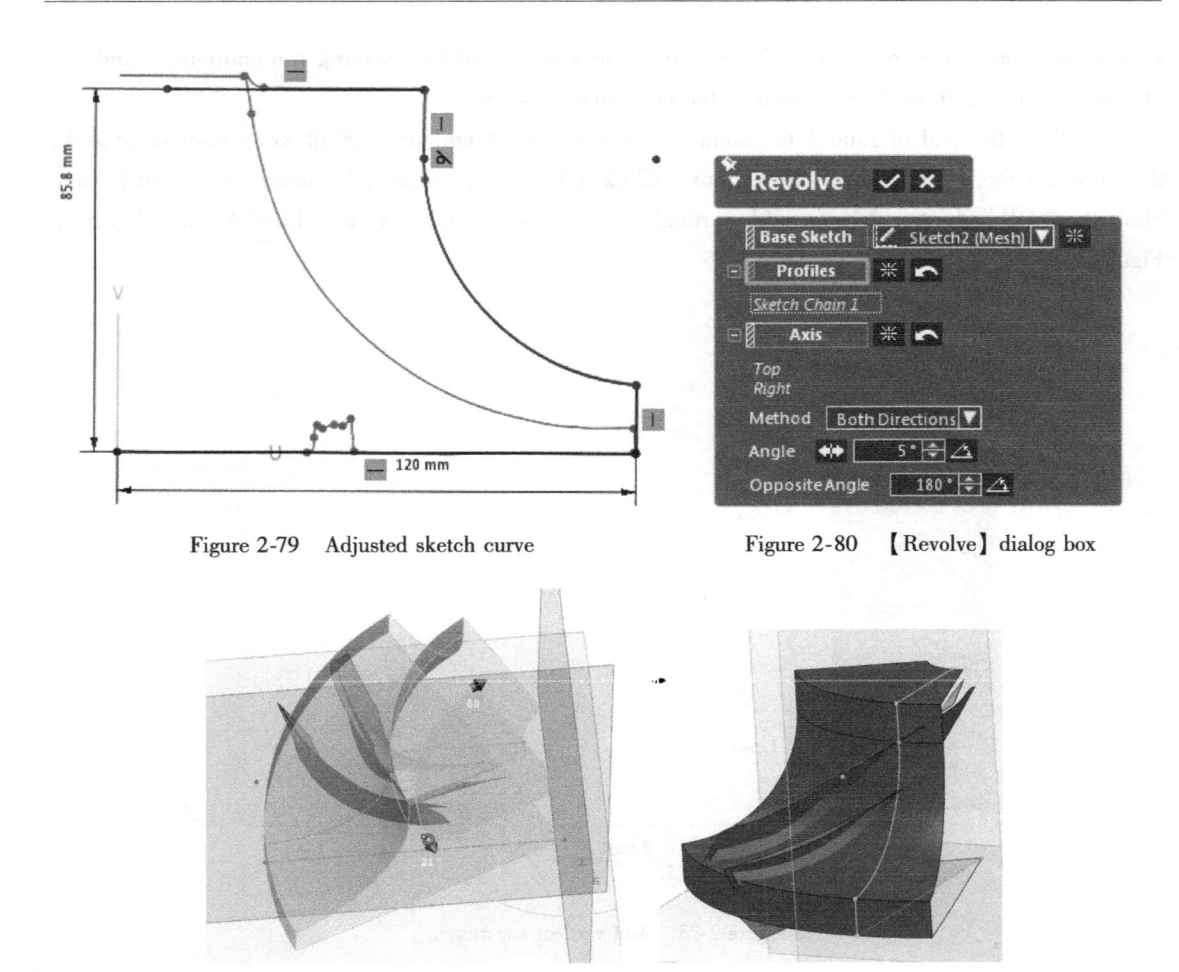

Figure 2-79 Adjusted sketch curve   Figure 2-80 【Revolve】dialog box

a)   b)

Figure 2-81 Create rotation surface

14) Click the 【Mesh Sketch】 command in the 【SKETCH】 module to pop up the【Mesh Sketch Setup】dialog box, as shown in Figure 2-82. Select【Rotational Method】. Click【Central Axis】, and select "Upper Plane" and "Right Plane", taking the intersection of these two planes as the central axis. Click【Base Plane】and select "Right". Adjust the two parameter values of【Silhouette Range】and【Offset Angle From Base Plane】until the section line of the upper end of the small blade is projected onto the sketch plane, as shown in Figure 2-83a.

15) Click the 【Line】command in the 【SKETCH】 module to pop up the 【Line】 dialog box. Select the line indicated by the arrow in Figure 2-83b, then click the button ☑ in the

Figure 2-82 Mesh sketch setup

▶ 62

dialog box. The line is created automatically. Drag the two ends of the line to extend it, as shown in Figure 2-83c. Click the "exit" button ⬚ in the upper left corner of the interface, or the "exit" button ⬚ in the lower right corner of the interface to exit the sketch.

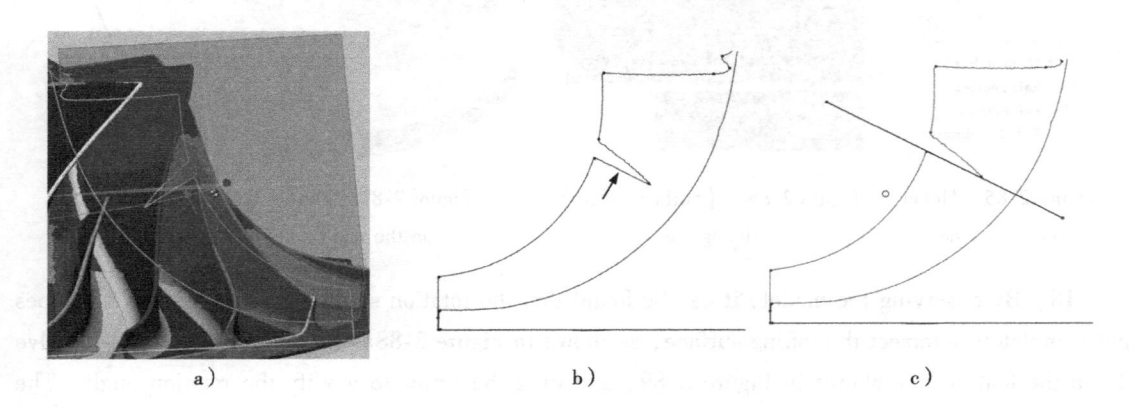

Figure 2-83　Create the section line at the top of the small blade

16) Click the 【Revolve】 command in the 【Create Surface】 group in the 【MODEL】 module to pop up the 【Revolve】 dialog box shown in Figure 2-80. Set 【Method】 to 【Both Directions】 and drag the two arrows shown in Figure 2-84a to enable the created rotation surface to intersect with the small blades. Click the box in front of the "Surface Bodies" in the model feature tree shown in Figure 2-85. An eye icon 👁 appears and the surface created before is visible, as shown in Figure 2-84b. Click the "OK" button ✅ in the 【Revolve】 dialog box, and the result is shown in Figure 2-84c.

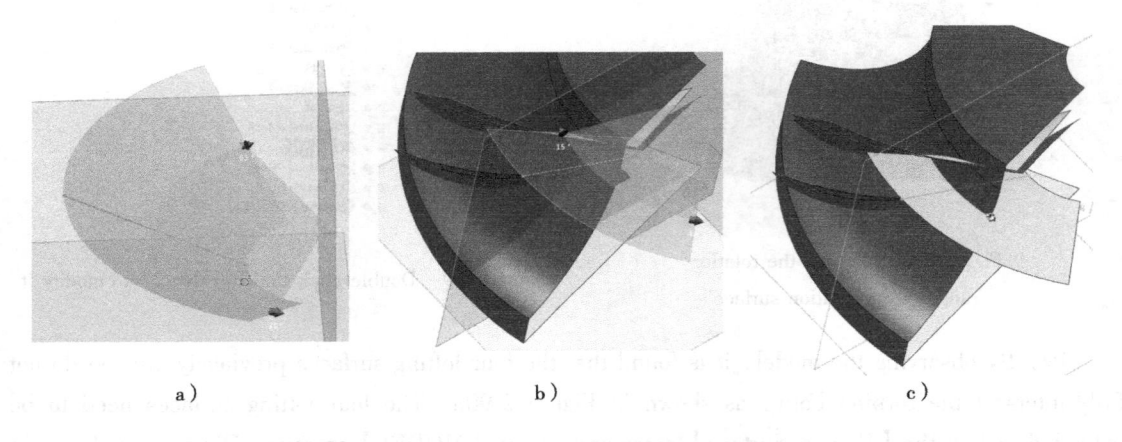

Figure 2-84　Create rotation surface

17) Click the 【Surface Offset】 command in the 【MODEL】 module to pop up the 【Surface Offset】 dialog box shown in Figure 2-86. Input 【Offset Distance】 to "0mm", select the three surfaces on the main body of the turbine, and click the "OK" button ✅. The result is shown in Figure 2-87.

3D Digital Design and Manufacturing

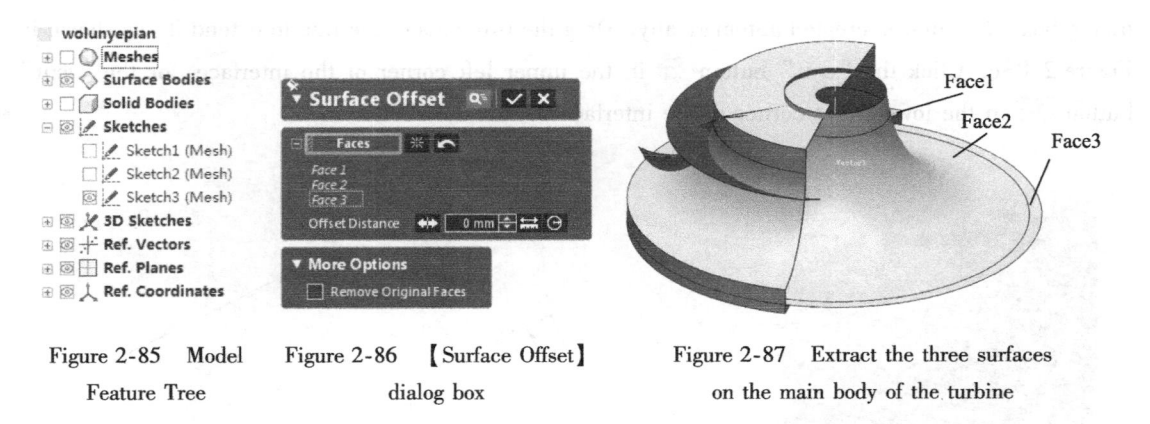

Figure 2-85 Model Feature Tree

Figure 2-86 【Surface Offset】 dialog box

Figure 2-87 Extract the three surfaces on the main body of the turbine

18) By observing the model, it can be found that the rotation surface created by Step 13) does not completely intersect the lofting surface, as shown in Figure 2-88a. So double-click the "Revolve 2" in the feature tree shown in Figure 2-89, and drag the arrow to modify the rotation angle. The result after the modification is shown in Figure 2-88b.

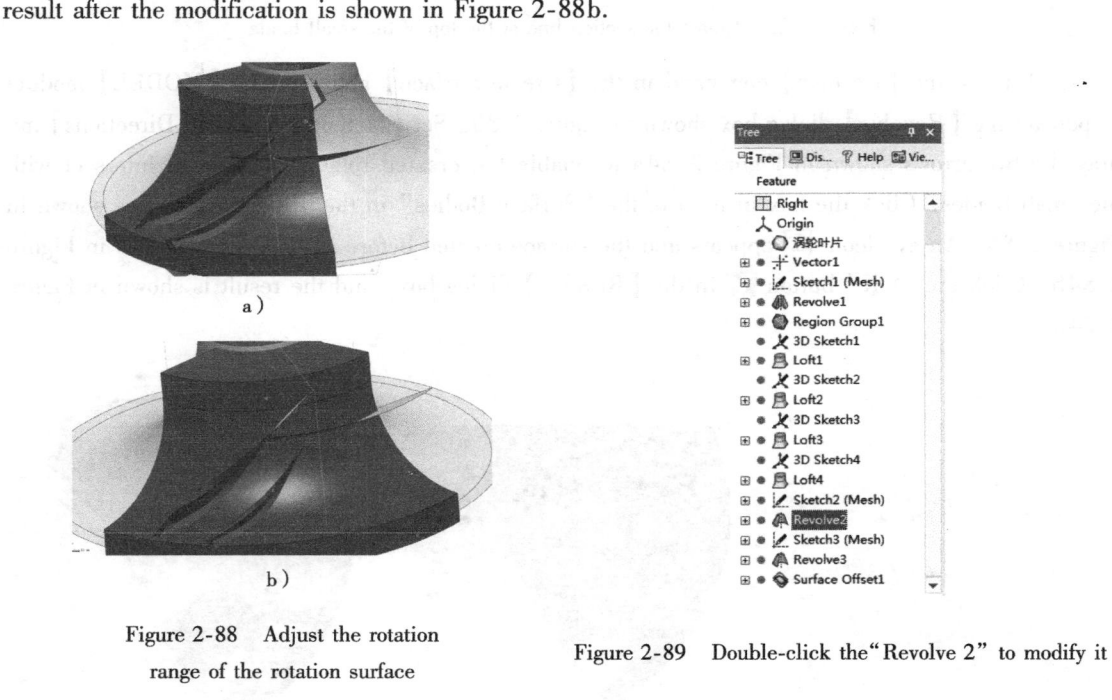

Figure 2-88 Adjust the rotation range of the rotation surface

Figure 2-89 Double-click the "Revolve 2" to modify it

19) By observing the model, it is found that the four lofting surfaces previously created do not fully intersect the turbine body, as shown in Figure 2-90a. The four lofting surfaces need to be extended. Click the 【Extend Surface】 command in the 【MODEL】 module. Dialog box shown in Figure 2-91 appears. Set the 【Distance】 to "10mm", and select the lofting surface indicated in Figure 2-90a. Preview effect shown in Figure 2-90b appears. When the lofting surface and the turbine body completely intersect, click the "OK" button ☑. The other three lofting surfaces are extended by the same method.

▶ 64

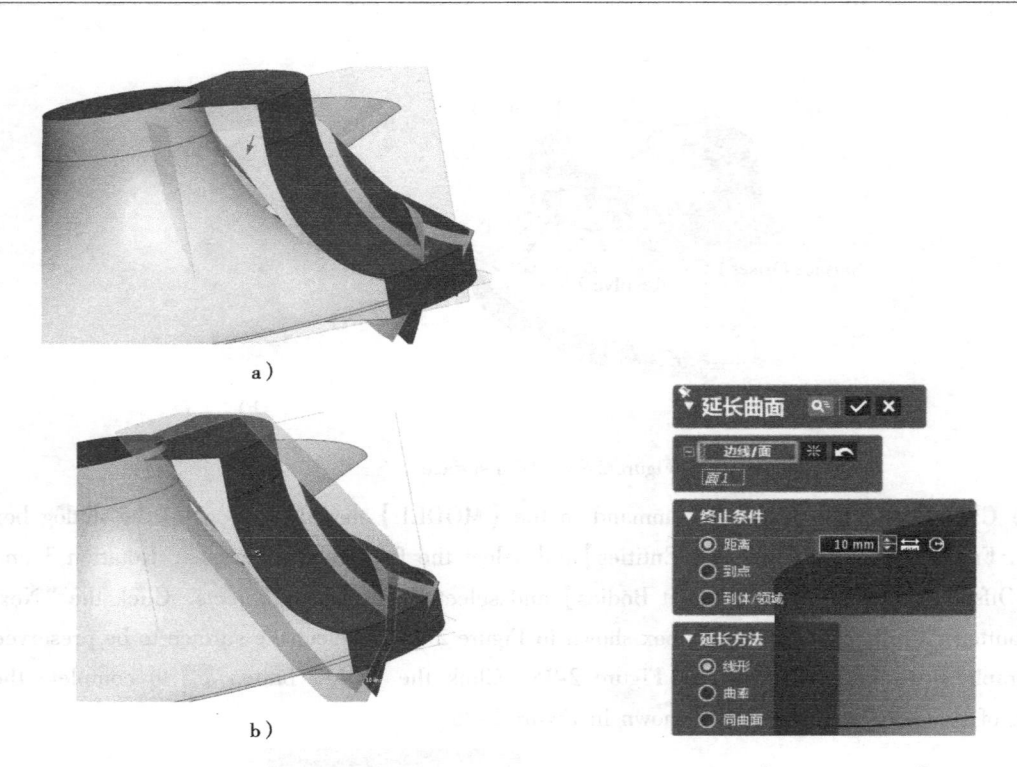

Figure 2-90　Extend the lofting surface　　　　Figure 2-91　【Extend Surface】dialog box

20）Click the 【Trim Surface】command in the 【MODEL】module to pop up the dialog box shown in Figure 2-92a. Click 【Tool Entities】, and select the three surfaces shown in Figure 2-93a as tools. Click 【Target Bodies】, and select the four lofting surfaces as objects. Click the "Next Stage" button ⭢ to show the dialog box shown in Figure 2-92b. Select the surface to be preserved in the graphics window, as shown in Figure 2-93. Click the "OK" button ✓ to complete the trimming of the four lofting surfaces.

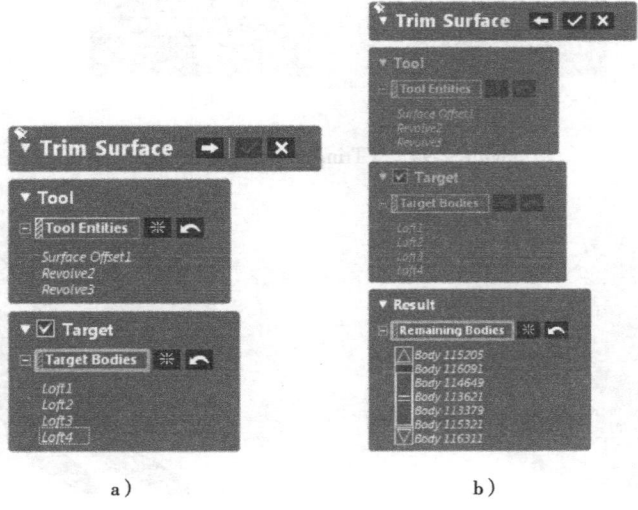

Figure 2-92　【Trim Surface】dialog box

3D Digital Design and Manufacturing

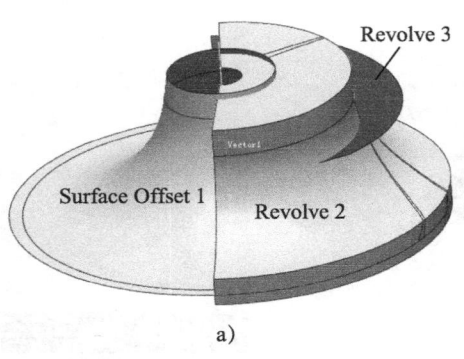
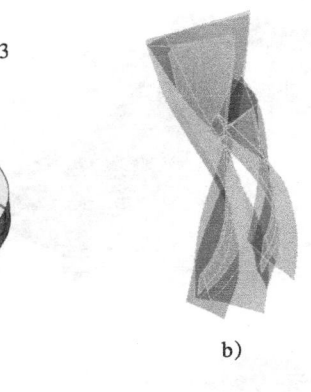

a)
b)

Figure 2-93　Trim surface

21) Click the【Trim Surface】command in the【MODEL】module to pop up the dialog box shown in Figure 2-94a. Click【Tool Entities】and select the four lofting surfaces, Rotation 3 and Surface Offset 1 as tools. Click【Target Bodies】and select Revolve 2 as objects. Click the "Next Stage" button 🔄 to pop up the dialog box shown in Figure 2-94b. Select the surface to be preserved in the graphics window, as shown in Figure 2-95. Click the "OK" button ✅ to complete the trimming of Revolve 2. The result is shown in Figure 2-96.

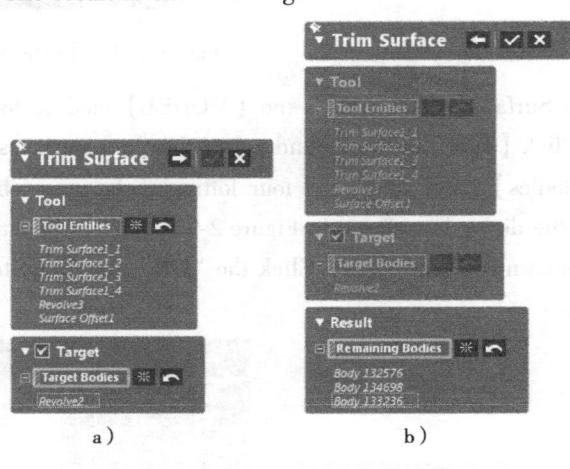

a)
b)

Figure 2-94　【Trim Surface】dialog box

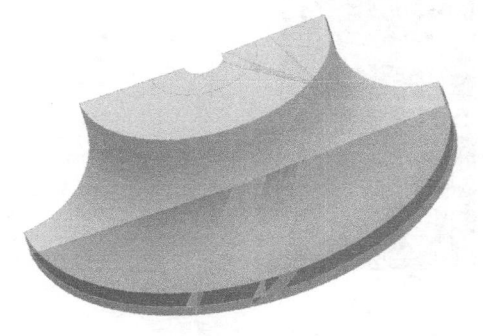

Figure 2-95　Select the surface to be preserved

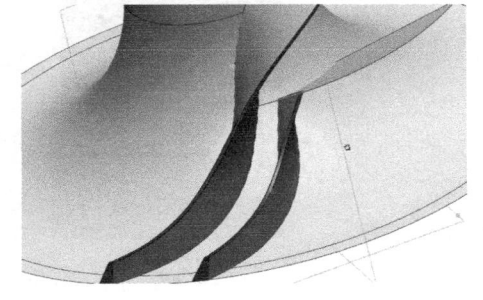

Figure 2-96　Effect after surface trimming

▶ 66

Project 2   Turbine Wheel

22) Click the 【Trim Surface】command in the 【MODEL】 module to pop up the dialog box shown in Figure 2-97a. Click 【Tool Entities】and select the upper plane as the tool. Click 【Target Bodies】 and select Surface Offset 1 as the object. Click the "Next" button ➡ to pop up the dialog box shown in Figure 2-97b. Select the surface you want to keep in the graphics window and click the "OK" button ✓, then the three surfaces extracted from the main body are cut off half, as shown in Figure 2-98.

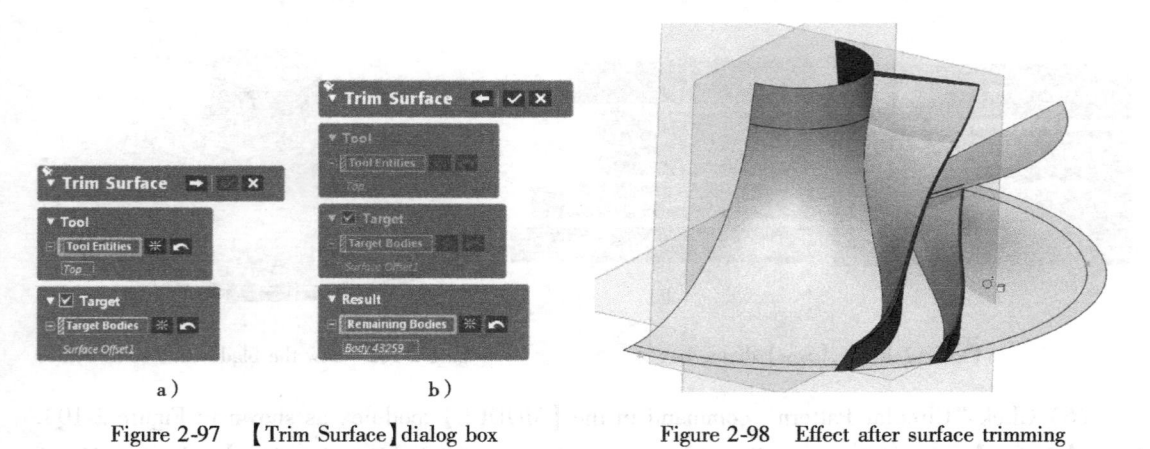

a)   b)

Figure 2-97   【Trim Surface】dialog box   Figure 2-98   Effect after surface trimming

23) The same method is used to trim the surface of Revolve 3. The result is shown in Figure 2-99b.

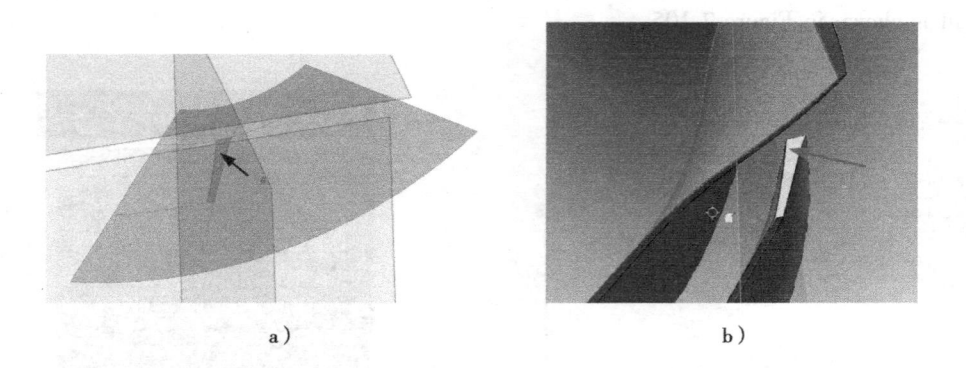

a)   b)

Figure 2-99   Effect after surface trimming Revolve 3

24) Continue trimming the surfaces in the same way, the resullt is shown in Figure 2-100.

a) Before trimming surface   b) After trimming surface

Figure 2-100   Effect after surface trimming

67 ◀

3D Digital Design and Manufacturing

25) Click the 【Sew】 command in the 【MODEL】 module to pop up the dialog box shown in Figure 2-101a. Select all the facets that make up the blade. Click the "Next" button ➡ to pop up the dialog box shown in Figure 2-101b. Click the "OK" button ✓ to make the meshes that make up the blade sewn together into an solid body. Sew the large blade in the same way, and the results after sewing are shown in Figure 2-102.

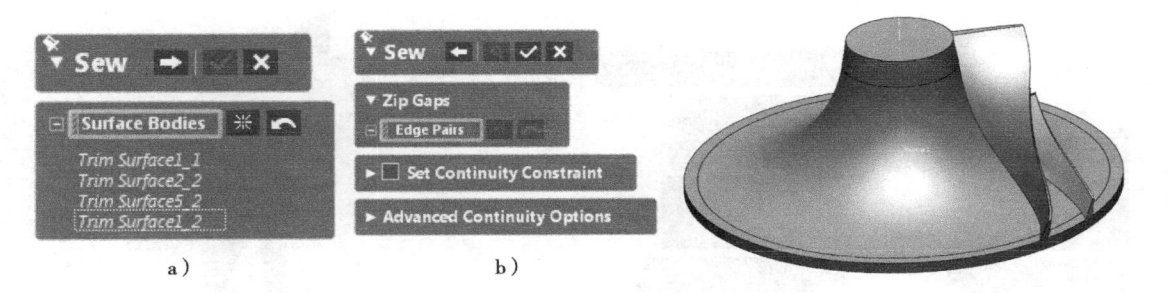

a)   b)

Figure 2-101   【Sew】 dialog box   Figure 2-102   Sew the blades into solid bodies

26) Click "Circular Pattern" command in the 【MODEL】 module, as shown in Figure 2-103. Click 【Bodies】and select the small and large sewn blades. Click 【Rotation Axis】and select Vector 1, and set 【No. of Instances】 to "10", as shown in Figure 2-104. Click the "OK" button ✓, and the result is shown in Figure 2-105.

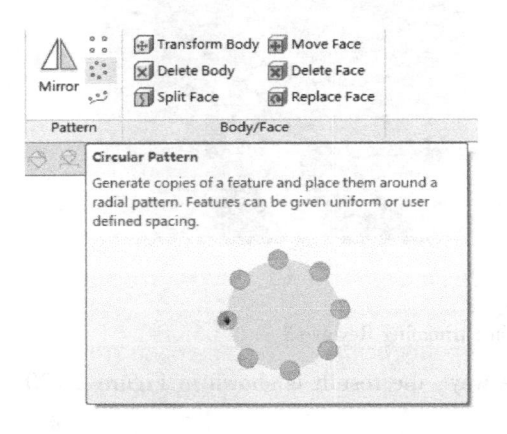

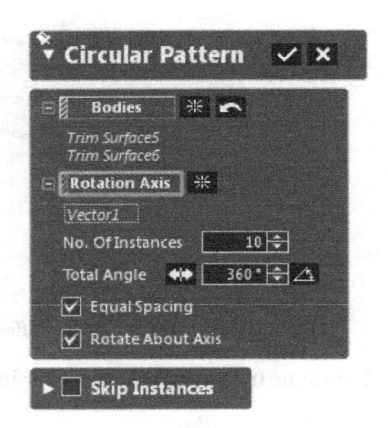

Figure 2-103   "Circular Pattern" command   Figure 2-104   【Circular Pattern】dialog box

27) Click the 【Boolean】command in the 【MODEL】 module to pop up the dialog box as shown in Figure 2-106. Select 【Operation Method】 as 【Merge】, select all Solid Bodies, and click the "OK" button ✓.

28) Click the 【Fillet】 command and select 【Variable Fillet】. The radius of the inner fillet on the side of the large impeller blade is 3 mm and the outer fillet is 1 mm. On the other side, the radius of the outer fillet is 0.3mm and the inner fillet is 1.5mm, as shown in Figure 2-107.

▶ 68

Project 2　Turbine Wheel

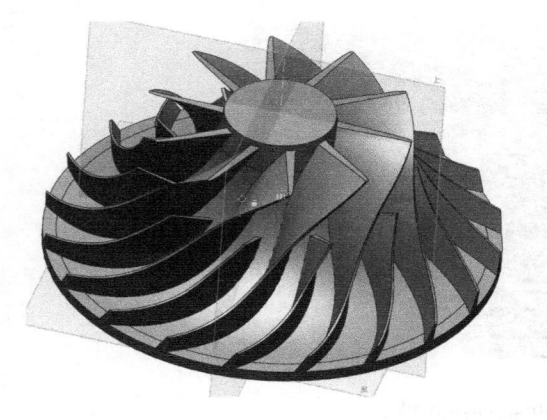

Figure 2-105　Effect after circular pattern

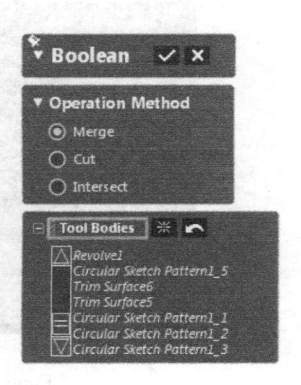

Figure 2-106　【Boolean】dialog box

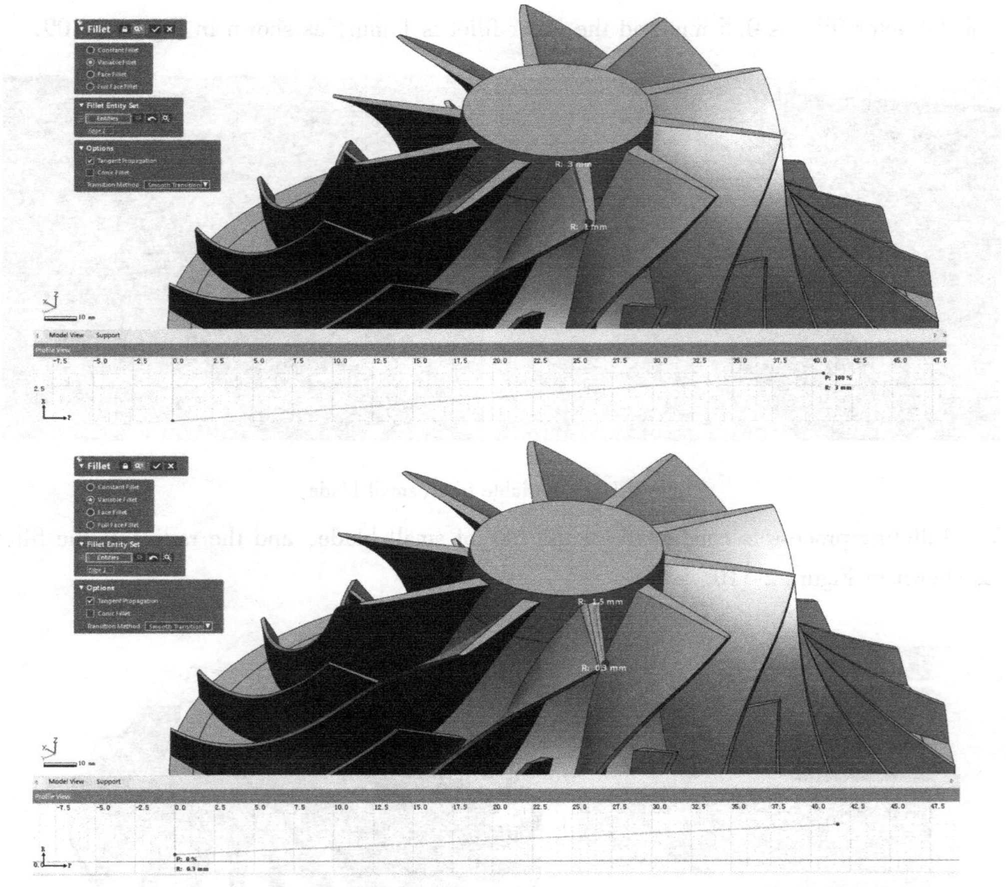

Figure 2-107　Variable fillet (large blade)

29) Filleting process is conducted for the root of large blade, and the radius of the fillet is 4 mm, as shown in Figure 2-108.

69 ◀

3D Digital Design and Manufacturing

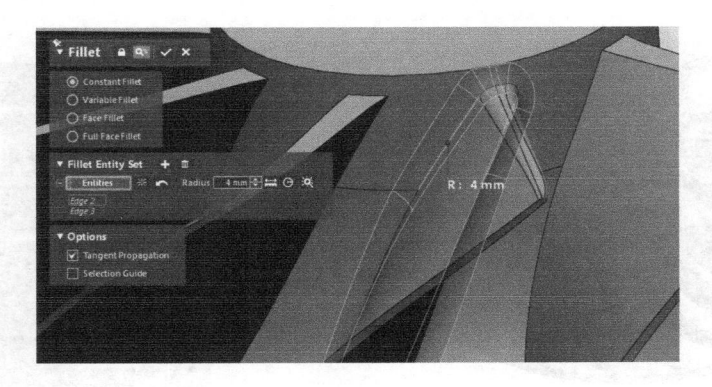

Figure 2-108　Filleting(large blade)

30) Click the 【Fillet】 command and select 【Variable Fillet】. The radius of the inner fillet on the side of the small impeller blade is 2 mm and the outer fillet is 1 mm. On the other side, the radius of the outer fillet is 0.5 mm and the inner fillet is 1 mm, as shown in Figure 2-109.

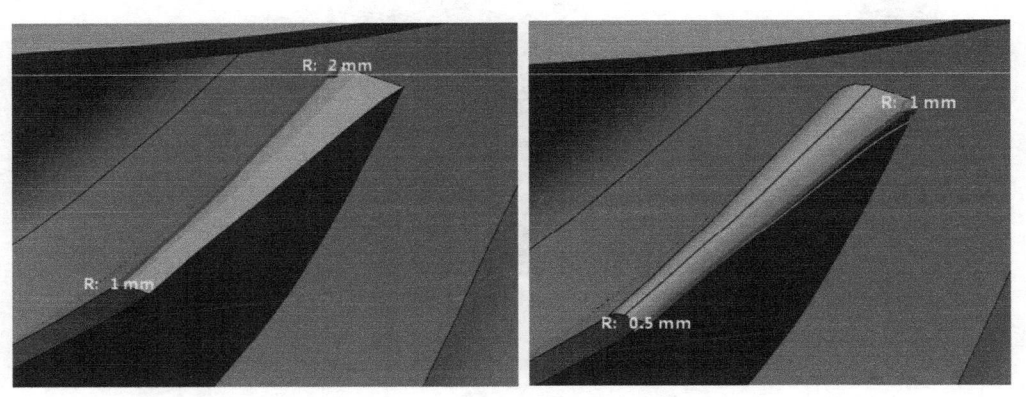

Figure 2-109　Variable fillet(small blade)

31) Filleting process is conducted for the root of small blade, and the radius of the fillet is 4 mm, as shown in Figure 2-110.

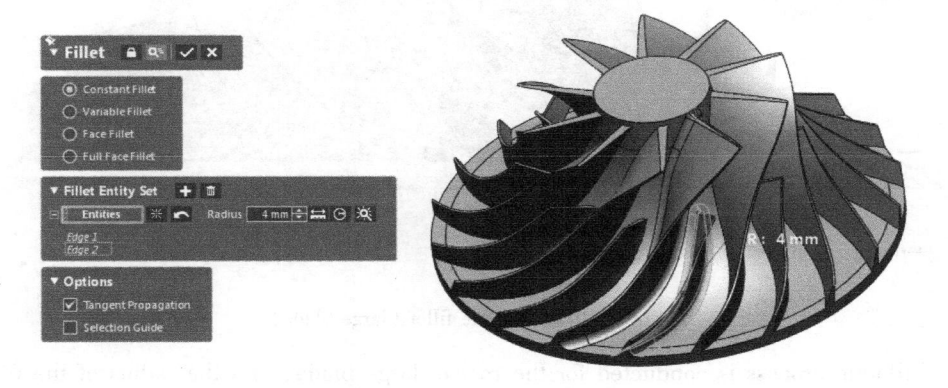

Figure 2-110　Filleting(small blade)

▶ 70

Project 2　Turbine Wheel

32) Filleting process is conducted for all blades and the model is derived, as shown in Figure 2-111.

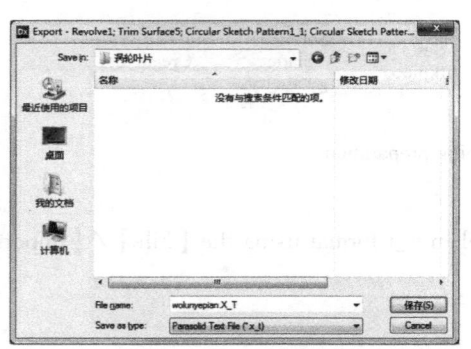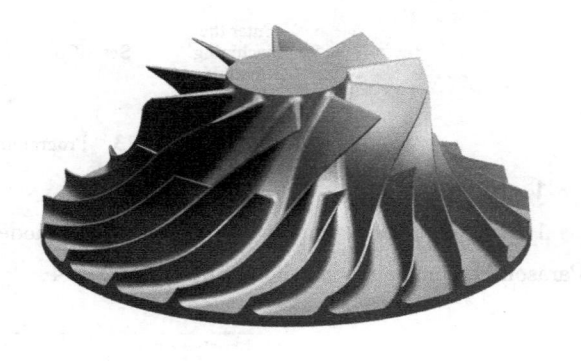

Figure 2-111　Filleting completed

Task 4　Machining Programming

2.4.1　Machining Analysis

The shape of the turbine wheel is complex and the requirement for machining precision is high. The blades are thinwalled parts, and they are easy to be deformed and interfered during processing.

In order to reduce the machining workload of five-axis milling, the blank of the turbine wheel is turned on a numerical control lathe to obtain the basic shape, as shown in Figure 2-112.

The machining process of the turbine wheel is shown in Table 2-1.

Figure 2-112　Blank of the turbine wheel

Table 2-1　The machining process of the turbine wheel

No.	Operation	Type	Operation Subtype	Tool
1	blade roughing		multi_blade_roughing	D8R4
2	main blade finishing		blade_finishing	D8R4
3	splitter blade finishing	mill_multi_blade	blade_finishing 工	D8R4
4	hub_finishing		hub_finishing	D8R4
5	blend_finishing		blend_finishing	D6R3

2.4.2　Preparation for Programming

Before creating the machining program for the turbine wheel, it is needed to do the preparatory work as shown in Figure 2-113.

71

3D Digital Design and Manufacturing

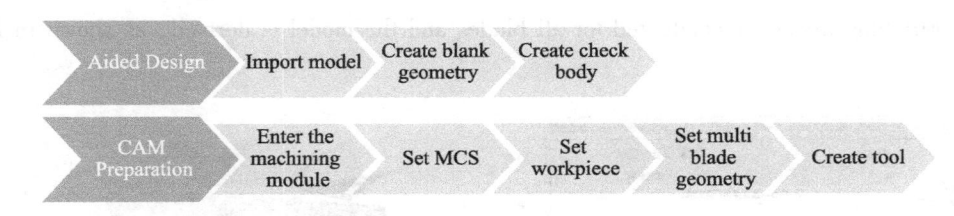

Figure 2-113　Programming preparation

1. Aided Design

1) Import model. Import the turbine wheel model in x_t format using the 【File】 / 【Import】 / 【Parasolid】 command, as shown in Figure 2-114.

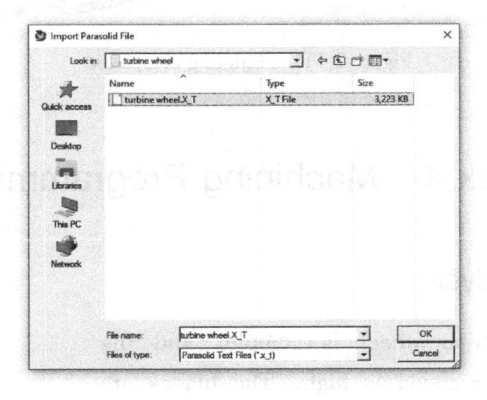

Figure 2-114　Import model

2) Create blank geometry. The blade edge line is selected and the turbine wheel axis center is selected by vector with the 【Revolve】 command. Create a cladding surface, as shown in Figure 2-115.

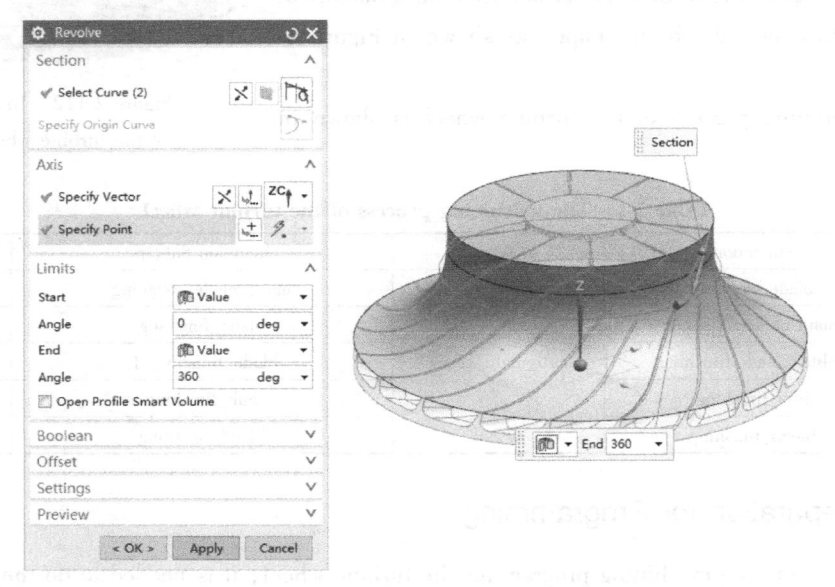

Figure 2-115　Create a cladding surface

▶ 72

Project 2 Turbine Wheel

The edge line is selected for extruding under the command of 【Extrude】. At the end of the extruding, select 【Until Extended】. Select the top surface of the turbine wheel, and select 【Unite】 for Boolean to unite it and the created cladding surface, as shown in Figure 2-116.

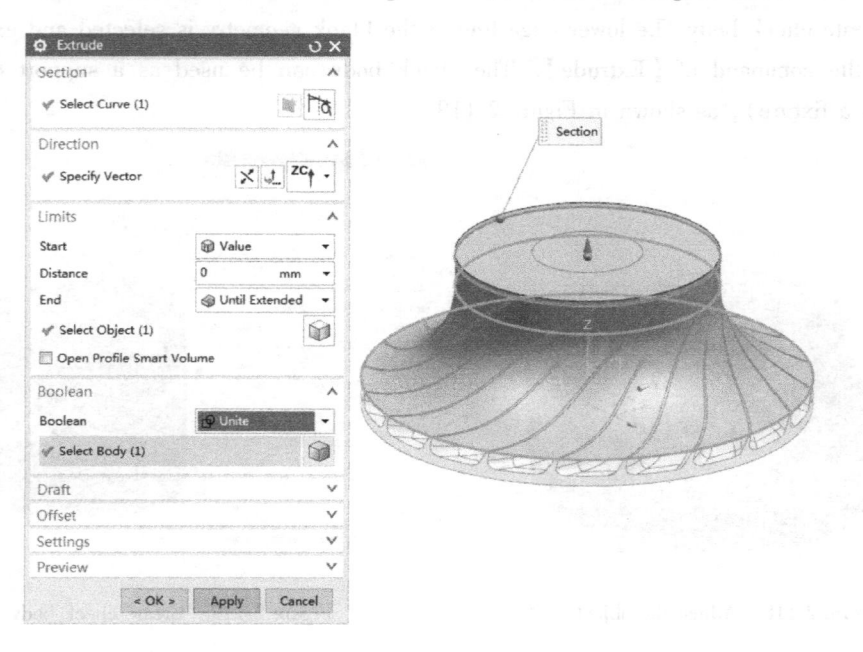

Figure 2-116   Extrude

Similarly, the【Extrude】command is used to select the lower edge line to extrude downward. The end distance of extruding is set to 25 mm. Select 【Unite】 for Boolean as shown in Figure 2-117. The creation of blank geometry is completed at present.

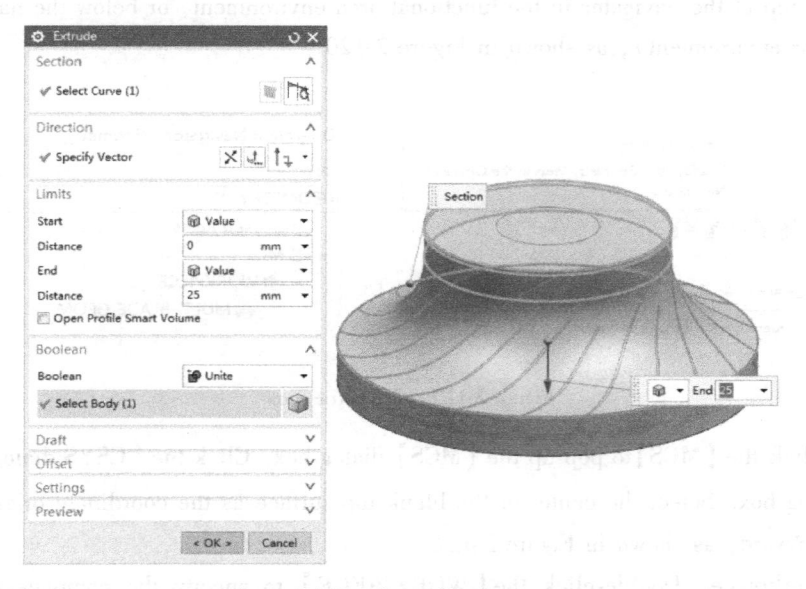

Figure 2-117   Create blank geometry

73

3D Digital Design and Manufacturing

For easy observation, the color and transparency of the blank geometry can be changed appropriately by clicking the 【Edit Object Display】 command or directly using the shortcut key < Ctrl + J > , as shown in Figure 2-118.

3) Create check body The lower edge line of the blank geometry is selected and extruded 250 mm under the command of 【Extrude】. The check body can be used as a support column (or regarded as a fixture), as shown in Figure 2-119.

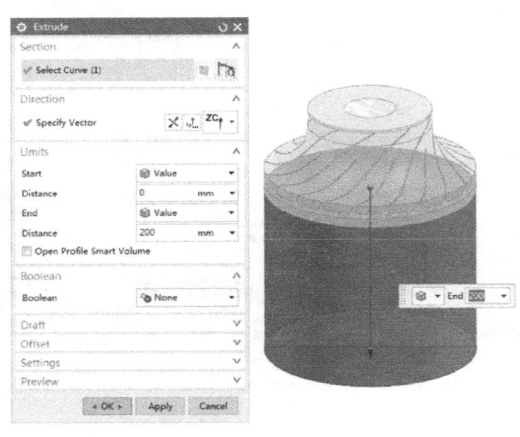

Figure 2-118　Adjust the object

Figure 2-119　Create check body

2. CAM Preparation

1) Enter the machining module and select 【cam_general】 and 【mill_multi_blade】 (multi-axis milling blade template) in the setting of pop-up machining environment.

2) Set up machine tool coordinates. After entering the machining module, switch to 【Geometry View】 (at the top of the navigator in the functional area environment, or below the navigator in the classical toolbar environment), as shown in Figure 2-120.

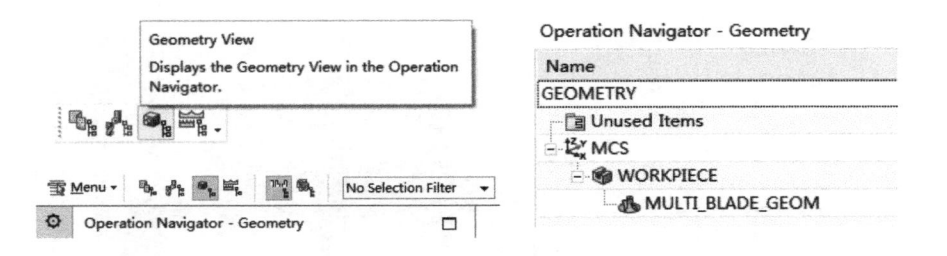

Figure 2-120　Geometric view

Double-click the 【MCS】to pop up the 【MCS】 dialog box. Click the "CSYS dialog box" button in the dialog box. Select the center of the blank top surface as the coordinate origin, and make the Axis ZM upward, as shown in Figure 2-121.

3) Set workpiece. Double-click the 【WORKPIECE】 to specify the component, blank and check body in turn, as shown in Figure 2-122.

▶ 74

Project 2　Turbine Wheel

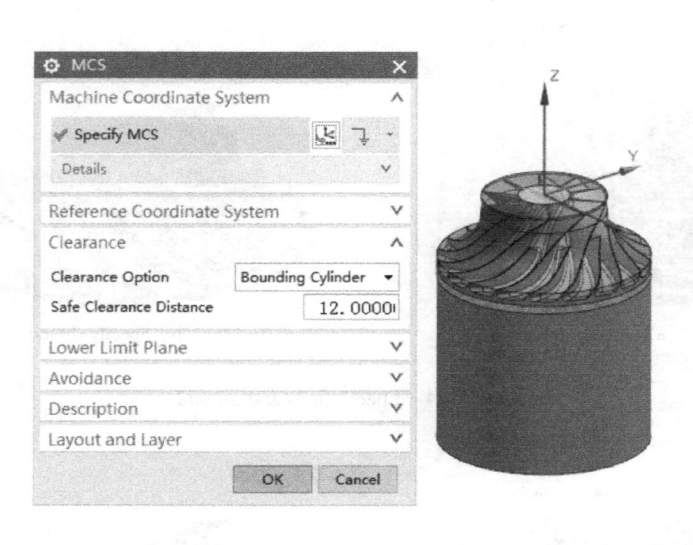

Figure 2-121　MCS setting

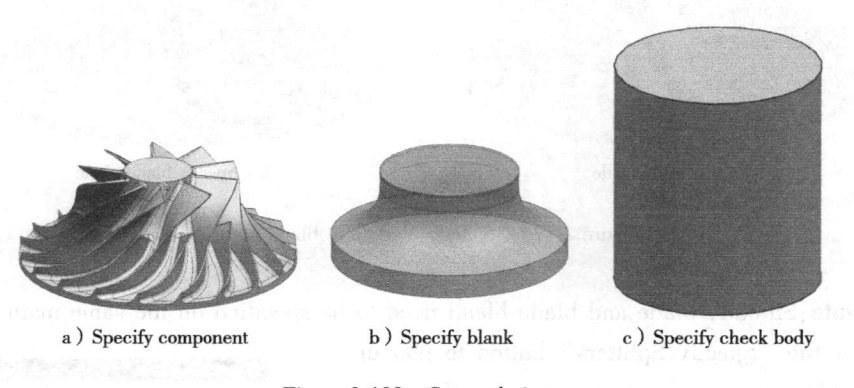

a）Specify component　　　　b）Specify blank　　　　c）Specify check body

Figure 2-122　Set workpiece

4）Set multi-blade geometry. Double-click the 【MULTI_BLADE_GEOM】 in the operation navigator geometry view to pop up the 【Multi Blade Geom】 dialog box, as shown in Figure 2-123.

① Click the "Specify Hub" button and select the inner large surface, as shown in Figure 2-124a.

② Click the "Specify Shroud" button and select the cladding surface of the main blade, as shown in Figure 2-124b.

③ Click the "Specify Blade" button and select a continuous surface on the main blade, as shown in Figure 2-125a.

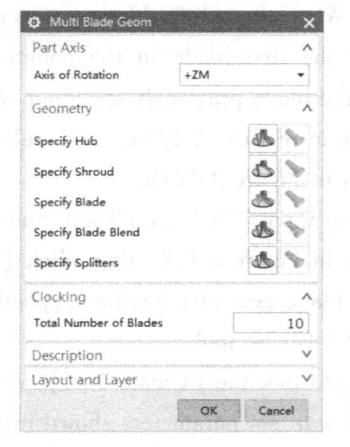

Figure 2-123　【Multi Blade Geom】 dialog box

④ Click the "Specify Blade Blend" button and select blend faces on the main blade, as shown in Figure 2-125b.

75

3D Digital Design and Manufacturing

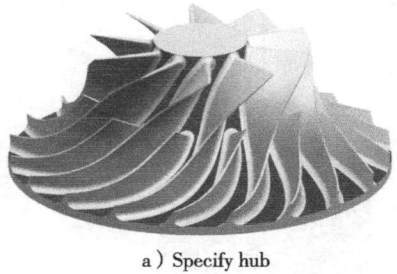 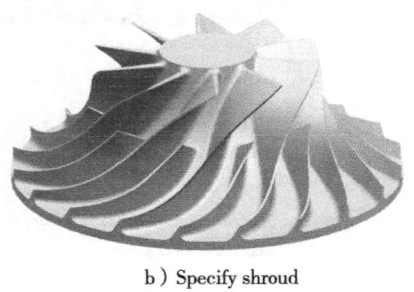

a) Specify hub   b) Specify shroud

Figure 2-124   Specify hub and shroud

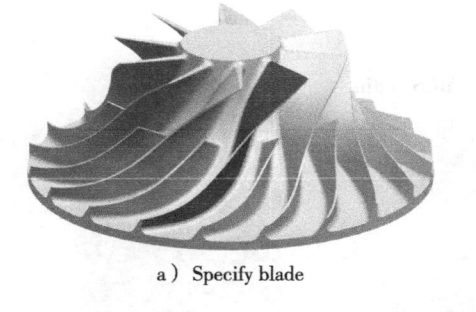 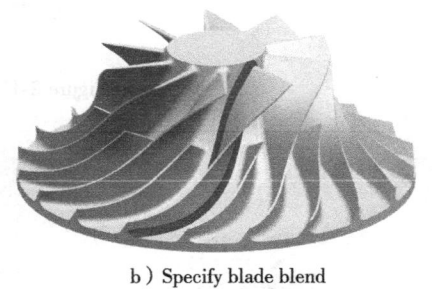

a) Specify blade   b) Specify blade blend

Figure 2-125   Specify blade and blade blend

Note: Shroud, blade and blade blend need to be specified on the same main blade.

⑤ Click the "Specify Splitters" button to pop up the 【Splitter Geometry】 dialog box as shown in Figure 2-126. When the splitter blade is specified, it is needed to select the first blade in the counterclockwise direction from the blade previously specified. Select wall faces, as shown in Figure 2-127a, and select blend faces, as shown in Figure 2-127b.

⑥ Input "10" in 【Total Number of Blades】, as shown in Figure 2-128, then click 【OK】. By now, the multi blade geometry has been specified completely.

5) Create tool.

① Click the 【Create Tool】 command to create tool D8R4 with set parameters shown in Figure 2-129.

Figure 2-126   【Splitter Geometry】
dialog box

② Click the 【Create Tool】 command to create tool D6R3 with set parameters shown in Figure 2-130.

76

Project 2　Turbine Wheel

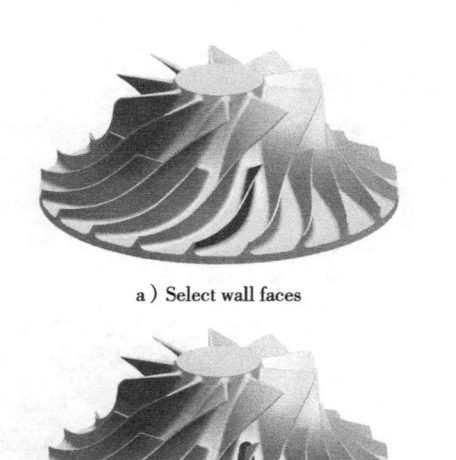

a) Select wall faces

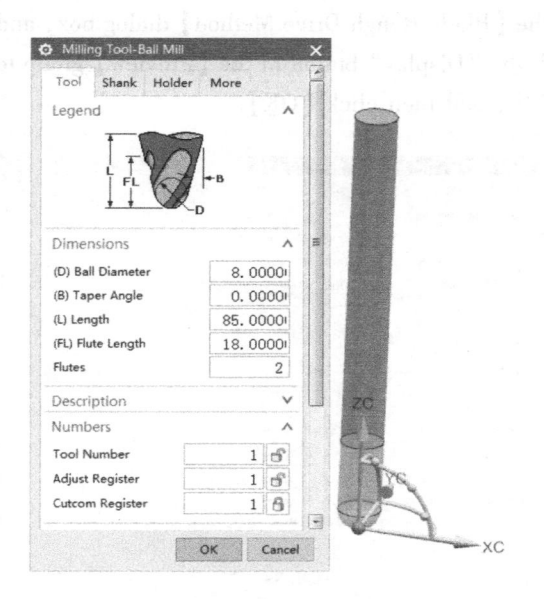

b) Select blend faces

Figure 2-127　Specify splitters

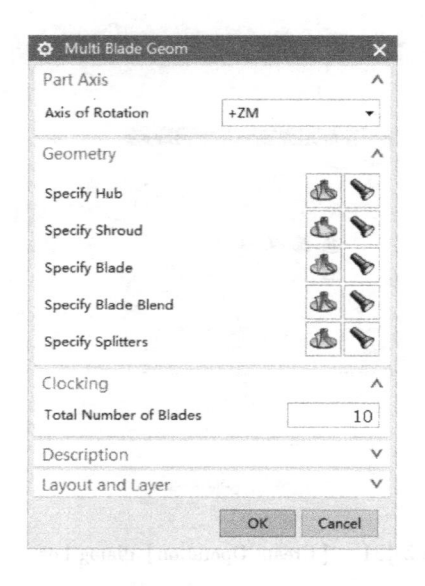

Figure 2-128　Total number of blades

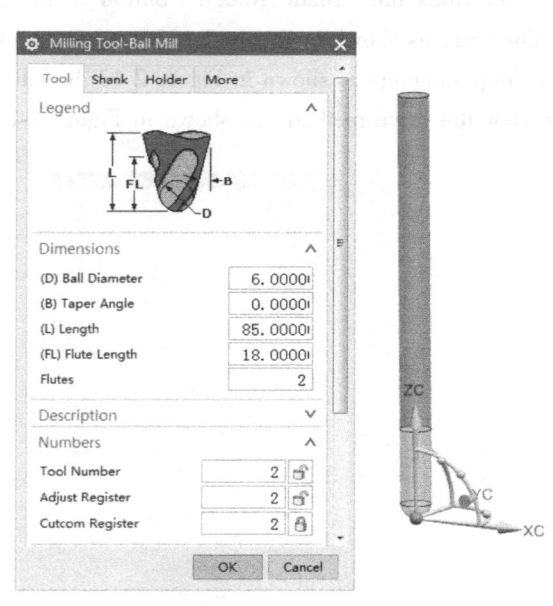

Figure 2-129　Create tool D8R4　　　　Figure 2-130　Create tool D6R3

2.4.3　Multi-blade Roughing

1) Click the 【Create Operation】 command to pop up the dialog box. Select 【Type】 as "mill_multi_blade", select 【Operation Subtype】 as the "Multi Blade Rough", and set the remaining parameters as shown in Figure 2-131. Click 【OK】.

2) Pop up the 【Multi Blade Rough】 dialog box shown in Figure 2-132a. The geometry has been set up. Click the "Display" button 🔖 to view the corresponding hub, shroud, blade and other geometries, as shown in Figure 2-132b.

77 ◀

3D Digital Design and Manufacturing

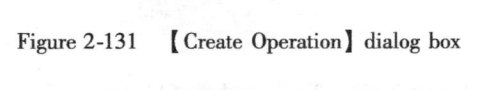

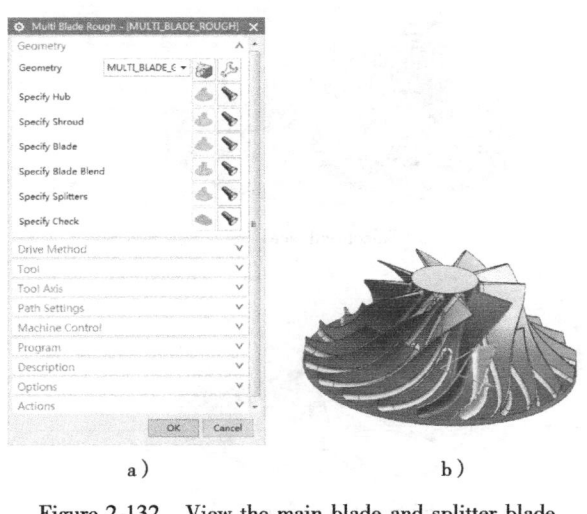

Figure 2-131　【Create Operation】dialog box

Figure 2-132　View the main blade and splitter blade

3) Click the "Blade Rough" button in the 【Drive Method】 group of 【Multi Blade Rough】 dialog box, as shown in Figure 2-133a, to pop up the 【Blade Rough Drive Method】 dialog box, and set the parameters as shown in Figure 2-133b. Click the "Display" button in the 【Preview】 group to preview the drive method, as shown in Figure 2-133b, and then click 【OK】.

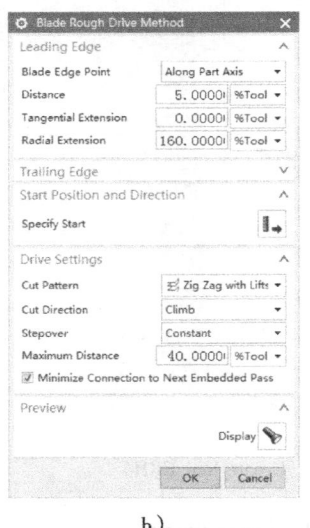

a)

b)

Figure 2-133　Drive method

4) Click the "Cut Levels" button 圖 in the【Path Settings】 group of 【Multi Blade Rough】 dialog box to pop up the 【Cut Levels】 dialog box. Select 【Depth Mode】 as 【Interpolate from Shroud to Hub】, 【Distance】 as constant "1mm", as shown in Figure 2-134.

5) Click the "Cutting Parameters" button 而 in the 【Path Settings】 group of 【Multi Blade Rough】 dialog box to pop up the 【Cutting Parameters】dialog box. Input 【Blade Stock】 as "0.6",

▶ 78

Project 2 Turbine Wheel

【Hub Stock】 as "0.4", 【Check Stock】 as "0.6", as shown in Figure 2-135.

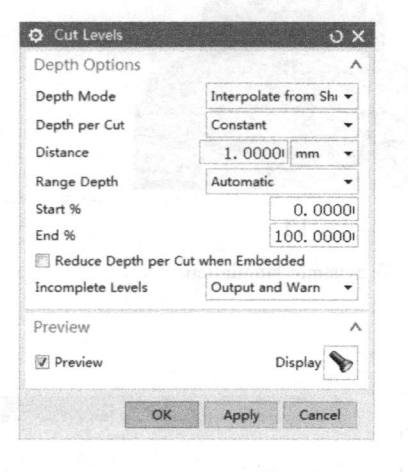 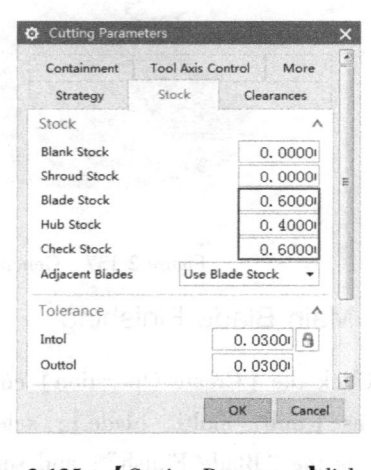

Figure 2-134 【Cut Levels】dialog box   Figure 2-135 【Cutting Parameters】dialog box

6) Click the "Feeds and Speeds" button in the 【Path Settings】 group of 【Multi Blade Rough】 dialog box. Input the spindle speed as "8000" and cutting feed rate as "4000", as shown in Figure 2-136a. Click the "Calculate Feeds and Speeds based on this value" button ▣ behind the "Spindle Speed" to calculate the surface speed and the feed per tooth, as shown in Figure 2-136b.

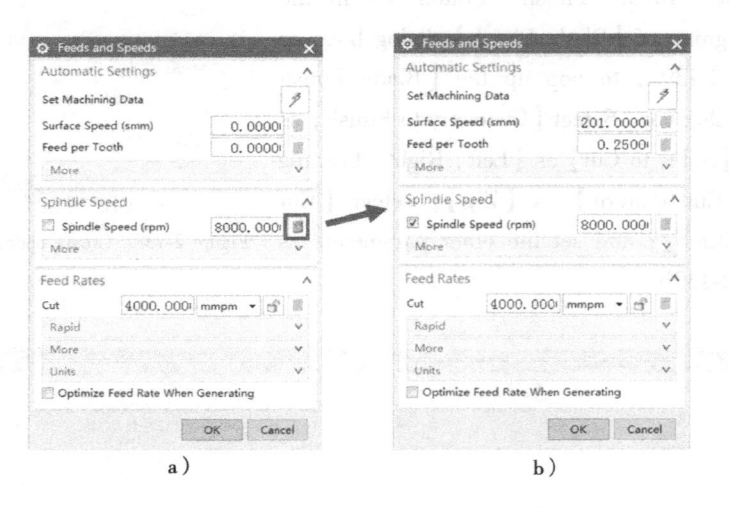

Figure 2-136 【Feeds and Speeds】 dialog box

7) Generate tool path. Click the "Generate" button ▶ in the 【Actions】 group of 【Multi Blade Rough】 dialog box, and the tool path of inter-blade roughing is as shown in Figure 2-137a. Click the "OK" button 🔧 to pop up the 【Tool Path Visualization】 dialog box. Click the 【2D Dynamic】 tab, and then click the "Play" button ▶ to start the cutting simulation. The simulation result is as shown in Figure 2-137b.

8) After clicking the 【OK】 button in the 【Multi Blade Rough】 dialog box, the 【MULTI_BLADE_ROUGH】 program appears in the program order view of the operation navigator.

79

3D Digital Design and Manufacturing

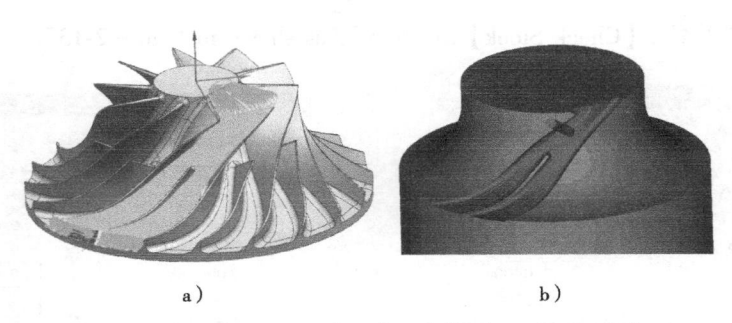

a)   b)

Figure 2-137   Generate tool path and 2D dynamic simulation

2.4.4   Main Blade Finishing

1) Click the 【Create Operation】 command. Select
【Type】 as 【mill _ multi _ blade】; select 【Operation
Subtype】 as the "Blade Finish"; and set the remaining
parameters as shown in Figure 2-138. Click 【OK】.

2) Pop up the 【Blade Finish】 dialog box. The
geometry has been set up. There is no need to set it
here.

3) Click the "Blade _ Finish" button  in the
【Drive Method】 group of 【Blade Finish】 dialog box, as
shown in Figure 2-139a, to pop up the 【Blade Finish
Drive Method】 dialog box. Select 【Geometry to Finish】 as
【Blade】; select 【Sides to Cut】 as 【Left, Right, Leading
Edge】; select 【Cut Pattern】 as 【Zig】; select 【Cut
Direction】 as 【Climb】; and set the other parameters as
shown in Figure 2-139b.

Figure 2-138   Create operation(blade_finish)

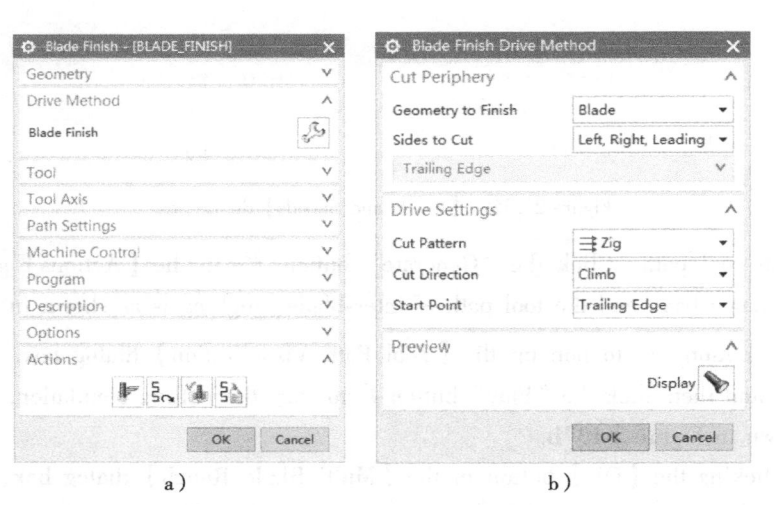

a)   b)

Figure 2-139   Drive method

▶ 80

Project 2　Turbine Wheel

4) Click the "Cut Levels" button ☰ in the 【Path Settings】 group of 【Blade Finish】 dialog box to pop up the 【Cut Levels】 dialog box. Select 【Depth Mode】 as【Interpolate from Shroud to Hub】; select 【Depth per Cut】 as 【Scallop】; enter【Scallop Height】 as "0. 02"; and set the other parameters as shown in Figure 2-140.

5) Click the "Feeds and Speeds" button in the 【Path Settings】 group of 【Blade Finish】 dialog box. Input the spindle speed as "8000" and cutting feed rate as "4000". Click the "Calculate Feeds and Speeds based on this value" button ▣ to calculate the surface speed and the feed per tooth.

6) Click the "Generate" button ⊮ in the 【Actions】 group of 【Blade Finish】 dialog box to view the tool path of main blade finishing, as shown in Figure 2-141a. The zoom-in view is shown in Figure 2-141b.

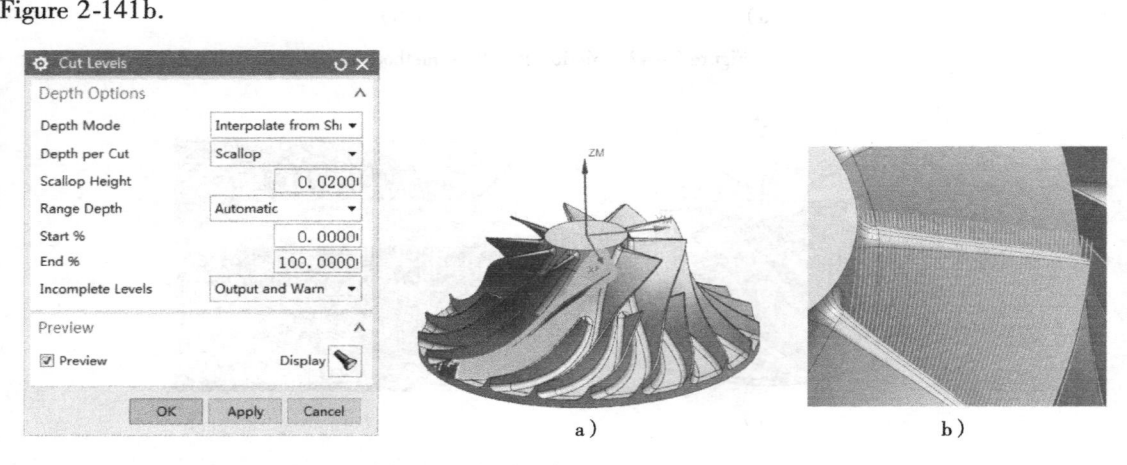

a)　　　　　　　　　　　　　　　　　　b)

Figure 2-140　【Cut Levels】dialog box　　　　　Figure 2-141　Generate the tool path of main blade finishing

2.4.5　Splitter Blade Finishing

The processing parameters of the splitter blade finishing are the same as those of the main blade finishing. It can be created by copying.

1) Select the 【BLADE_FINISH】 program in the program order view of the operation navigator, then click the right key and select 【Copy】 from the right-click menu that pops up. Right-click again and select 【Paste】 from the right-click menu that pops up. The 【BLADE_FINISH_COPY】 program will appear.

2) Double-click the【BLADE_FINISH_COPY】 program to pop up the 【Blade Finish】 dialog box, as shown in Figure 2-142a. Click the button ⚙ in the 【Drive Method】 group to pop up the 【Blade Finish Drive Method】 dialog box. Select【Geometry to Finish】 as 【Splitter 1】, and click 【OK】, as shown in Figure 2-142b.

3) Click the "Generate" button ⊮ in the 【Actions】 group of 【Blade Finish】 dialog box to view the tool path of splitter blade finishing, as shown in Figure 2-143a. The zoom-in view is shown in Figure 2-143b.

81

3D Digital Design and Manufacturing

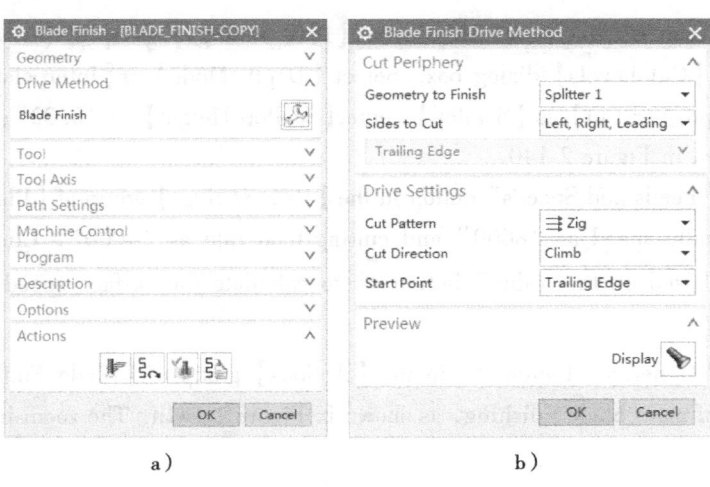

a)   b)

Figure 2-142   Modify the drive method

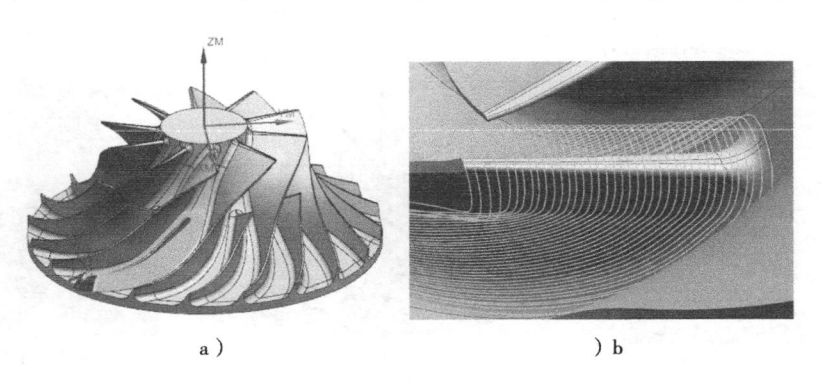

a)   ) b

Figure 2-143   Generate the tool path of splitter blade finishing

2.4.6   Hub Finishing

1) Click the 【Create Operation】command; select 【Type】 as 【mill _ multi _ blade 】; select 【Operation Subtype】 as the " Hub Finish"; and set the remaining parameters as shown in Figure 2-144. Click 【OK】.

2) Pop up the【Hub Finish】dialog box, in which the setting of geometry is completed. No setting is required.

3) Click the "Hub Finish" button  in the 【Drive Method】 group of 【Hub Finish】 dialog box, and pop up the【Hub Finish Drive Method】dialog box. Set the【Radial Extension】of the leading edge to "50". The other parameters are as shown in Figure 2-145a. The effects of the radial extension of the leading edge set to "0" and "50" are shown in Figure 2-145b and c.

4) Click the "Feeds and Speeds" button in the 【Path Settings】

Figure 2-144   Create operation
(hub_finish)

82

Project 2　Turbine Wheel

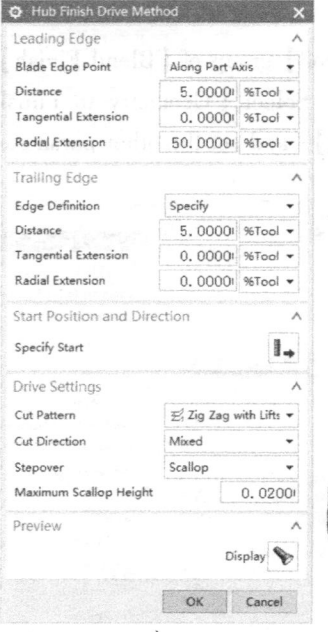

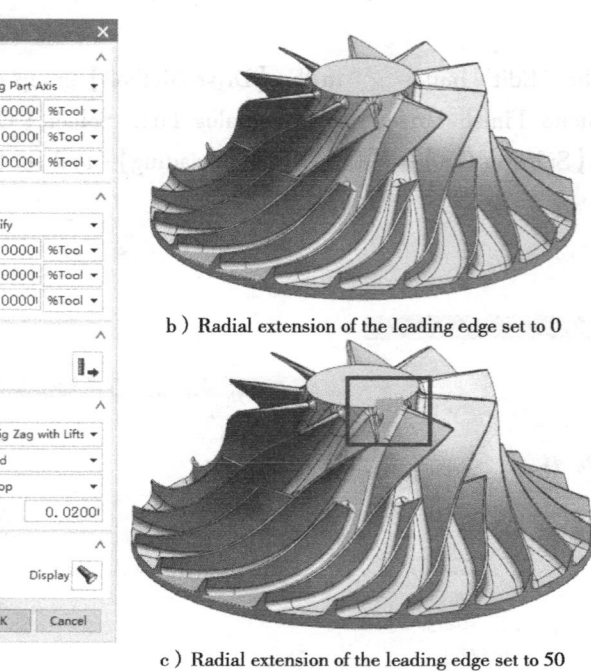

b) Radial extension of the leading edge set to 0

c) Radial extension of the leading edge set to 50

a)

Figure 2-145　Drive method of hub finishing

group of 【Hub Finish】 dialog box. Input the spindle speed as "8000" and cutting feed rate as "4000". Click the "Calculate Feeds and Speeds based on this value" button 🖩 to calculate the surface speed and the feed per tooth.

5) Click the "Generate" button ⧫ in the 【Actions】 group of 【Hub Finish】 dialog box to view the tool path of hub finishing, as shown in Figure 2-146.

6) After clicking the 【OK】 button in the 【Hub Finish】 dialog box, the 【HUB_FINISH】 program appears in the program order view of the operation navigator.

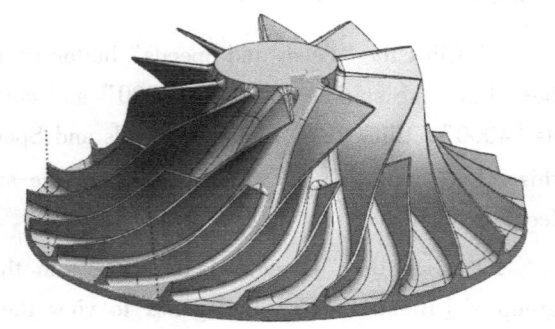

Figure 2-146　Generate the tool path of hub finishing

2.4.7　Blend Machining

Blend machining includes blend finishing of main blades and blend finishing of splitter blades. Firstly, create the blend finishing program of the main blades, and then create the blend finishing program of the splitter blades by copying and pasting. Specific steps are described below.

1) Click the 【Create Operation】 command; select 【Type】 as 【mill_multi_blade】; select 【Operation Subtype】 as the "Blend Finish"; and set the remaining parameters as shown in Figure 2-147. Click 【OK】.

83

3D Digital Design and Manufacturing

2) Pop up the【Blend Finish】dialog box. The geometry has been set up. There is no need to set it here.

3) Click the "Edit" button in the【Drive Method】group of【Blend Finish】dialog box to pop up the【Blend Finish Drive Method】dialog box. Select【Geometry to Finish】as【Blade Blend】; select【Sides to Cut】as【Left, Right, Leading】; and set the other parameters as shown in Figure 2-148.

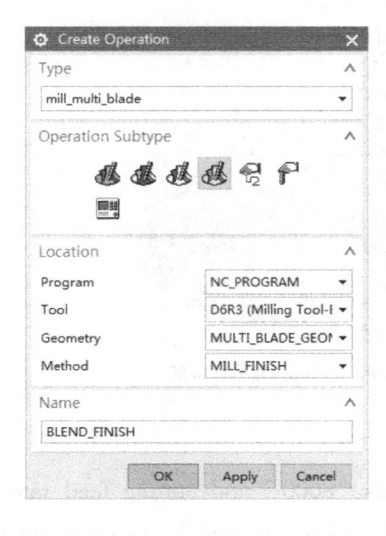 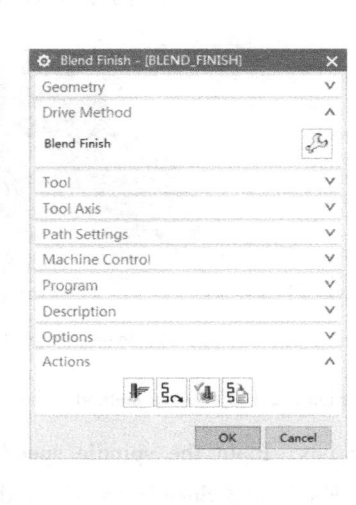 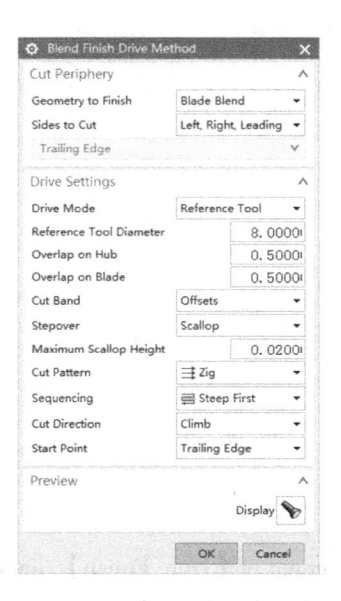

Figure 2-147　Create blend finishing process　　　　Figure 2-148　Drive method of blend finishing

4) Click the "Feeds and Speeds" button in the【Path Settings】group of【Blend Finish】dialog box. Input the spindle speed as "8000" and cutting feed rate as "4000". Click the "Calculate Feeds and Speeds based on this value" button to calculate the surface speed and the feed per tooth.

5) Click the "Generate" button in the【Actions】group of【Blend Finish】dialog box to view the tool path of blend finishing of main blades, as shown in Figure 2-149.

6) After clicking the【OK】button in the【Blend Finish】dialog box, the【BLEND_FINISH】program appears in the program order view of the operation navigator.

Figure 2-149　Create the tool path of blend finishing of main blades

7) Select the【BLEND_FINISH】program in the program order view of the operation navigator, then click the right key and select【Copy】from the right-click menu that pops up. Right-click again and select【Paste】from the right-click menu that pops up. The【BLEND_FINISH_COPY】program will appear.

8) Double-click【BLEND_FINISH_COPY】program to pop up the【Blend Finish】dialog box

▶ 84

Project 2 Turbine Wheel

as shown in Figure 2-150a. Click the button 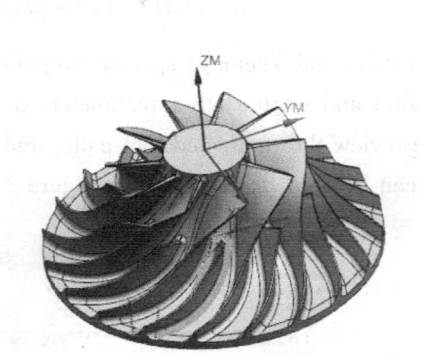 in the 【Drive Method】group to pop up the 【Blend Finish Drive Method】 dialog box. Select 【Geometry to Finish】 as 【Splitter 1 Blend】 and click 【OK】, as shown in Figure 2-150b.

9) Click the "Generate" button 〖 in the 【Actions】group of 【Blend Finish】 dialog box to view the tool path of splitter blade blend finishing, as shown in Figure 2-151.

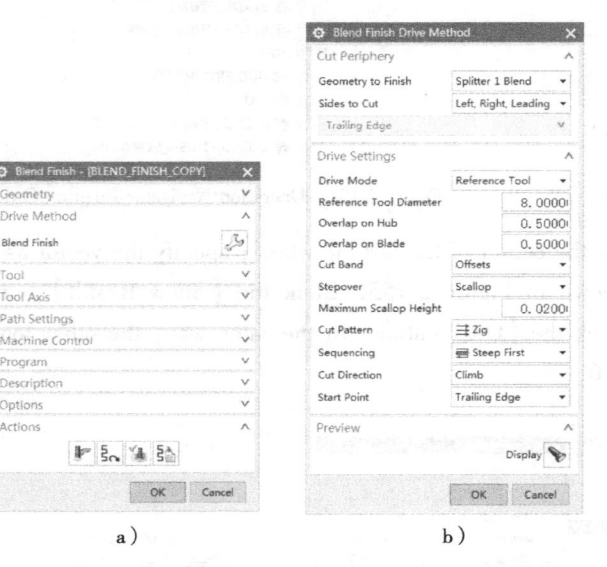

a)   b)

Figure 2-150   Drive method of splitter 1 blend

Figure 2-151   Generate tool path

2.4.8 Copy Program

1) Click the 【Create Program】 command as shown in Figure 2-152 to pop up the 【Create Program】 dialog box with set parameters as shown in Figure 2-153a. After clicking 【OK】button, pop up the【Program】 dialog box shown in Figure 2-153b. After clicking 【OK】button, ROUGH will appear in the operation navigator. Drag the 【MULTI_BLADE_ROUGH】 program below 【ROUGH】.

Figure 2-152   Create program command

2) Create 【BLADE】, 【HUB】 and 【BLEND】 programs in the same way. Drag the【BLADE_FINISH】 and 【BLADE_FINISH_COPY】 programs below 【BLADE】; drag the 【HUB_FINISH】 program below 【HUB】; drag the【BLEND_FINISH】 and 【BLEND_FINISH_COPY】 programs below 【BLEND】, as shown in Figure 2-154.

3) Select the 【MULTI_BLADE_ROUGH】 program (blade rough), then click the right key and select the 【Object】/【Transformations】 command from the pop-up right-click menu. In the 【Transformations】 dialog box, select 【Type】 as 【Rotate About a Line】; select 【Line Method】 as

85 ◀

3D Digital Design and Manufacturing

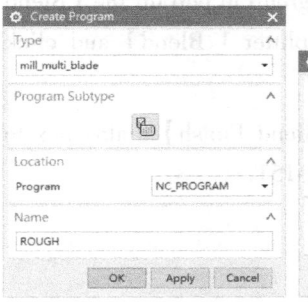

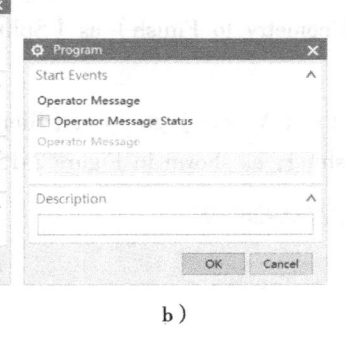

a)   b)

Figure 2-153   Create program

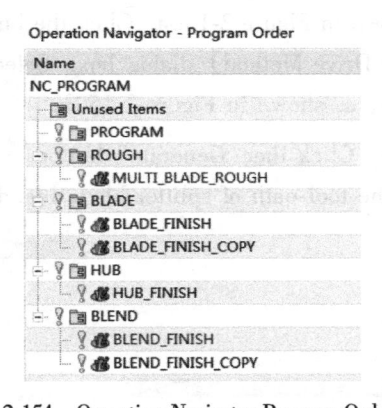

Figure 2-154   Operation Navigator-Program Order view

【Point and Vector】; specify the point as the center of the tarbine wheel; specify the vector as the + ZC; and set the other parameters as shown in Figure 2-155. Click the 【Show Result】 button to preview the replicated tool path, and click the 【OK】 button. In the same way, the other tool path can be copied, as shown in Figure 2-156.

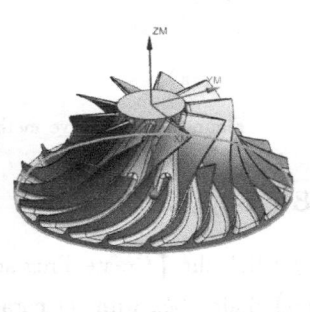

Figure 2-155   Copy program by 【Transformations】

Figure 2-156   Copy tool path

▶ 86

Project 2　Turbine Wheel

2.4.9　Machining Simulation

In NX software, there are two ways of machining simulation: one is to transfer the simulation machine tool for simulation, the other is to convert the generated program directly to 2D or 3D for simulation. The machining simulation is shown in Figure 2-157.

In NX, the corresponding tool paths of the turbine wheel's runner surfaces and blades are generated by the processing module, and the actual movement of NC machine tool is simulated by using the simulation environment of five-axis machining center. The actual cutting verification is reduced by simulating the actual action of NC machine tool. The collision problem between tool, workpiece and machine tool components and fixture is solved, and the correctness of NC processing program and post-processor is verified, so as to shorten the production cycle and reduce the cost.

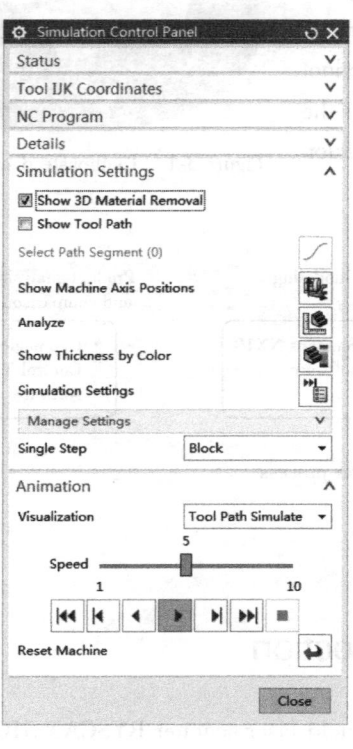

Figure 2-157　Machining simulation

87 ◀

Project 3　Toothpaste Dispenser

This project takes the toothpaste dispenser (Figure 3-1) as the carrier. Data collection is carried out by hand-held laser scanner BYSCAN 510; data processing is carried out by the Geomagic Design X 2016 software; product reverse modeling is carried out by the Siemens NX 10 software; and product analysis, comparison and report exporting are carried out by the Geomagic Control X 2018 software. Project implementation process is shown in Figure 3-2.

Figure 3-1　Toothpaste dispenser

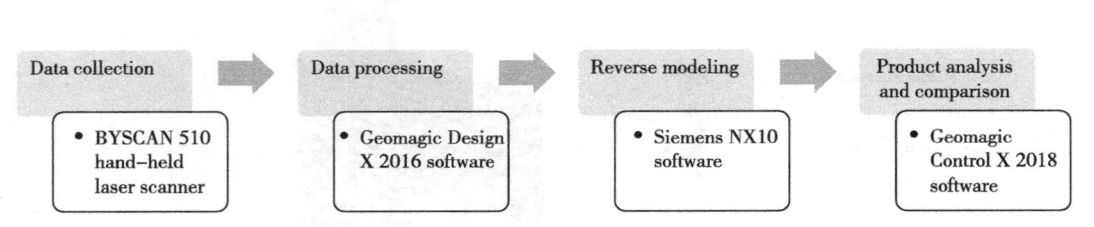

Figure 3-2　Project implementation process

Task 1　Data Collection

The data collection of this project is based on the hand-held laser scanner BYSCAN 510, which is equipped with the scanning software ScanViewer. Refer to the turbine wheel scanning for the specific scanning steps and processes.

3.1.1　Stick Mark Points

Stick mark points on the front and back sides of the toothpaste dispenser before the scanning starts, as shown in Figure 3-3. Because there are many stiffeners in the interior of the toothpaste dispenser, it is necessary to pay attention not to make the mark points too close to the stiffeners or edges.

Figure 3-3　Stick mark points

Project 3　Toothpaste Dispenser

3.1.2　Scan Mark Points

1）Open the supporting scanning software ScanViewer of the hand-held laser scanner BYSCAN 510.

2）In the scanning control panel, set the【Scanning Resolution Settings】to 1mm; set the 【Exposure Parameter Settings】to 1ms; select the【Mark Point】option; and then click the【Start】 button, as shown in Figure 3-4.

3）As shown in Figure 3-5, the scanner is facing the toothpaste dispenser. Press the scan key on the scanner to start scanning. Software interface in the process of scanning is shown in Figure 3-6. Save the data after scanning.

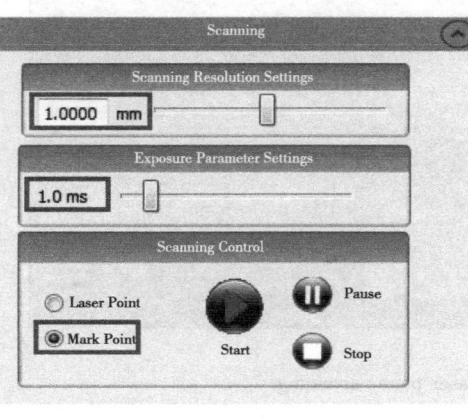

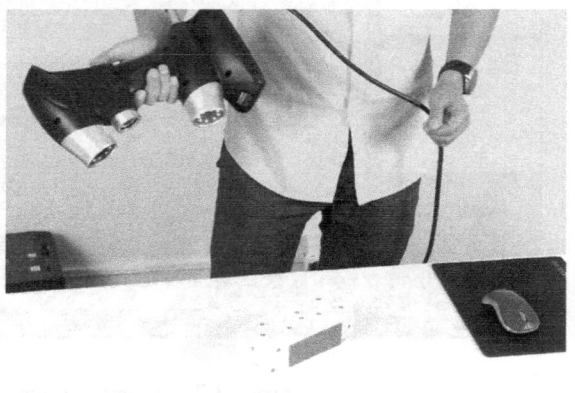

Figure 3-4　Parameter settings　　　　　Figure 3-5　Scanning of the toothpaste dispenser

Figure 3-6　Software interface when scanning mark points

89 ◀

3.1.3 Scan Laser Points

1) In the scanning control panel, set the 【Scanning Resolution Settings】 to 0.3mm; set the 【Exposure Parameter Settings】 to 2.1ms; and click the 【Start】 button to start scanning, as shown in Figure 3-7.

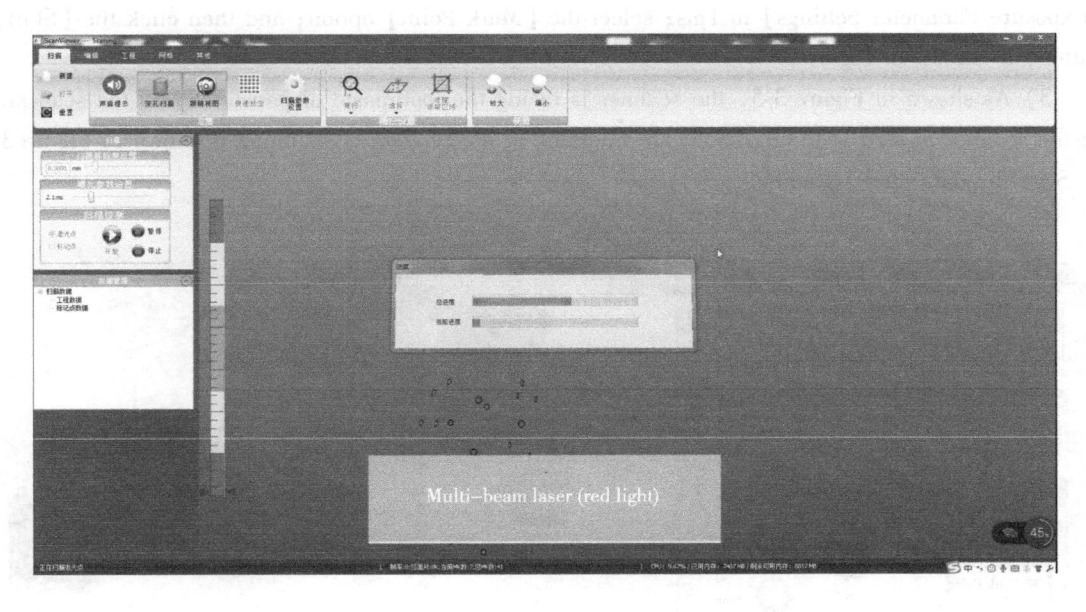

Figure 3-7   Parameter settings of laser points scanning

2) Press the scan key on the scanner to start scanning. Software interface in the process of scanning is shown in Figure 3-8.

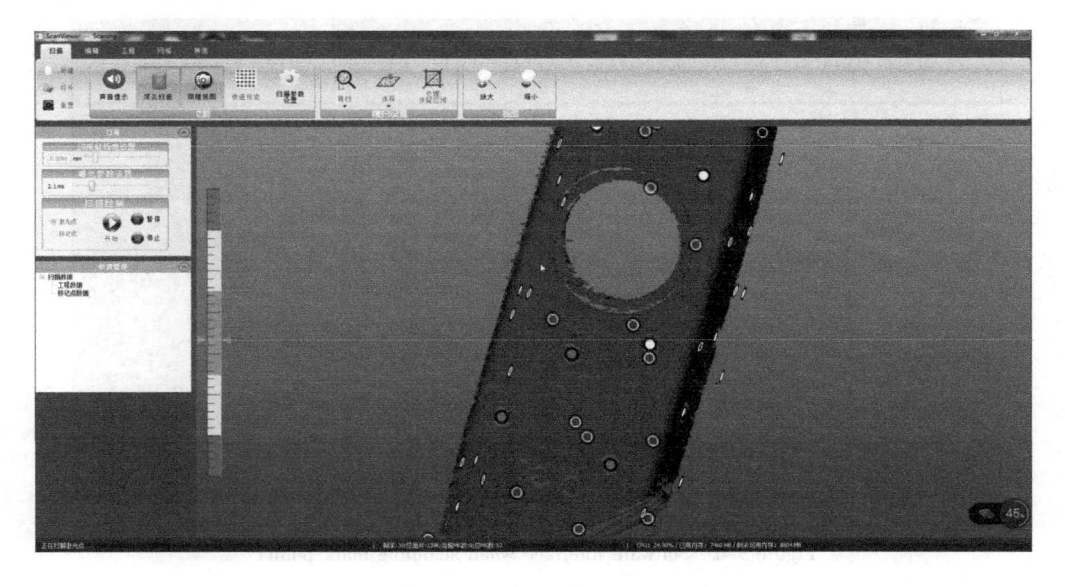

Figure 3-8   Software interface when scanning laser points

Project 3　Toothpaste Dispenser

Because the workpiece is small, it can be held in hand and switched to single-beam laser scanning mode. The alternating use of single-beam and multi-beam laser helps to scan small areas, as shown in Figure 3-9 and Figure 3-10.

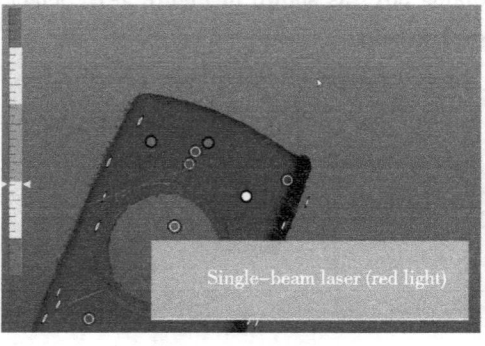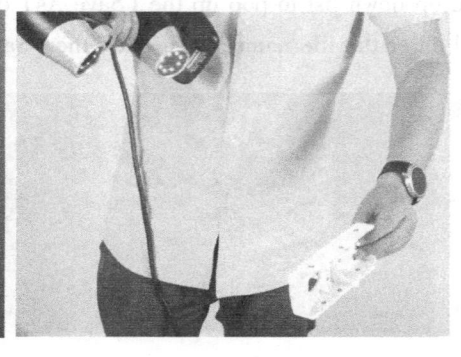

Figure 3-9　Single-beam laser scanning

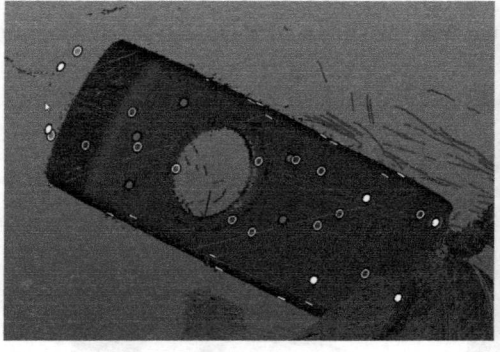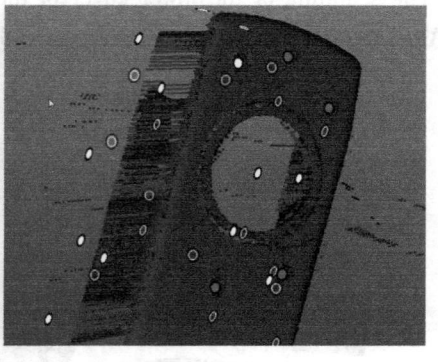

Figure 3-10　Multi-beam laser and single-beam laser scanning

3) After the scanning is completed, press the scan key on the scanner to stop scanning, then click the 【Pause】 button in the scanning software. Use the lasso tool to select the scanned irrelevant data and press the < Delete > key on the keyboard to delete it, as shown in Figure 3-11.

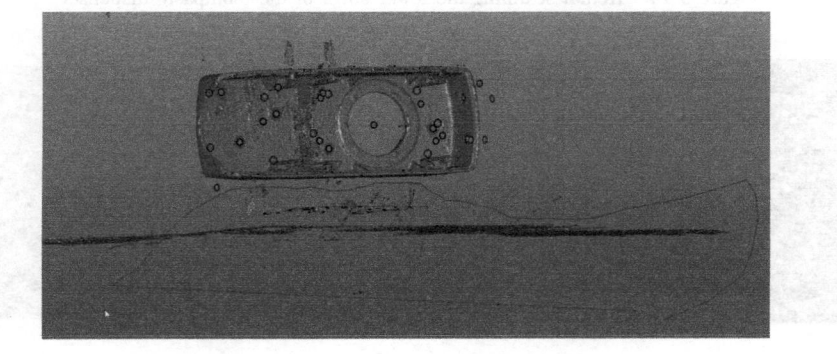

Figure 3-11　Select and delete the irrelevant data

3D Digital Design and Manufacturing

4) After the scanning is completed, click the 【Generate Mesh】 command in the 【Work】 tab, and the system starts to generate the grid and displays the progress bar, as shown in Figure 3-12.

5) Click the 【Save】command in the 【Mesh】 tab and select the 【Mesh File (* STL)】 in the pop-up drop-down list to pop up the 【Save As】 dialog box, as shown in Figure 3-3. Select the save path and enter the file name; then click the 【Save】 button.

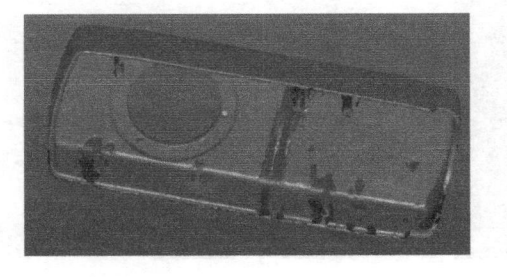

Figure 3-12　Generate the mesh model of the toothpaste dispenser　　　　Figure 3-13　Save as STL File

6) The scanning of the lower cover of the toothpaste dispenser is similar to that of the upper cover. After the scanning is completed, save the data, as shown in Figure 3-14 and Figure 3-15.

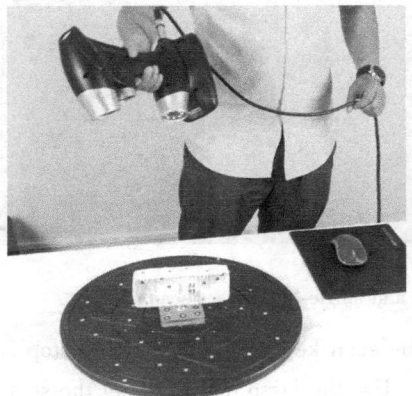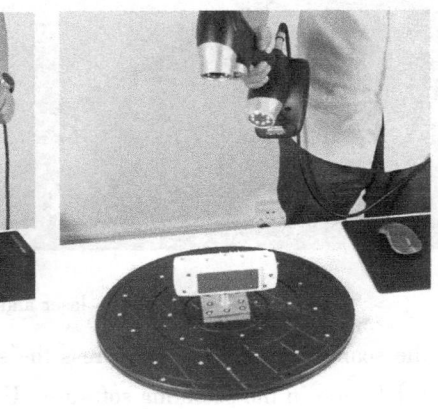

Figure 3-14　Actual scanning the lower cover of the toothpaste dispenser

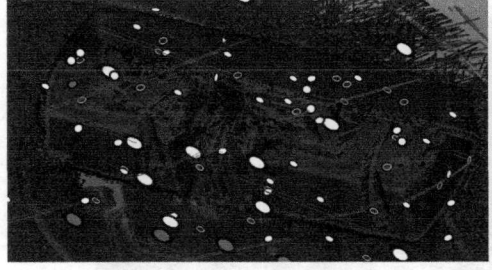

Figure 3-15　Software interface for scanning the lower cover of the toothpaste dispenser

92

Project 3　Toothpaste Dispenser

Task 2　Data Processing

3.2.1　Import Model

1) Open the software. Click the Geomagic Design X 2016 program on the "Start" menu, or double-click the icon **DX** on the desktop, to start the Geomagic Design X 2016 application software.

2) Import the model data. Click the "Import" button 📂 on the top left of the interface and select the model data generated by scanning to import the data into the Geomagic Design X 2016 software.

3.2.2　Scanning Data Processing of the Upper Cover

1) Click the 【Healing Wizard】command in the 【POLYGONS】 module to pop up the 【Healing Wizard】 dialog box. The software will automatically retrieve various defects in the mesh model, such as non-manifold vertices, overlapping poly-faces, suspended poly-faces, crossing poly-faces, etc. , as shown in Figure 3-16. After clicking the "OK" button ✅, the software will automatically repair the defects retrieved.

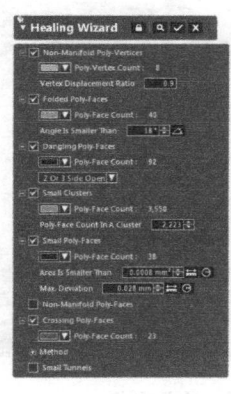

Figure 3-16　Repair mesh defects automatically with 【Healing Wizard】

2) Observation is done by rotating, translating, zooming-in and zooming-out the model, to find out whether there are defects. One of the defects is shown in Figure 3-17a.

Select the defective area as shown in Figure 3-17b and press the < Delete > key to delete the selected data as shown in Figure 3-17c. Click the 【Fill Holes】 command in the 【POLYGONS】 module to pop up the dialog box as shown in Figure 3-18. Select the hole as shown in Figure 3-17d. The parameters are set by default, and click the "OK" button ✅ to complete the operation. The effect is shown in Figure 3-17e.

Fill the holes in other defects by 【Fill Holes】command, as shown in Figure 3-19.

93 ◀

3D Digital Design and Manufacturing

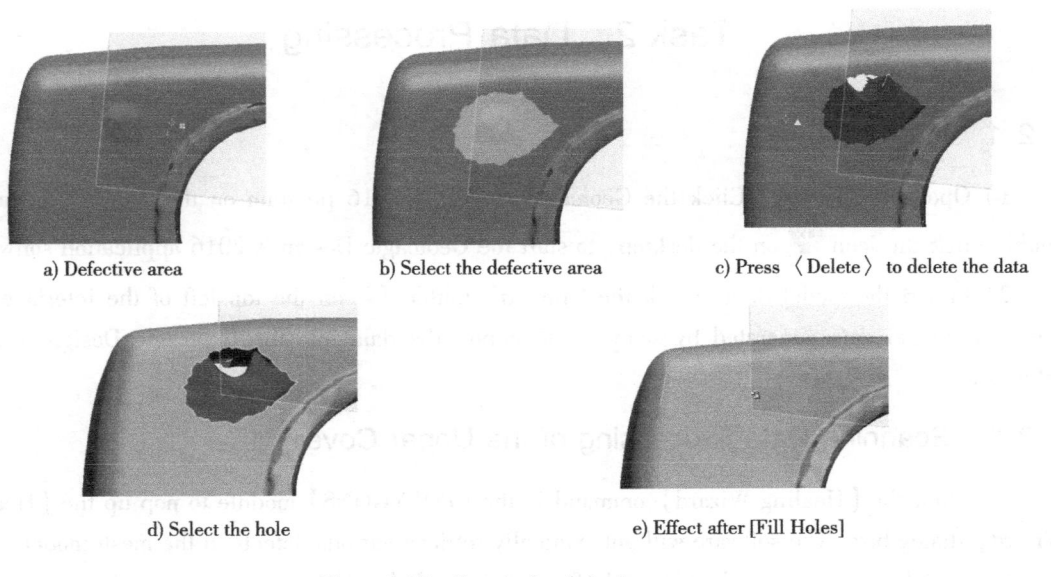

a) Defective area   b) Select the defective area   c) Press 〈Delete〉 to delete the data

d) Select the hole   e) Effect after [Fill Holes]

Figure 3-17   Repair the defective area

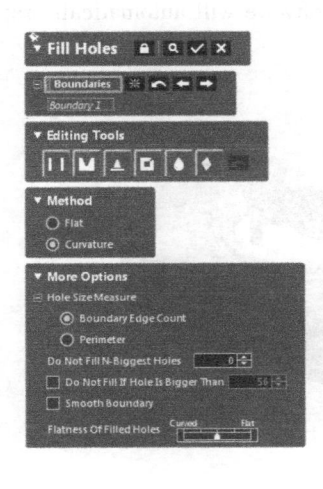

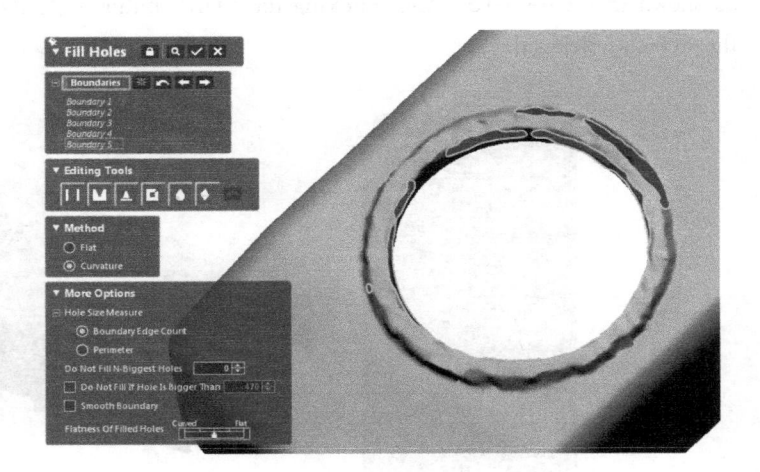

Figure 3-18   【Fill Holes】 dialog box   Figure 3-19   Holes filling

3) Click the 【Enhance Shape】command in the 【POLYGONS】 module to pop up the dialog box shown in Figure 3-20. Set all the three options at their default values, and then click the "OK" button ✓ to complete the operation. Enhancing shape is used to sharpen the sharp areas (edges and corners) on the mesh and smooth the plane or cylindrical areas to improve the quality of the mesh.

4) Select the upper cover of the toothpaste dispenser; select the 【Menu】/【File】/【Export】command to pop up the 【Export】 dialog box, as shown in Figure 3-21. Click the "OK" button ✓ to export the data, as shown in Figure 3-21b.

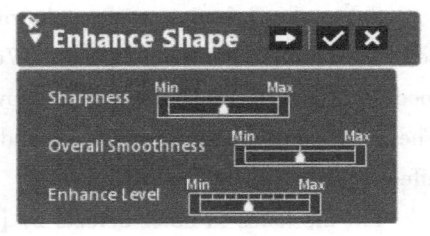

Figure 3-20   【Enhance Shape】 dialog box

▶ 94

Project 3　Toothpaste Dispenser

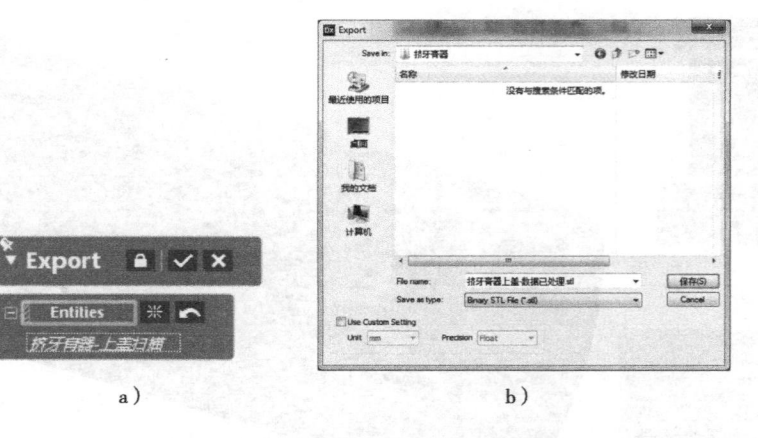

a)　　　　　　　　　　　　　b)

Figure 3-21　Export the data of the upper cover of the toothpaste dispenser

3.2.3　Scanning Data Processing of the Lower Cover

1) After clicking the【Healing Wizard】command in the【POLYGONS】module and clicking the "OK"button, the software will automatically repair the retrieved defects, as shown in Figure 3-22.

Figure 3-22　Repair mesh defects automatically with【Healing Wizard】(low cover)

2) Click the【Fill Holes】command in the【POLYGONS】module; select the hole you want to fill; set the parameters by default; and click the "OK" button ✓ to complete the operation, as shown in Figure 3-23.

3) Click the【Enhance Shape】command in the【POLYGONS】module, and keep the default values for the parameters to enhance the shape.

4) Select the lower cover of the toothpaste dispenser; select the【Menu】/【File】/【Export】command to pop up the【Export】dialog box, as shown in Figure 3-24a. Click the "OK" button ✓ to export the data, as shown in Figure 3-24b.

95

3D Digital Design and Manufacturing

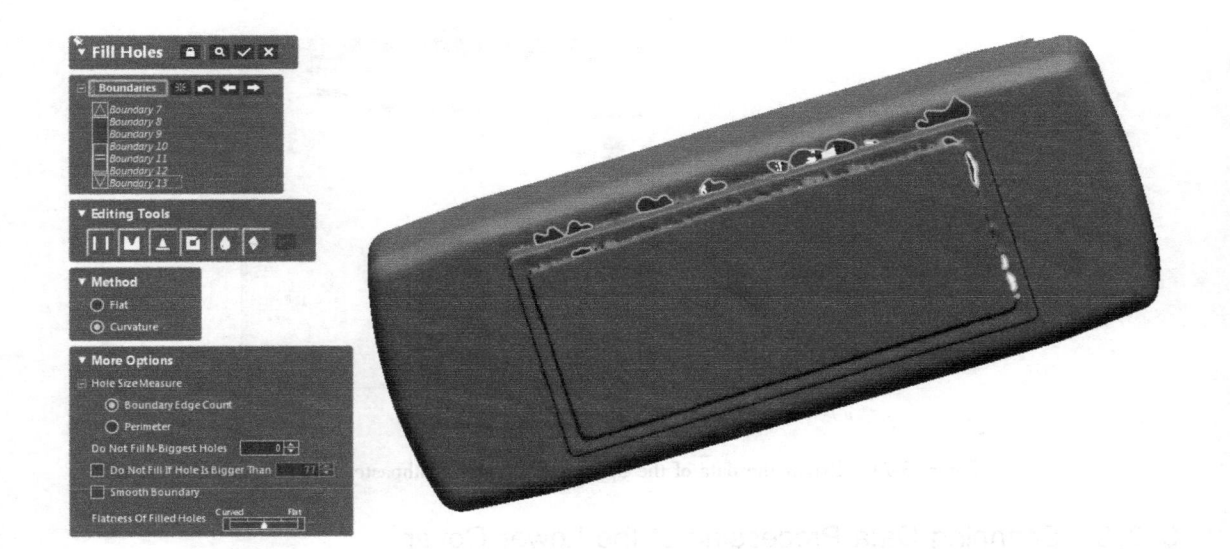

Figure 3-23   Holes filling(low cover)

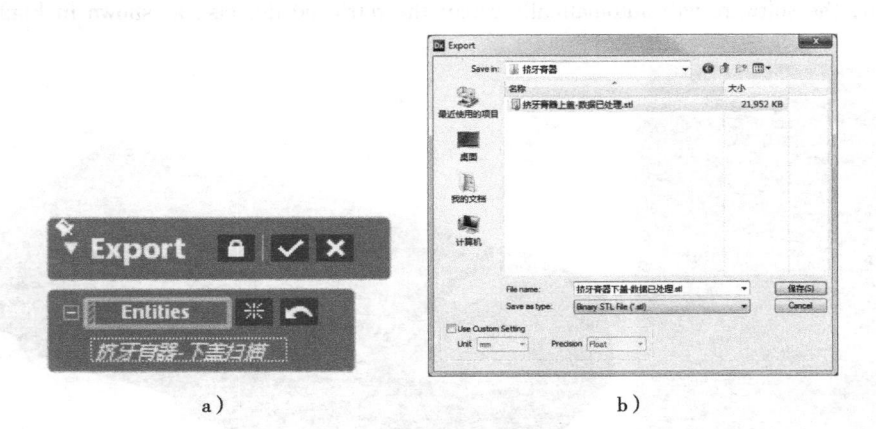

a)   b)

Figure 3-24   Export the data of the lower cover of the toothpaste dispenser

Task 3   Reverse Modeling

3.3.1   Import Model

1) The scanning files of the upper and lower cover of the toothpaste dispenser are imported respectively. Change the color of a scanned model to make it easier to distinguish between the upper and lower cover, as shown in Figure 3-25.

2) According to personal habits, the color of the sheet can be changed to make the subsequent created sheet blue for easy observation, as shown in Figure 3-26.

▶ 96

Project 3 Toothpaste Dispenser

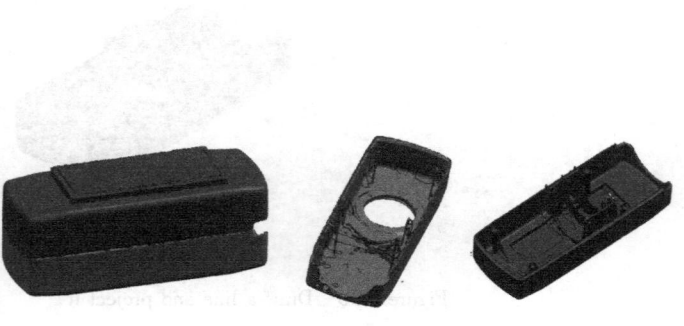

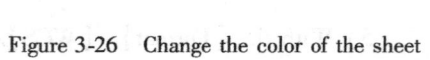

Figure 3-25 Import scanning files

Figure 3-26 Change the color of the sheet

3.3.2 Determine Coordinates

1) Through the function of 【Line】in the 【Basic Curves】command, select the 【Point on Face】to create two lines, as shown in Figure 3-27.

2) Through the 【Swept】command and the two lines created, the surface is generated, and it is adjusted under the 【Enlarge】command, as shown in Figure 3-28.

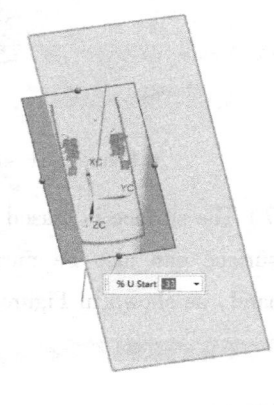

Figure 3-27 Create two lines

Figure 3-28 Adjust the surface size after sweeping

3) The tentatively created plane is the horizontal plane. Through the 【Format】/ 【WCS】/ 【Orient】command, 【Type】is selected as 【Inferred】and capture points should be selected as 【Point on Face】. Then click three points that are roughly aligned in 90°, as shown in Figure 3-29.

4) Draw a line on the side plane of the lower cover by using the 【Line】function in the 【Basic Curves】command. After completion, the line is projected to the XC-YC plane by the 【Project Curve】command, as shown in Figure 3-30.

97 ◀

3D Digital Design and Manufacturing

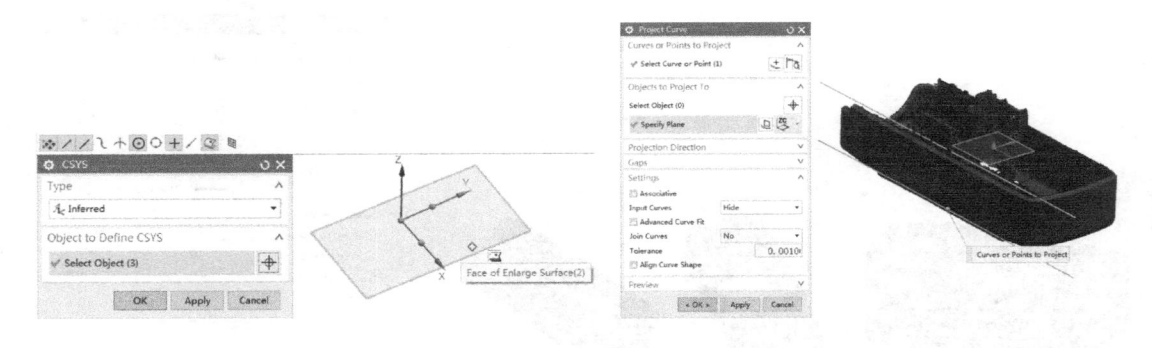

Figure 3-29   Orient

Figure 3-30   Draw a line and project it

5) With the 【Format】/【WCS】/【Orient】command, first click the Axis X, then click the projection line, so that the direction of the Axis X is parallel to the direction of the projection line, as shown in Figure 3-31.

6) With the 【Extrude】command, extrude the line created before, and then measure the distance from the other side, as shown in Figure 3-32.

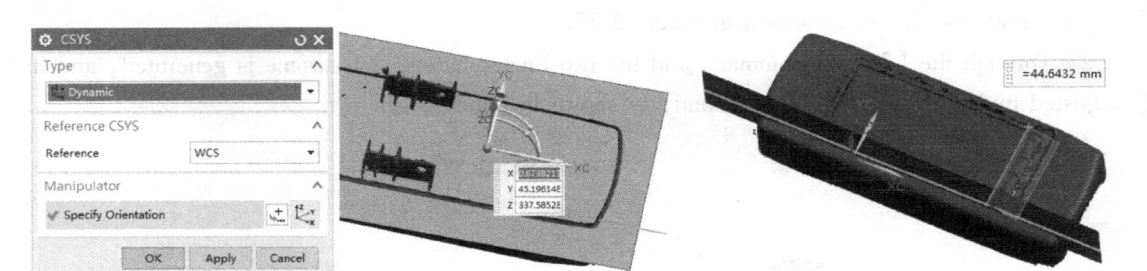

Figure 3-31   Axis X orient

Figure 3-32   Extrude and measure the distance

7) The surface is biased by half of the distance measured under the 【Offset Surface】command. The surface can also be mirrored to the other side for observation under the 【Transformations】command, as shown in Figure 3-33 and Figure 3-34.

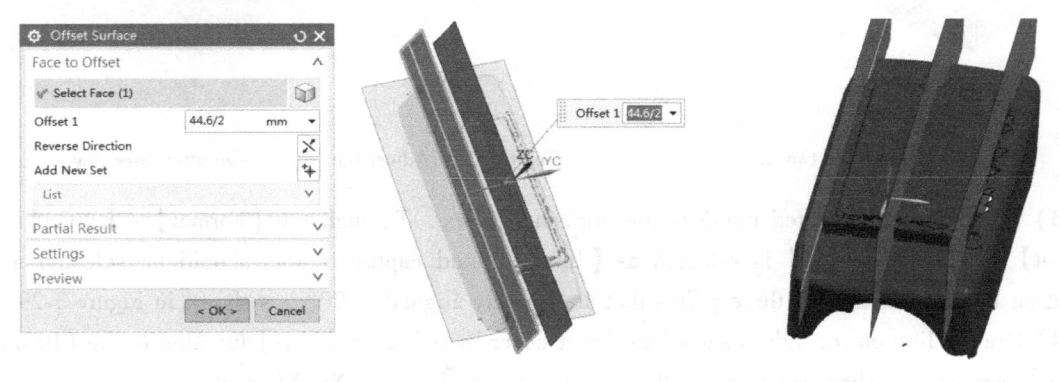

Figure 3-33   Offset the surface

Figure 3-34   Mirror the surface

8) By using the 【Arc】function in the 【Basic Curves】command, two point on the surface are

▶ 98

Project 3　Toothpaste Dispenser

selected to create an arc, and the arc is extended under the 【Curve Length】 command, as shown in Figure 3-35.

9) Extrude the arc with the 【Law Extension】 command, and trim the surface by the Plane XC-ZC, as shown in Figure 3-36.

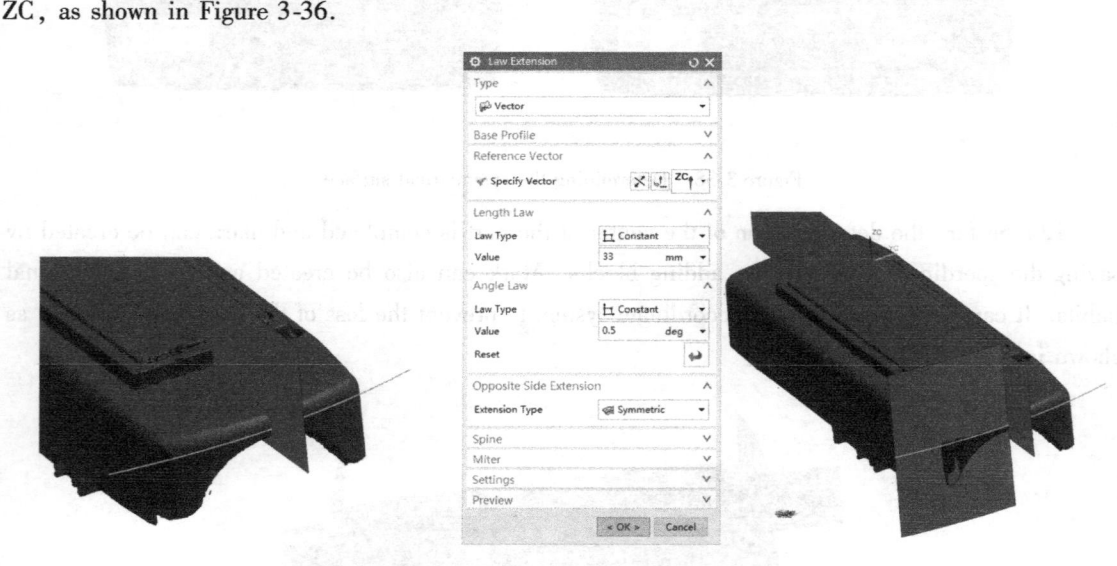

Figure 3-35　Create curve　　　　　　　　Figure 3-36　Law extension and trimming

10) Mirror the surface to the other side. By using the 【Arc】 function in the 【Basic Curves】 command, the two ends are selected as the starting point and the end point, and the arc is redrawn to ensure it is symmetric about Plane XC-ZC. In this way, extruding is carried out under the 【Law Extension】 command, and the size of the surface is adjusted appropriately, as shown in Figure 3-37.

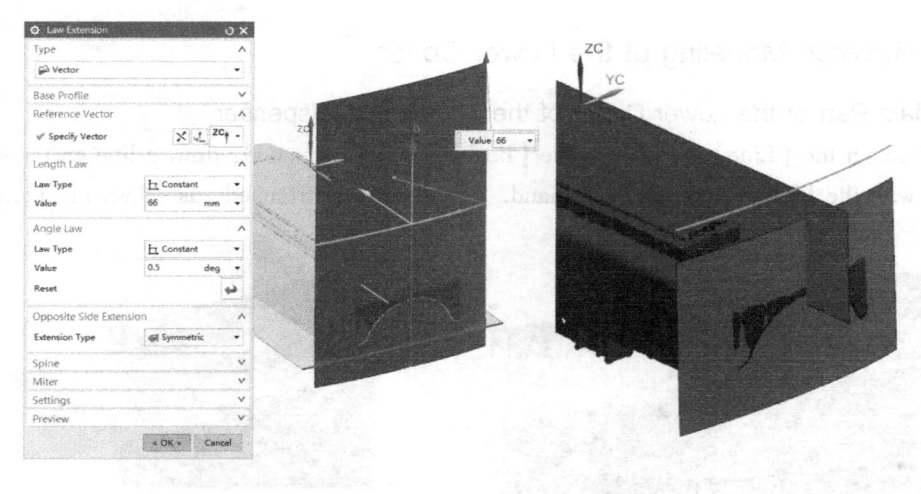

Figure 3-37　Law extension and trimming

11) Use the 【Transformations】 command to mirror the surface to the other side. If there is a deviation, measure the distance. Then click the Axis X and move half of the measured distance for adjustment step by step until it meets the requirements, as shown in Figure 3-38.

99 ◀

3D Digital Design and Manufacturing

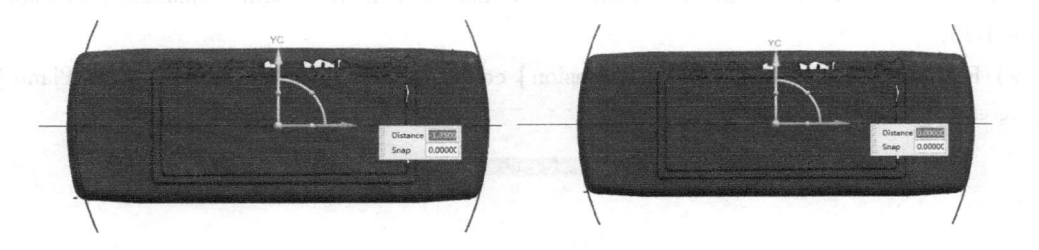

Figure 3-38　Determining the symmetrical surface

12) So far, the determination of the center of the part is completed and mark can be created by saving the coordinate system or by adding blocks. Mark can also be created according to personal habits. It can be used to mark the coordinate system to prevent the loss of the coordinate system, as shown in Figure 3-39.

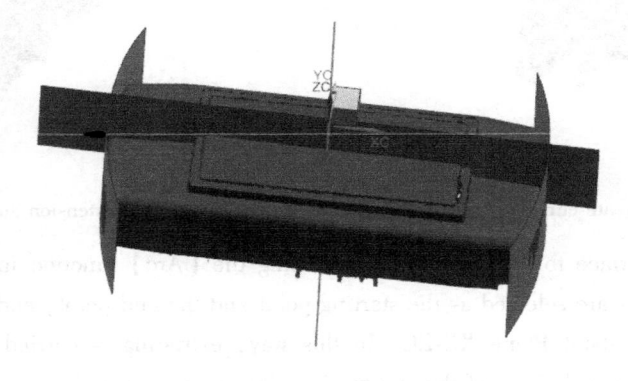

Figure 3-39　Mark coordinate system

3.3.3　Reverse Modeling of the Lower Cover

1. Main Part of the Lower Cover of the Toothpaste Dispenser

1) Through the 【Line】 function of the 【Basic Curves】 command, draw a line on the side and extrude it with the 【Law Extension】 command. Two sheets are trimmed, as shown in Figure 3-40.

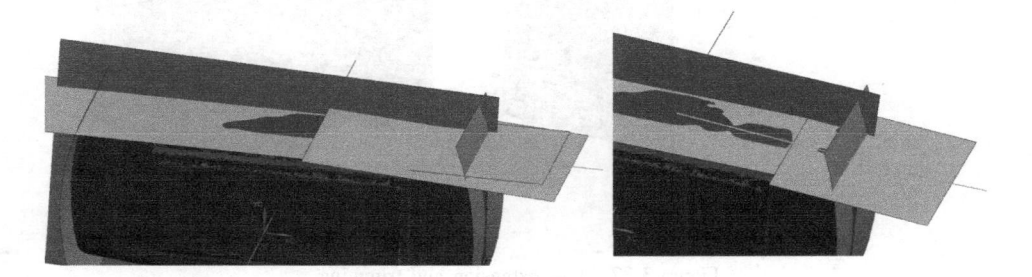

Figure 3-40　Create a sheet and trim it

2) Mirror the sheet by means of the 【Transformations】 command for observation. There are some deviations in the sheets, so equalize them, as shown in Figure 3-41.

▶ 100

Project 3　Toothpaste Dispenser

3）The original surface is offset inward by 0.15mm through the function of the【Offset Surface】 command, and the mirroring is performed for observation again, as shown in Figure 3-42 and Figure 3-43.

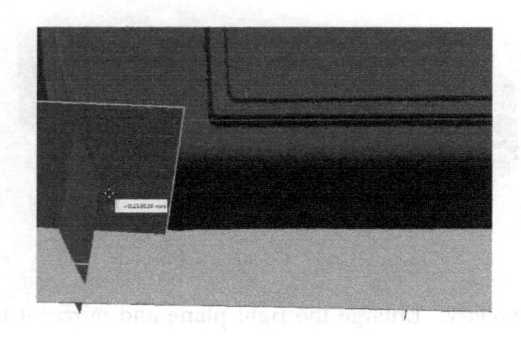

Figure 3-41　It is observed that there are some deviations in the sheets

Figure 3-42　Offset the original surface

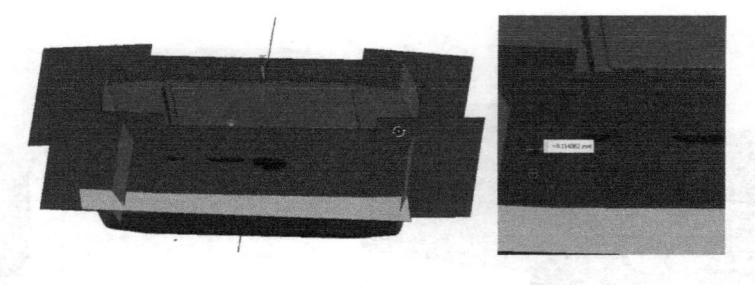

Figure 3-43　Re-mirror the surface

4）Observe the parallelism between the top surface and the bottom surface. After roughly measuring the distance, the bottom surface is offset to a certain distance under the【Offset Surface】 command, as shown in Figure 3-44.

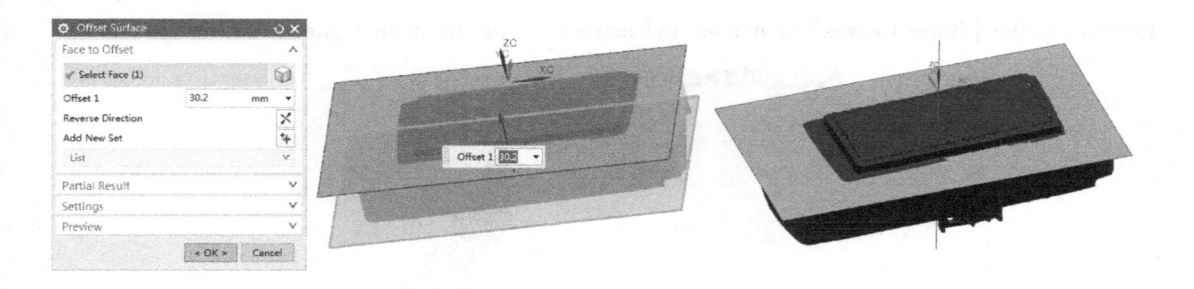

Figure 3-44　Offset the bottom surface

5）Two lines parallel to the YC direction are created at the two slopes respectively under the 【Line】 function in the【Basic Curves】command and are extruded under the【Law Extension】 command, as shown in Figure 3-45.

6）Trimming is done under the commands of【Trim Body】and【Trim Sheet】. If the length is

101

3D Digital Design and Manufacturing

insufficient, it can be extended under the 【Extension Sheet】 command. Till the final trimming is completed, it is sewed into a solid body under the 【Sew】 command, as shown in Figure 3-46.

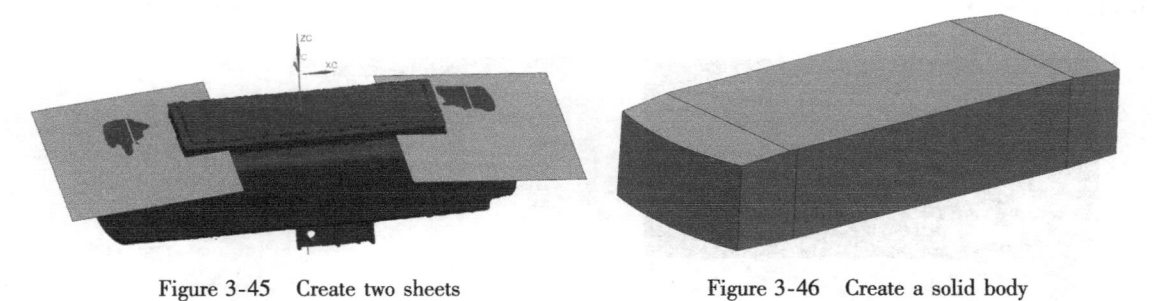

Figure 3-45　Create two sheets　　　　　　　　　Figure 3-46　Create a solid body

7) Observe the actual product and judge its symmetry. Enlarge the right plane and mirror it to the left and observe the result. The measurement results show that the deviation is about 0.1 mm, which meets the requirements. Substitute the plane by 【Replace Face】 command, as shown in Figure 3-47.

Figure 3-47　Replace the plane

8) Next, the boss features are drawn. Draw a line parallel to the Axis YC by the 【Line】 function in the 【Basic Curves】 command and extrude it, as shown in Figure 3-48.

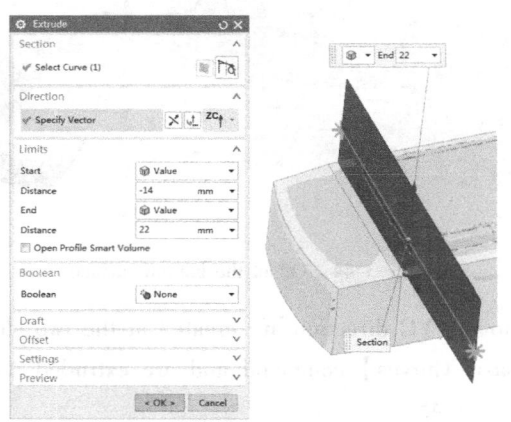

Figure 3-48　Create a line and extrude it

▶ 102

Project 3　Toothpaste Dispenser

9) Mirror the surface to the other side, which is actually used to stick the toothpaste dispenser on the wall. Therefore, it is symmetric about the coordinate system. Offset adjustments are made under the 【Offset Surface】 command, as shown in Figure 3-49.

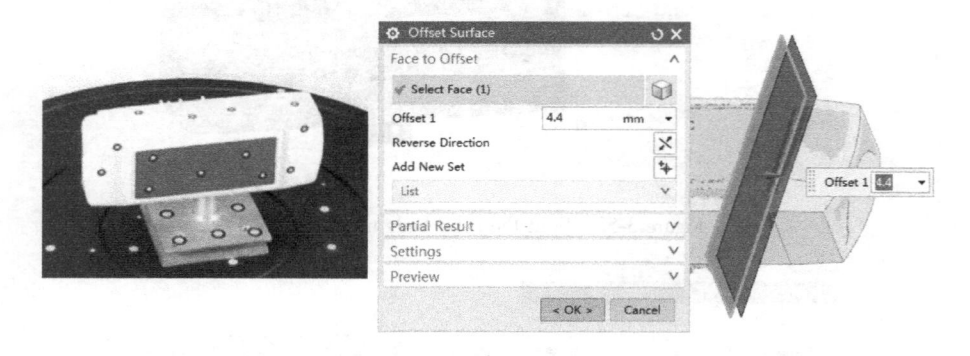

Figure 3-49　Offset the surface

10) After measuring the distance of the top of the boss, the offset is made under the command of the 【Offset Surface】, as shown in Figure 3-50.

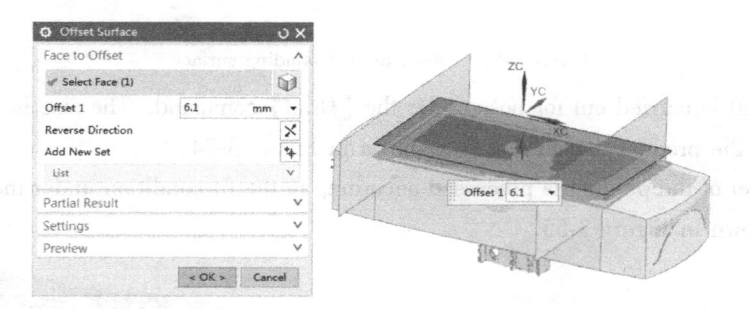

Figure 3-50　Offset the surface after measuring the distance

11) Thicken the boss under the 【Thicken】 command, then replace the surface under the 【Replace Face】 command, as shown in Figure 3-51.

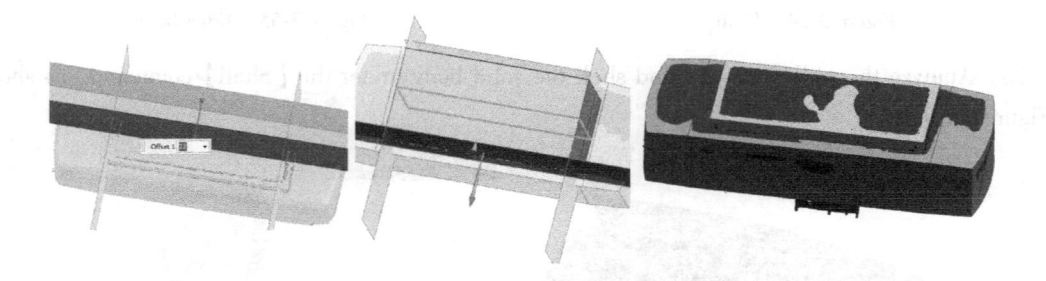

Figure 3-51　Create a boss

12) The top-down offset is 1 mm under the 【Offset Face】 command, and then the solid body is split under the 【Split Body】 command, as shown in Figure 3-52.

13) The distance can be roughly measured by the【Distance Measurement】command, and the surrounding surface can be offset under the 【Offset Face】 command, as shown in Figure 3-53.

103 ◀

3D Digital Design and Manufacturing

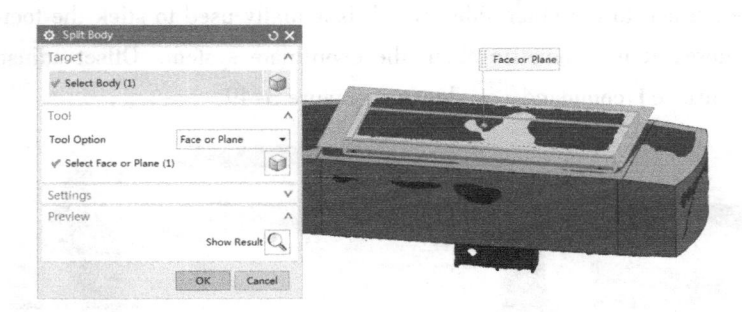

Figure 3-52　Split the solid body after offset

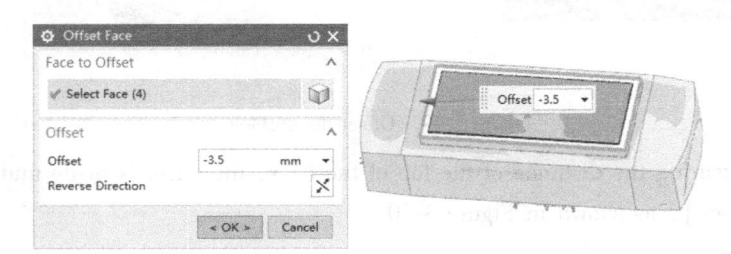

Figure 3-53　Offset the surrounding surface

14）The draft is carried out for boss under the 【Draft】command. The size is about 3°, which is convenient for the product demoulding, as shown in Figure 3-54.

15）The fillet of the product is processed according to the fit condition under the 【Edge Blend】 command, as shown in Figure 3-55.

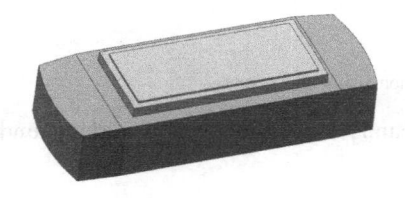

Figure 3-54　Draft

Figure 3-55　Edge blend

16）Analyze the wall thickness and shell the solid body under the 【Shell】command, as shown in Figure 3-56.

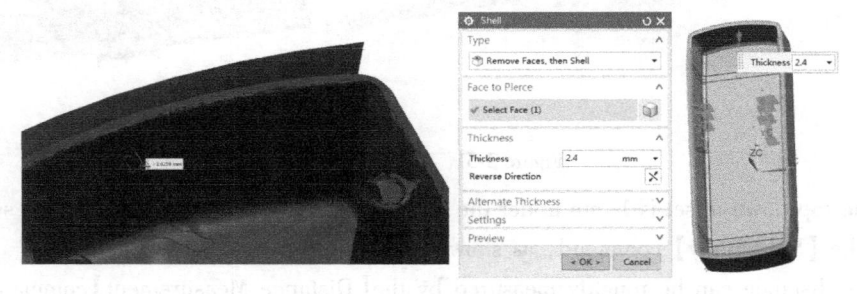

Figure 3-56　Analyze the wall thickness and shell the solid body

▶ 104

Project 3　Toothpaste Dispenser

17) Create two lines by using the 【Line】 function in the 【Basic Curves】 command and extrude them into two sheets, as shown in Figure 3-57.

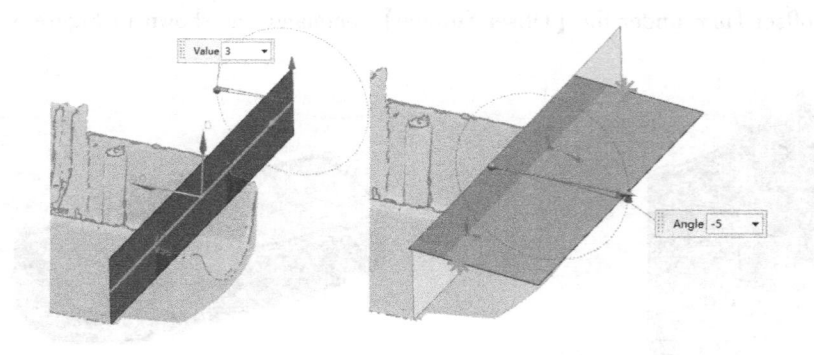

Figure 3-57　Create two sheets

18) The two sheets are trimmed and sewn under the 【Trim Body】 command. Split the solid body under the 【Split Body】 command and create fillets of 1.5mm, as shown in Figure 3-58.

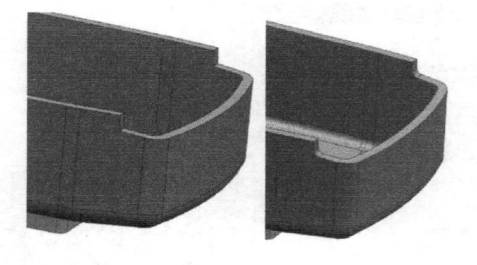

Figure 3-58　Split the body and create fillets

2. Internal Characteristics of the Lower Cover of the Toothpaste Dispenser

1) Create a circle using the 【Arc】 function in the 【Basic Curves】 command, which cancels the 【String Mode】 and selects the 【Full Circle】. Double-click the circle to observe the diameter, which is about 6.5mm, as shown in Figure 3-59.

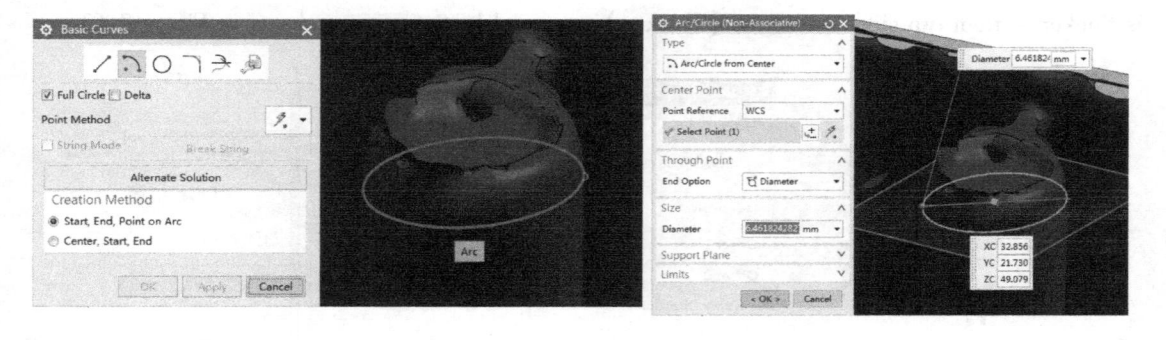

Figure 3-59　Create a circle

2) Through the 【Line】 function in the 【Basic Curves】 command, the center of the arc is captured, and a line passing through the center and parallel to Axis Z is drawn. Then create a tube

105 ◀

3D Digital Design and Manufacturing

with a diameter of 6.5mm under the 【Tube】 command, as shown in Figure 3-60.

3) The bottom plane is created under the 【Enlarge】 command. After measuring and analying, the surface is offset 1mm under the 【Offset Surface】 command, as shown in Figure 3-61.

Figure 3-60　Create a tube

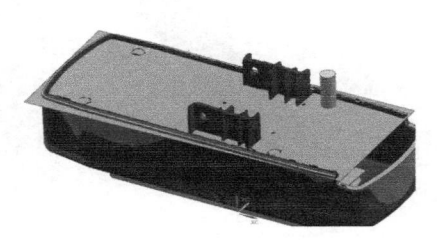

Figure 3-61　Offset the surface

4) Substitute the face under 【Replace Face】 command, as shown in Figure 3-62.

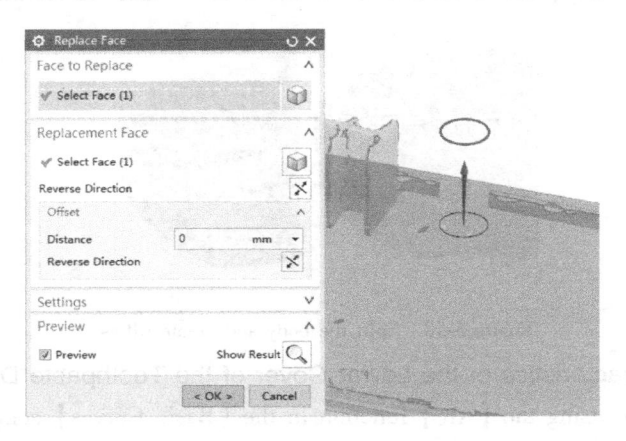

Figure 3-62　Substitute the face

5) The center line of the cylinder is extruded under the 【Extrude】 command, and then the sheet is thickened from two sides under the 【Thicken】 command by 0.5mm, as shown in Figure 3-63.

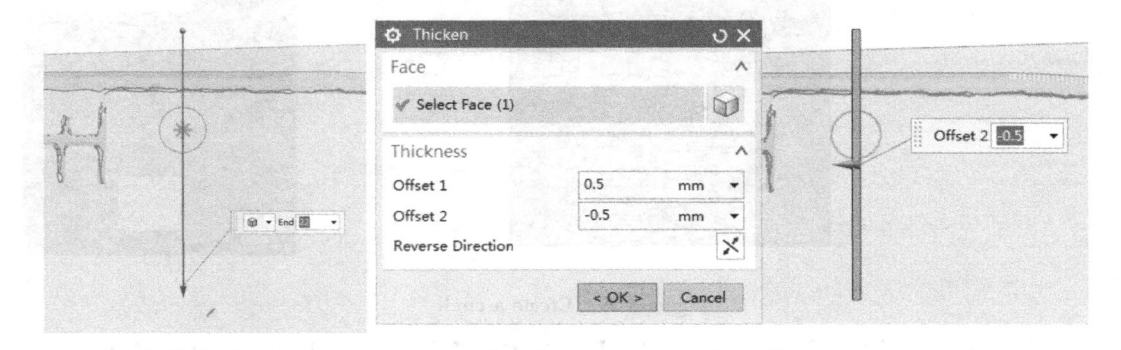

Figure 3-63　Extrude and thicken

▶ 106

Project 3   Toothpaste Dispenser

6) Multiple replacements are made through the 【Replace Face】 command. After the replacement is completed, add 0.3° draft, as shown in Figure 3-64 and Figure 3-65.

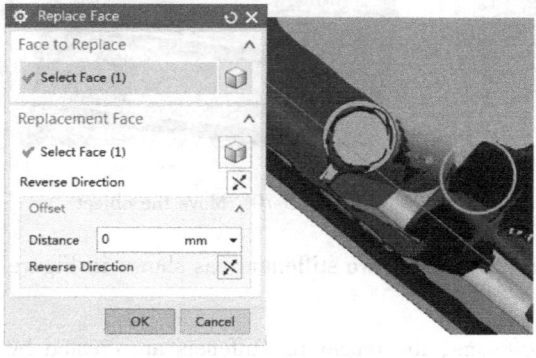
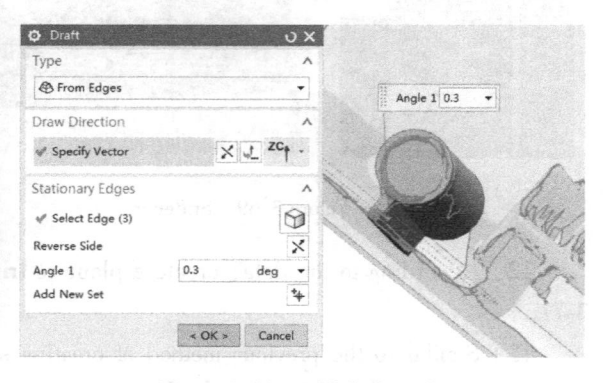

Figure 3-64   Replace the faces   Figure 3-65   Add draft angle

7) Through the 【Tube】 command, create a cylinder of 4mm in diameter, then offset a face 2 mm away from the top surface to trim the cylinder. Then subtract the two cylinders and add a certain draft angle, as shown in Figure 3-66.

8) Copy and move features by using the 【Move Object】 command. Then mirror the features to the other side under the 【Transformations】 command, as shown in Figure 3-67.

9) Through the 【Line】 Function of the

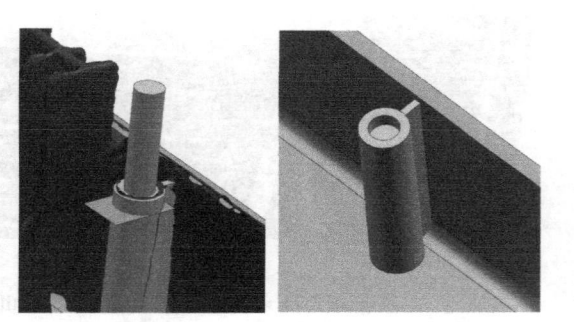

Figure 3-66   Feature creation

【Basic Curves】 command, create a line, and extrude it along Axis E with an angle of 0.1°. Thickening process is carried out, with the top closing to the scanning data and 0.1 draft adding to the other side, as shown in Figure 3-68.

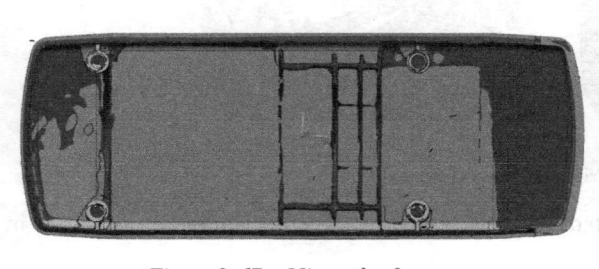
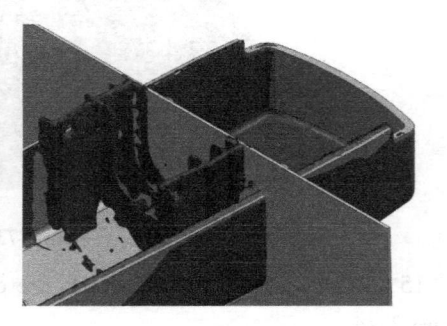

Figure 3-67   Mirror the feature   Figure 3-68   Thicken

10) Create the surface of the stiffener in other places for trimming; conduct subtraction; and conduct the fillet processing, as shown in Figure 3-69.

11) Move and copy the stiffener to the appropriate location by using the 【Move Object】 command, as shown in Figure 3-70.

107

3D Digital Design and Manufacturing

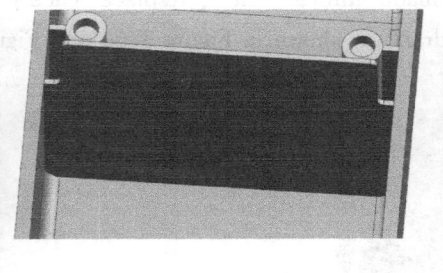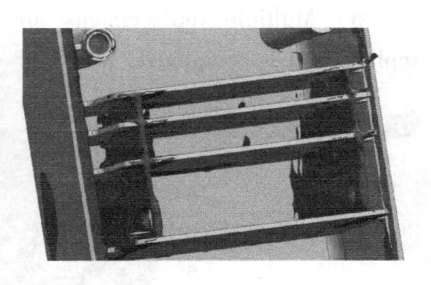

Figure 3-69   Stiffener

Figure 3-70   Move the object

12) According to the data, create a plane to trim the middle two stiffeners, as shown in Figure 3-71.

13) Similar to the previous method of creating stiffeners, the remaining stiffeners are created by commands such as 【Law Extension】, 【Thicken】 and 【Replace Face】, etc. , as shown in Figure 3-72.

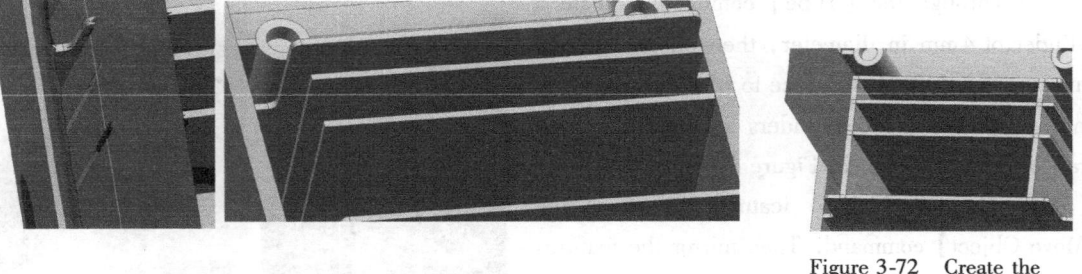

Figure 3-71   Trim the body (middle stiffeners)

Figure 3-72   Create the remaining stiffeners

14) After creating three sheets, make fillets of the sheets through the 【Face Blend】 command. After sewing, trim the three stiffeners under the 【Trim Body】 command, as shown in Figure 3-73.

Figure 3-73   Trim the three stiffeners

15) In the same way, the sheets are created and sewed to trim the other stiffener, as shown in Figure 3-74.

16) Through the 【Arc】 function in the 【Basic Curves】 command, create a full circle through three points. Check the size of the circle, and capture the center of the arc to create a line parallel to the direction of the Axis X. Then create a tube with a diameter of 4mm according to the previous analysis and measurement. Conduct subtraction in the end, as shown in Figure 3-75.

▶ 108

Project 3　Toothpaste Dispenser

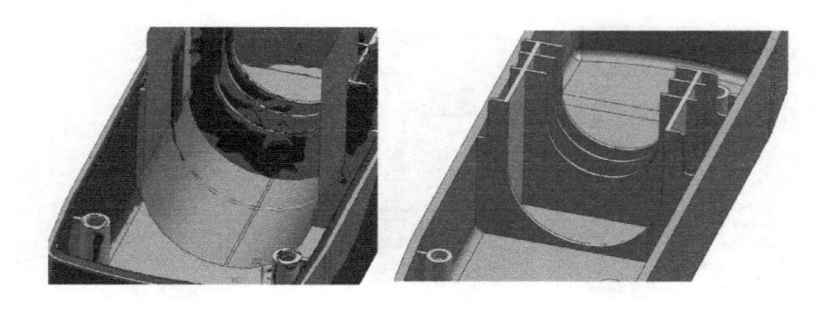

Figure 3-74　Trim the other stiffener

Figure 3-75　Make holes

17) The axis of the cylinders are extruded along the Axis Y to create two planes under the 【Extrude】 command. Obtain the intersection lines under the 【Intersection Curve】 command, as shown in Figure 3-76.

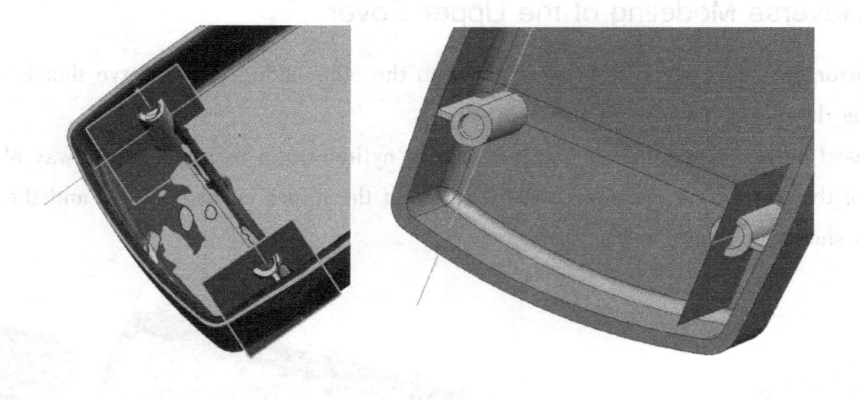

Figure 3-76　Extrude and obtain the intersection lines

18) Create a face under the 【Ruled】 command. Thickening is then carried out under the 【Thicken】 command, and the draft is made at 0.3°, as shown in Figure 3-77.

19) Create the sheets and sew them to trim the body, as shown in Figure 3-78.

20) Up to now, the reverse modeling of the lower cover of the toothpaste dispenser has been basically completed, and some features have been made, as shown in Figure 3-79.

109 ◀

3D Digital Design and Manufacturing

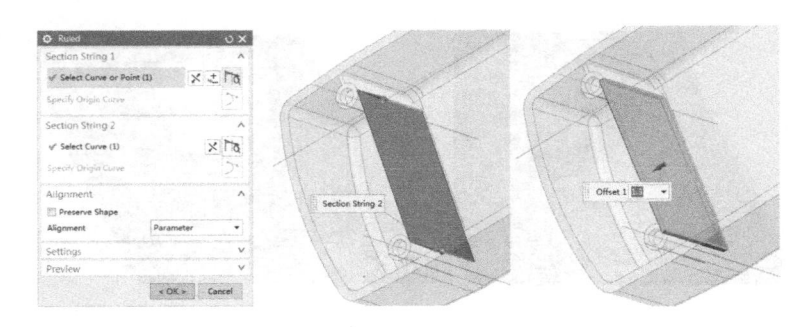

Figure 3-77　Create and thicken a face

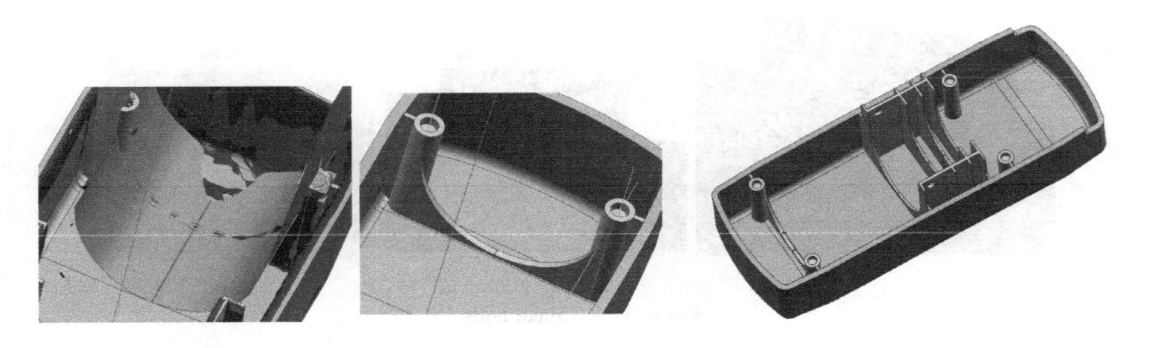

Figure 3-78　Trim the last stiffener

Figure 3-79　The lower cover
of toothpaste dispenser is basically completed

3.3.4　Reverse Modeling of the Upper Cover

1) Mirror the main part of the lower cover to the other side, and observe that it fits well with the scanning data, as shown in Figure 3-80.

2) Based on the center line of the lower cover cylinder and referring to the way of creating the same part of the previous lower cover, the features of the upper cover cylinder and the stiffener are created, as shown in Figure 3-81.

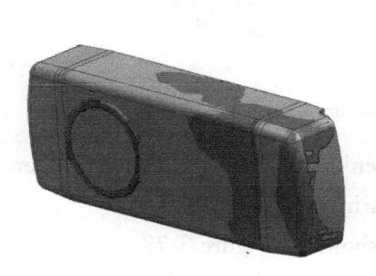

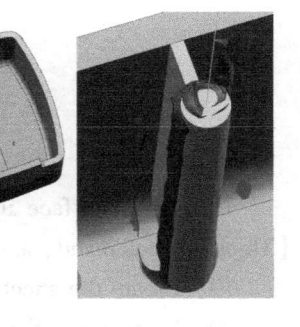

Figure 3-80　Observe the fit between
the mirrored lower cover and the upper cover

Figure 3-81　Create the features
of the upper cover cylinder and the stiffener

▶ 110

Project 3　Toothpaste Dispenser

3) To ensure the fitting, the top of the small cylinder should fit the bottom of the big cylinder, as shown in Figure 3-82.

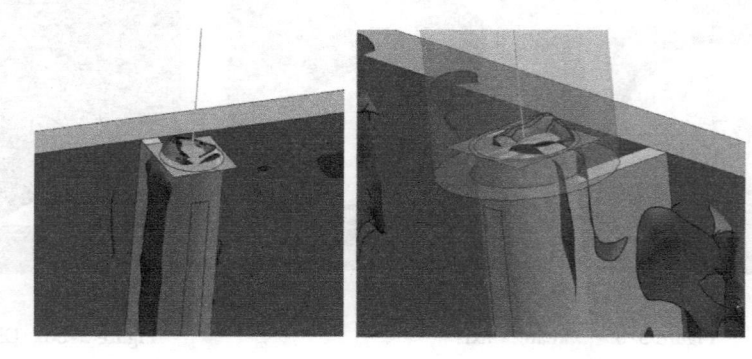

Figure 3-82　The top and bottom surfaces should be fitted

4) By 【Extract Geometry】command, extract three faces, as shown in Figure 3-83.

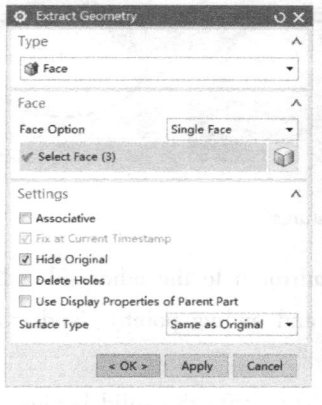

Figure 3-83　Extract three faces

5) The shape of this part is relatively simple and there is no complex surface, and the creation method is similar to the one of the lower cover features. Complete the creation of the internal features of the upper cover, as shown in Figure 3-84.

6) The function of 【Arc】 in the 【Basic Curves】 command is used. Capture the center of the circle and draw

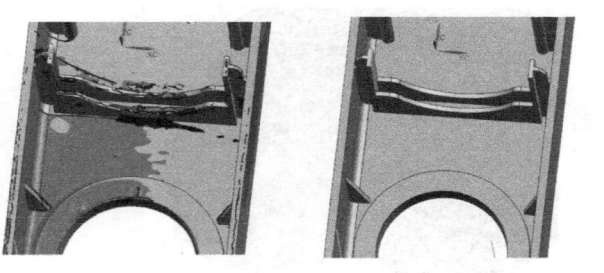

Figure 3-84　Create the internal features

a line parallel to the Axis Z. However, this axis cannot be guaranteed locating on the YC-ZC plane, so it needs further processing, as shown in Figure 3-85.

7) A line on the YC-ZC plane is generated by means of 【Intersection Curve】 or 【Project Curves】, as shown in Figure 3-86.

111 ◀

3D Digital Design and Manufacturing

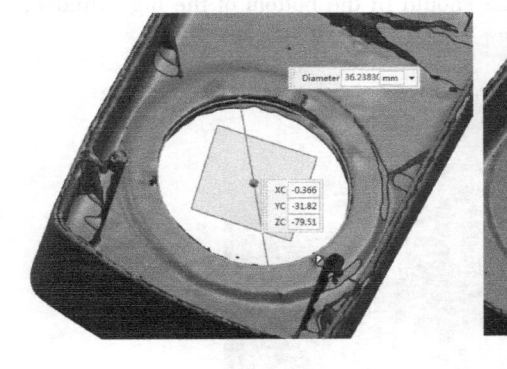  

Figure 3-85　Create a axis　　　　　Figure 3-86　Create a line

8) Next, refine the features, as shown in Figure 3-87.

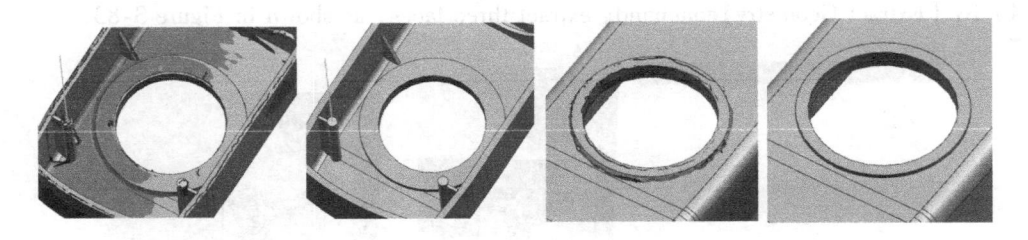

Figure 3-87　Refine the features

9) Create an arc, and half-cut after extruding; then mirror it to the other side. Draw an arc again through 3 points (select both ends for starting point and ending point), and extrude it, as shown in Figure 3-88.

10) The extruded sheet is mirrored to the other side. Then trim the solid bodies of the upper and lower covers with the sheets and make fillets, as shown in Figure 3-89.

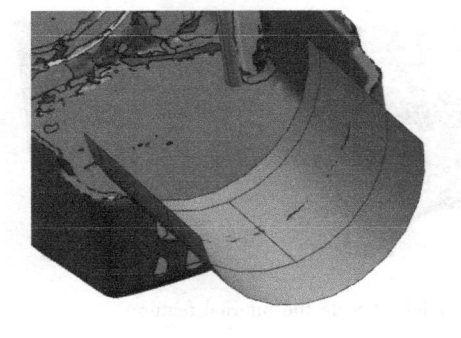　　

Figure 3-88　Create an arc　　　　Figure 3-89　Process the features of the toothpaste dispenser

3.3.5　Detail Processing

After roughly modeling the upper and lower covers of the toothpaste dispenser, the model is

▶ 112

Project 3　Toothpaste Dispenser

refined, such as filleting, shelling, face replacing, etc. Matching buckles are also made, as shown in Figure 3-90 and Figure 3-91.

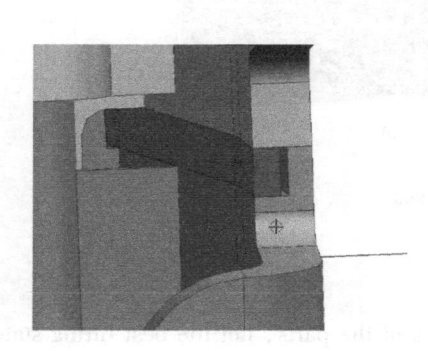

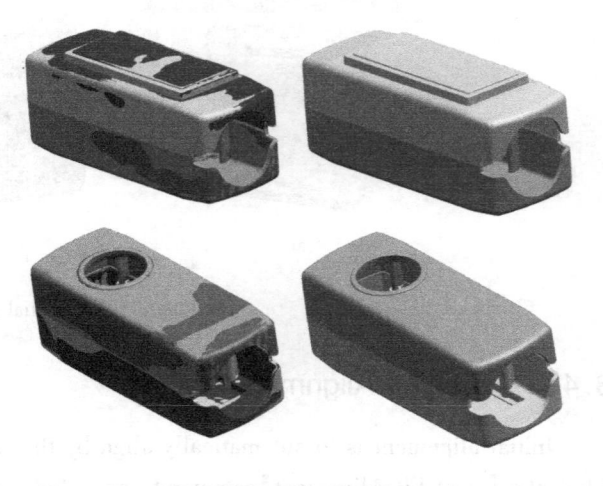

Figure 3-90　Buckle processing

Figure 3-91　Completion of the reverse
modeling of the toothpaste dispenser

Task 4　Product Analysis and Comparison

3.4.1　Data Import

Model and scanned data can be directly dragged to the Geomagic Control X software window or imported through the【Import】function in the toolbar, as shown in Figure 3-92.

Figure 3-92　Import the data

3.4.2　Initial Model Alignment

Click the【Initial Alignment】button _{Alignment}, and check the【Enhance Alignment Accuracy With Feature Recognition】, as shown in Figure 3-93a, then click the "OK" button ✓. Automatic alignment of the scanned data with the reverse modeling model is carried out through features, as shown in Figure 3-93b.

113 ◀

3D Digital Design and Manufacturing

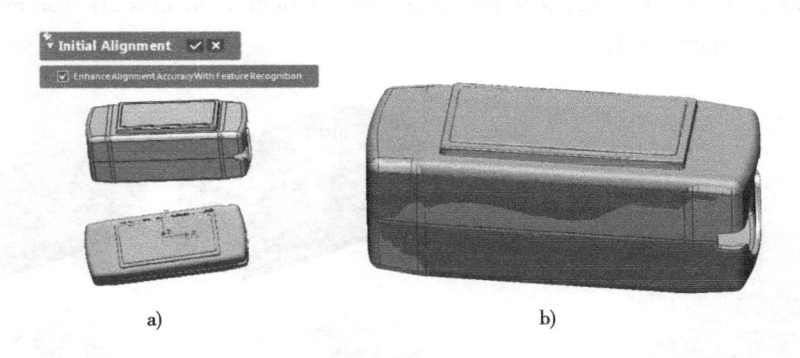

a)   b)

Figure 3-93　Initial alignment

3.4.3　Best Fit Alignment

Initial alignment is to automatically align by the features of the parts, not the best fitting state. Click the 【Best Fit Alignment】command, and click the "OK" button ☑ to align the scanned data with the reverse model to the best state, as shown in Figure 3-94.

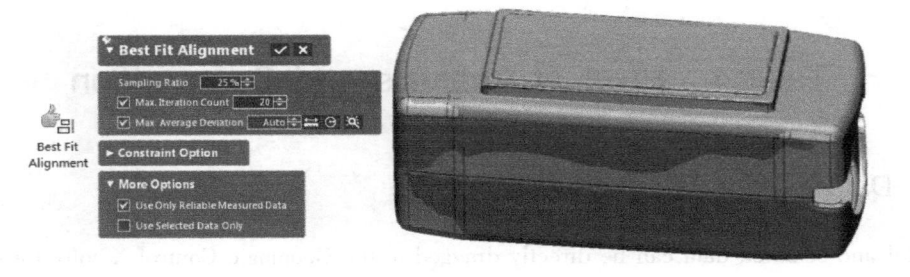

Figure 3-94　Best fit alignment

3.4.4　3D Comparison

Click 【3D Compare】command to compare the 3D data, as shown in Figure 3-95.

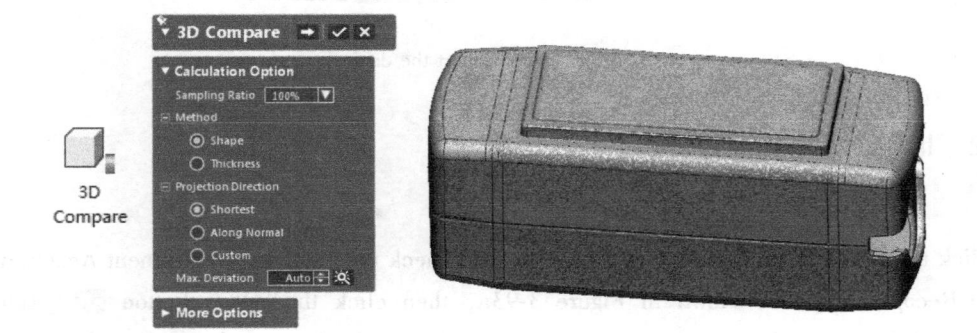

Figure 3-95　3D Comparison

▶ 114

Project 3　Toothpaste Dispenser

Click the "Next Stage" button ➡ and proceed to the next stage, as shown in Figure 3-96. In the 【3D Compare】 dialog box, modify 【Use Specific Tolerance】 to "±0.1", and click the "OK" button ✓ to confirm.

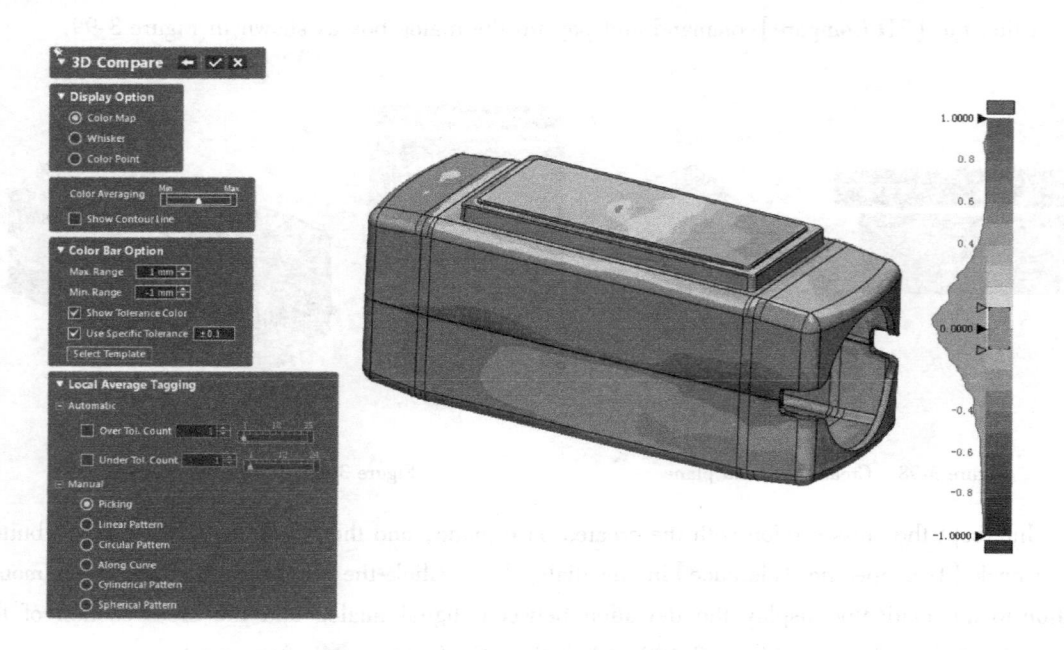

Figure 3-96　3D comparative color display

More detailed data can be viewed by clicking the 【Histogram】 button Histogram ▶ in the 3D comparison properties, as shown in Figure 3-97.

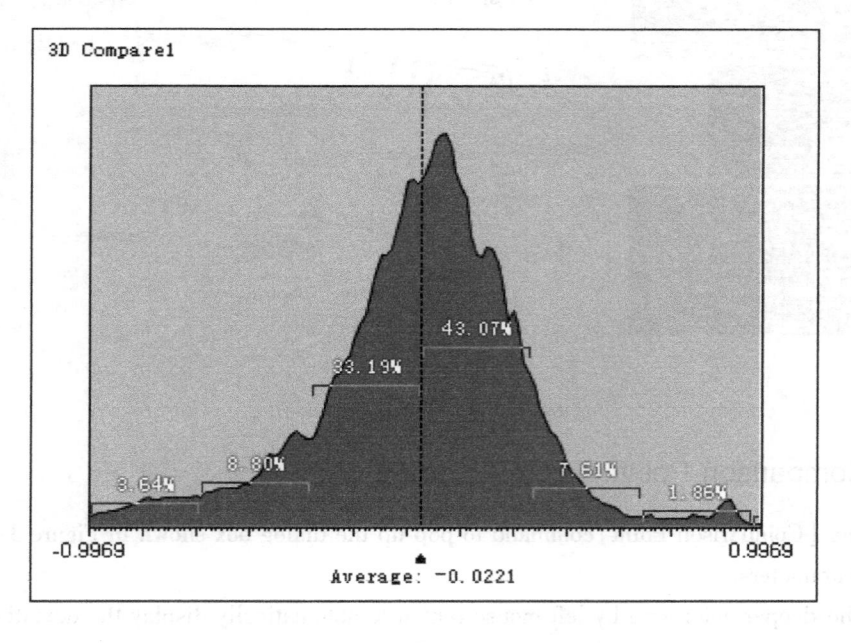

Figure 3-97　3D comparison histogram

115 ◀

3D Digital Design and Manufacturing

3.4.5 2D Comparison

Click the 【Plane】command to create a central plane, as shown in Figure 3-98.

Click the 【2D Compare】command and pop up the dialog box as shown in Figure 3-99.

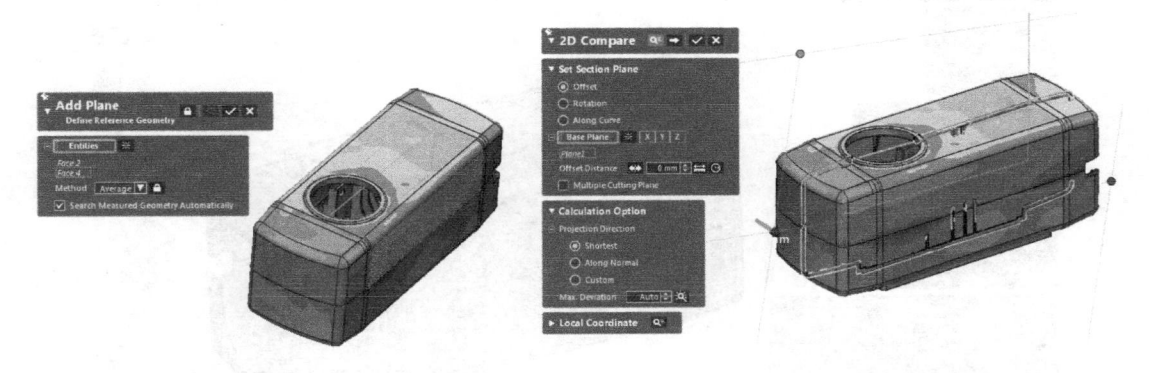

Figure 3-98 Create a central plane   Figure 3-99 2D Comparison

Intercept the cross section with the created base plane, and then click the "Next Stage" button ➡. Check 【Use Specific Tolerance】in the dialog box. Click the wireframe border by left mouse button to automatically display the deviation between digital analog and the cross section of the scanned data, as shown in Figure 3-100. Click the "OK" button ✅ after completion.

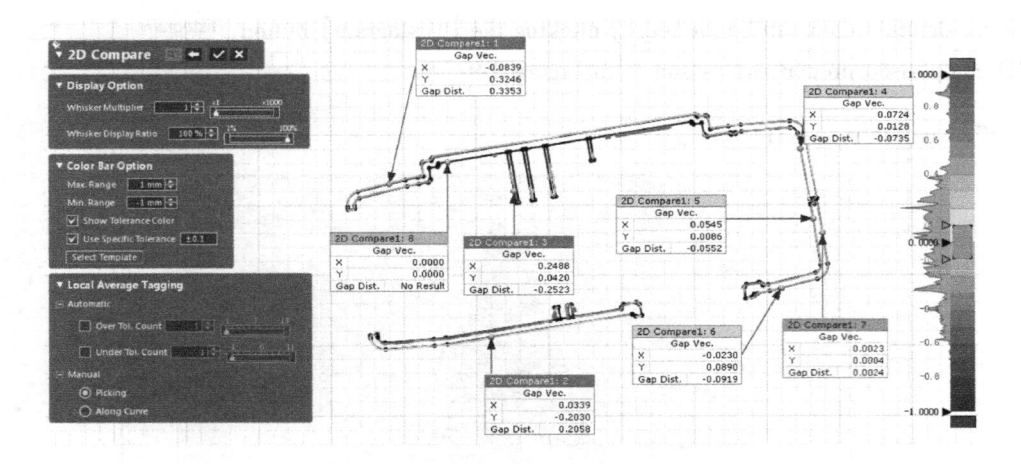

Figure 3-100 Deviation display

3.4.6 Comparison Point

Click the 【Comparison Point】command to pop up the dialog box shown in Figure 3-101. Select the default parameters.

Click the deeper color area by left mouse button to automatically display the deviation value, as shown in Figure 3-102. Click the "OK" button ✅ after confirming completion.

▶ 116

Project 3 Toothpaste Dispenser

Figure 3-101   Comparison point

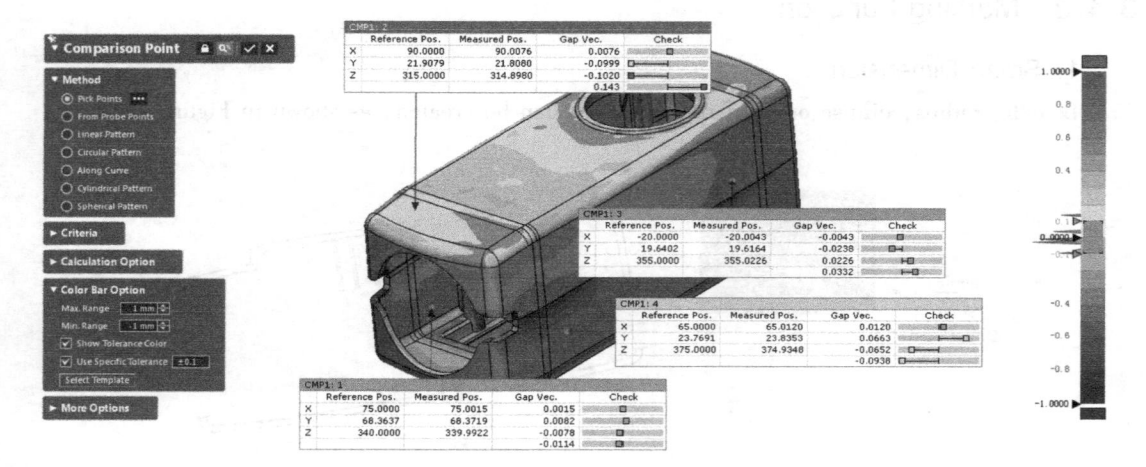

Figure 3-102   Display deviation value

3.4.7 Cross Section

Click 【Menu】/【Insert】/【Cross Section】command to pop up the【Cross Section】 dialog box. Then capture the section line with the plane created previously, as shown in Figure 3-103.

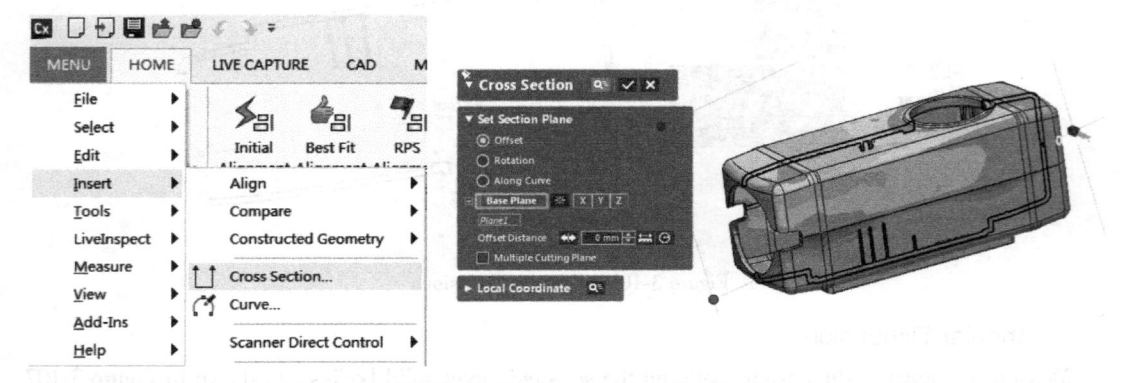

Figure 3-103   Cross section

117 ◀

3D Digital Design and Manufacturing

Click the "OK" button ☑ to automatically jump to the dimensioning interface, as shown in Figure 3-104.

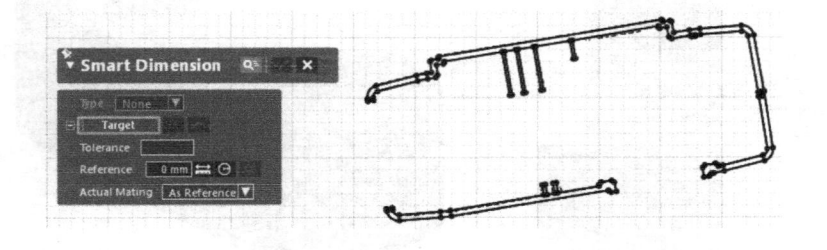

Figure 3-104 Dimensioning interface

3.4.8 Marking Function

1. Smart Dimension

Length, radius, ellipse or angular dimensions can be created, as shown in Figure 3-105.

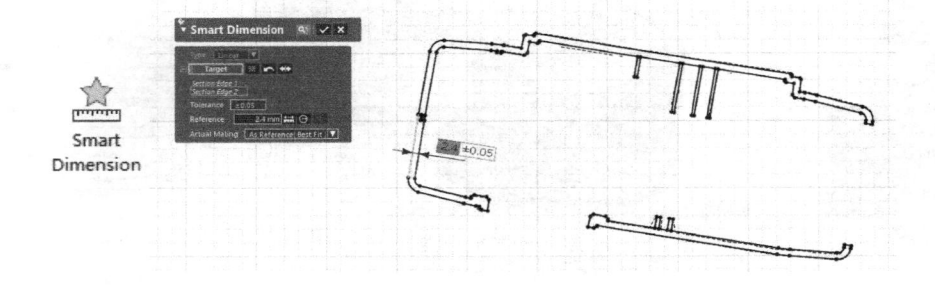

Figure 3-105 Smart dimensions

2. Linear Dimension

Measure the linear dimensions between the selected target solid bodies, as shown in Figure 3-106.

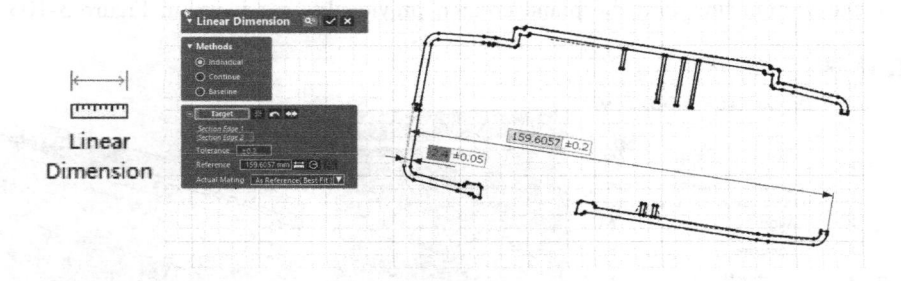

Figure 3-106 Linear dimensions

3. Angular Dimension

Measure the angular dimensions between the selected target solid bodies, as shown in Figure 3-107.

▶ 118

Project 3 Toothpaste Dispenser

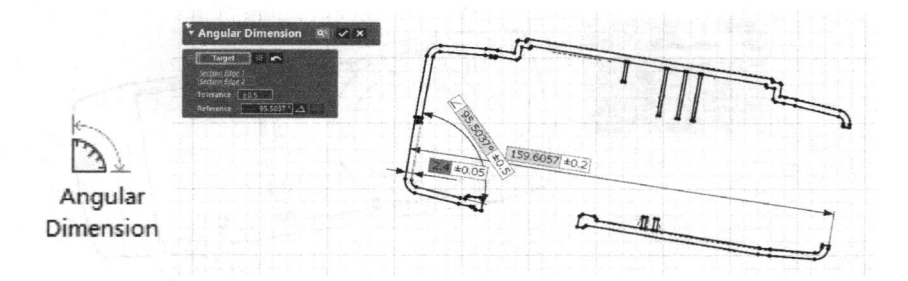

Figure 3-107   Angular dimensions

4. Datum

A datum is the position of the feature or a starting point of the geometry feature, as shown in Figure 3-108.

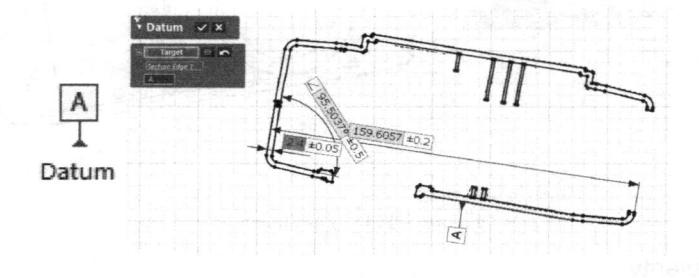

Figure 3-108   Datum

5. Straightness

Straightness is used to control the shape errors of planar or spatial lines, as shown in Figure 3-109.

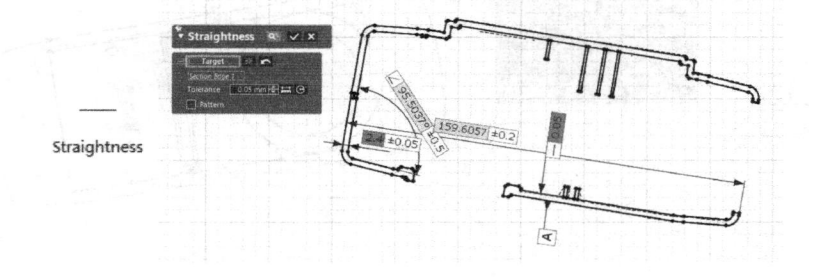

Figure 3-109   Straightness

6. Circularity

Circularity refers to the degree to which the cross section of a workpiece approaches the theoretical circle, as shown in Figure 3-110.

7. Parallelism

Parallelism refers to the degree to which two planes or two lines are parallel. Parallelism can be considered as a special case of angularity, as shown in Figure 3-111.

119 ◀

3D Digital Design and Manufacturing

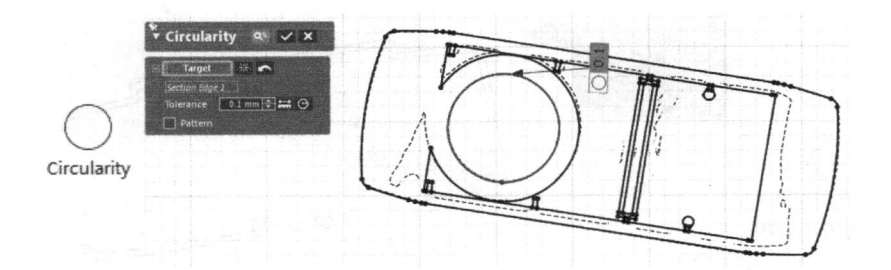

Figure 3-110　Circularity

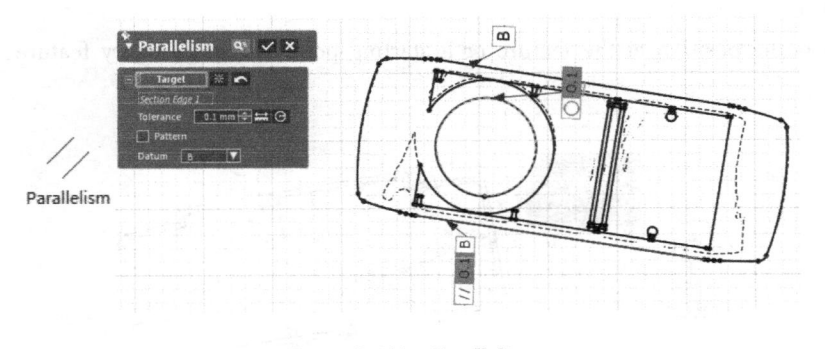

Figure 3-111　Parallelism

8. Perpendicularity

Perpendicularity is used to evaluate the vertical state between lines, planes, or between line and plane, as shown in Figure 3-112.

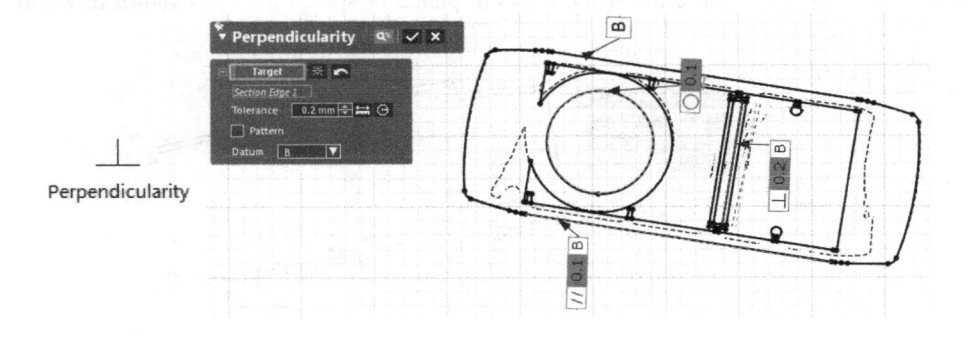

Figure 3-112　Perpendicularity

9. Angularity

According to the reference specification, angularity is used to control the deviation of a surface, axis or plane from a non-90° angle, as shown in Figure 3-113.

10. Position

Position is used to describe the exact position of a feature relative to a reference or other feature, as shown in Figure 3-114.

▶ 120

Project 3　Toothpaste Dispenser

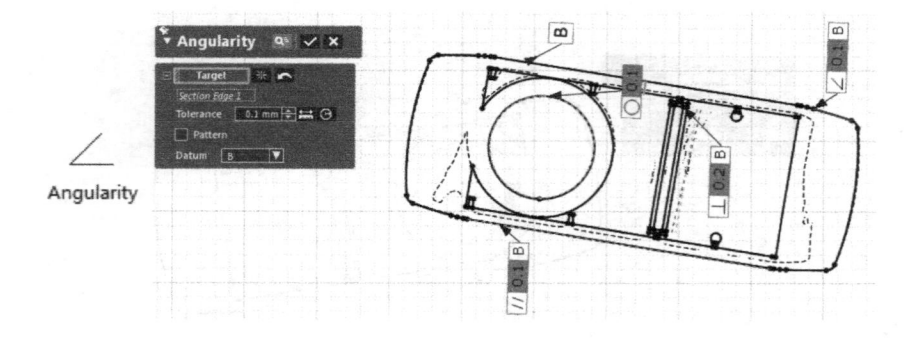

Figure 3-113　Angularity

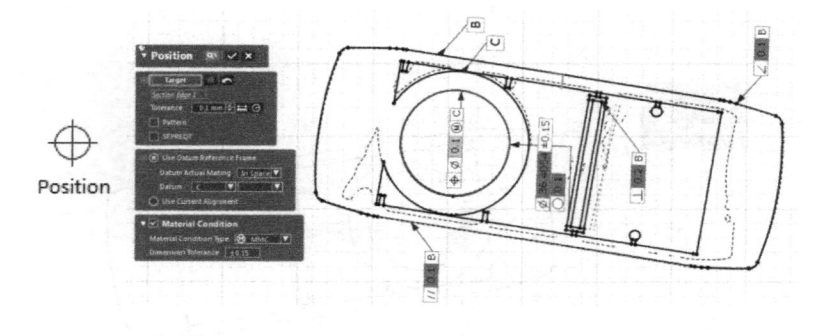

Figure 3-114　Position

11. Concentricity

Concentricity is used to control the deviation of the centers of several diameters from a particular datum, as shown in Figure 3-115.

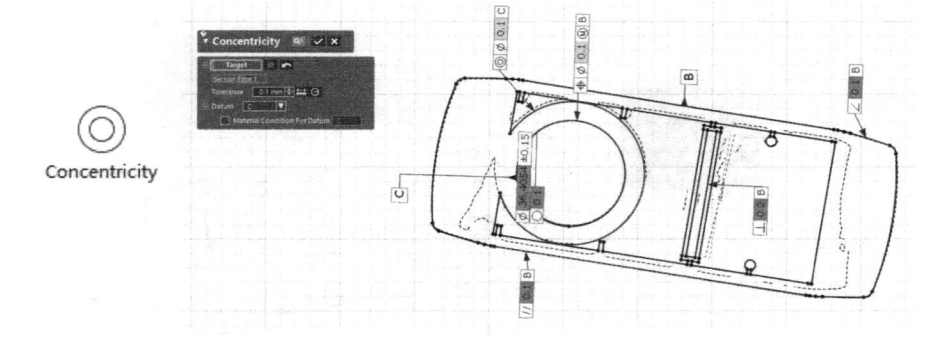

Figure 3-115　Concentricity

12. Symmetry

Symmetry is used to control the deviation of the midpoint of two features from a specific datum.

First, click the 【Vector】command to create a central line, as shown in Figure 3-116.

Set the central line as a datum, as shown in Figure 3-117.

121

3D Digital Design and Manufacturing

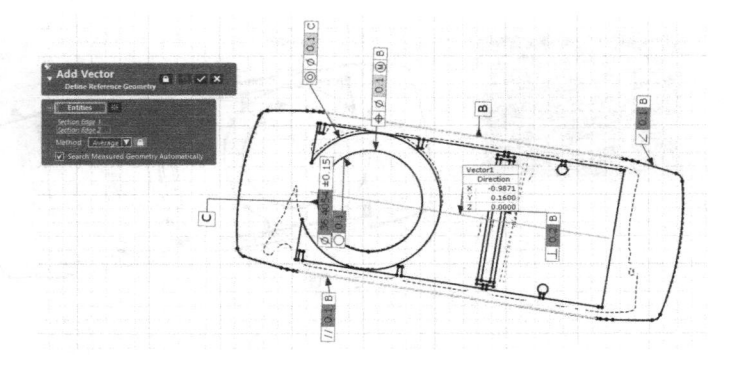

Figure 3-116   Central line

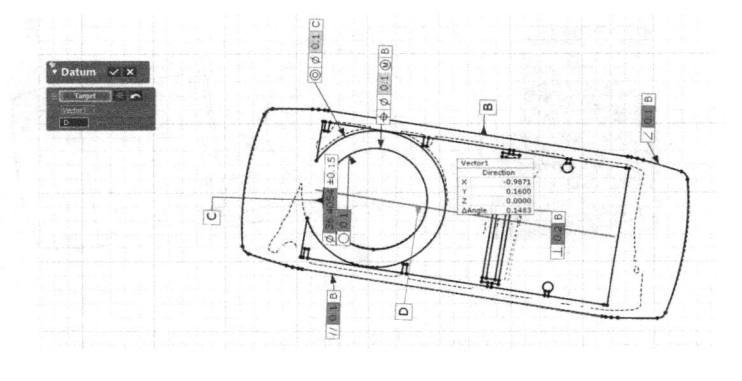

Figure 3-117   Set the central line as a datum

Select the object of Vector 1 and Vector 2, and analyze the symmetry of the two lines about the datum on D, as shown in Figure 3-118.

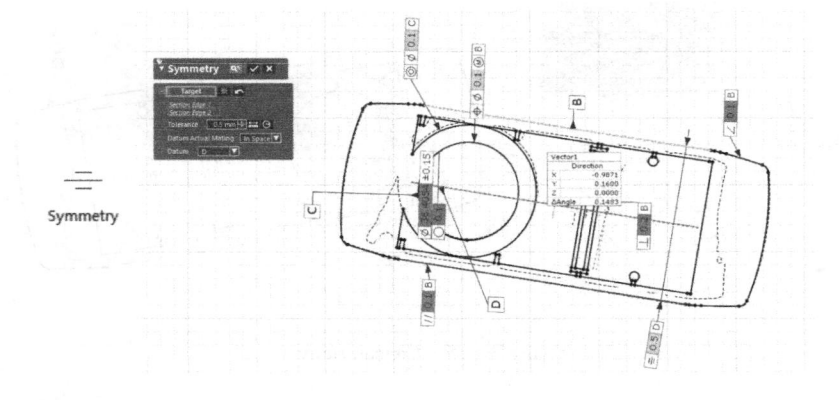

Figure 3-118   Symmetry

After completion, click the left button among the two buttons in the lower right corner of the screen to exit, as shown in Figure 3-119.

▶ 122

Project 3   Toothpaste Dispenser

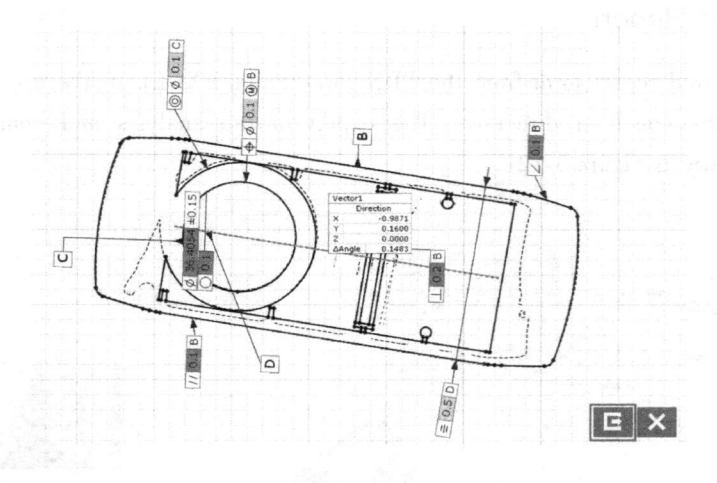

Figure 3-119   Exit after completion

3.4.9   Generate Report

Click the【Generate Report】command to pop up the dialog box shown in Figure 3-120, then click【Generate】.

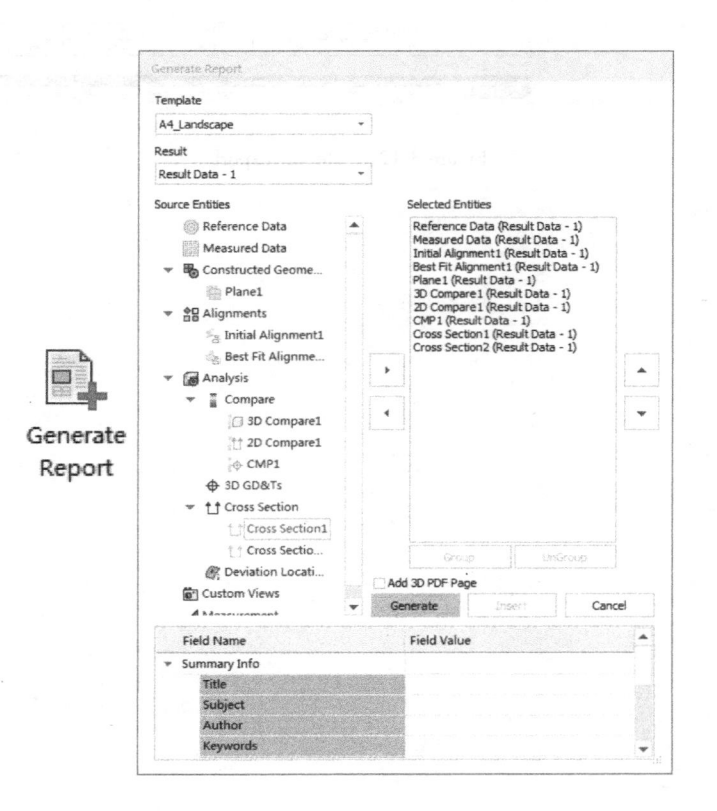

Figure 3-120   Generate report

123

3D Digital Design and Manufacturing

3.4.10　Export Report

　　Select the desired report form from the PDF/PowerPoint/EXCEL in the upper left corner and save the file to the specified directory. The export of the analysis and comparison report is completed, as shown in Figure 3-121.

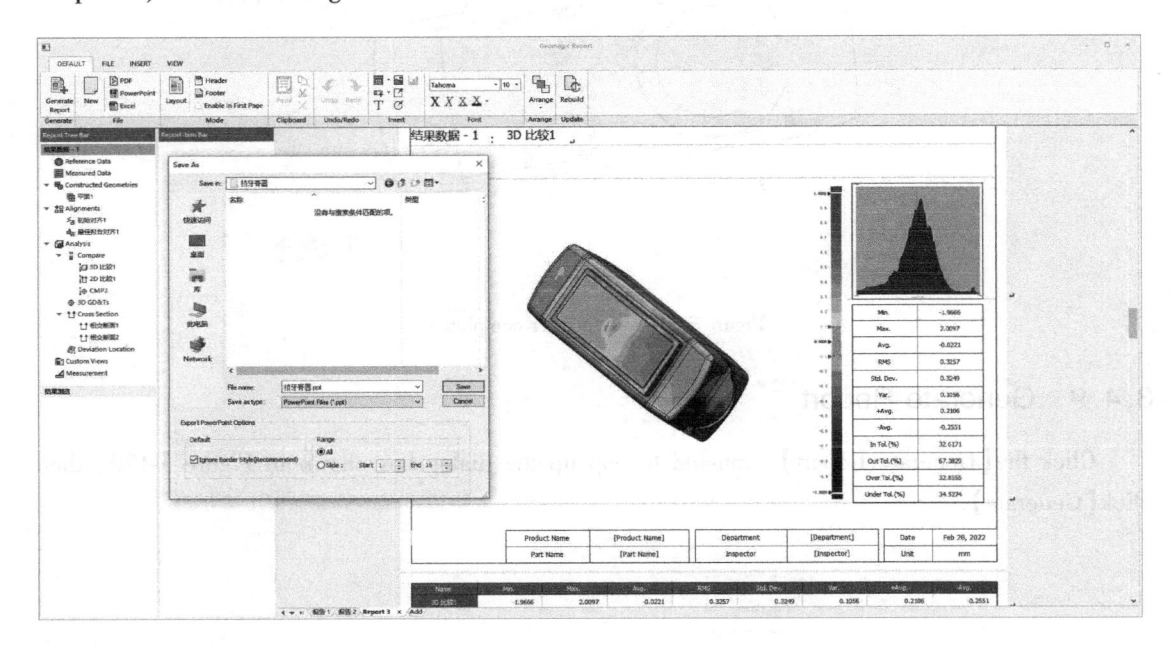

Figure 3-121　Report export

▶ 124

Project 4 Gamepad

This project takes the rocker gamepad shown in Figure 4-1 as the carrier. Win3DD monocular 3D scanner is used for data collection; Geomagic Wrap 2017 software is used for data processing; and finally the product reverse modeling and numerical control processing simulation are completed in Siemens NX 10 software. Project implementation process is shown in Figure 4-2.

Figure 4-1   The rocker gamepad

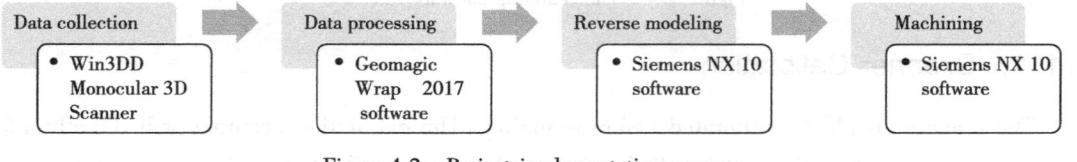

Figure 4-2   Project implementation process

Task 1   Data Collection

The model data of the gamepad is collected using Win3DD monocular 3D scanner. On the basis of technical advantages, Win3DD series products are greatly improved in appearance design, software functions and accessories configuration, etc. Beside high accuracy, the products also have the characteristics of easy learning and easy use. The hardware system of Win3DD monocular 3D scanner consists of scanning head, platform and tripod, as shown in Figure 4-3. The scanning software interface is shown in Figure 4-4.

125

3D Digital Design and Manufacturing

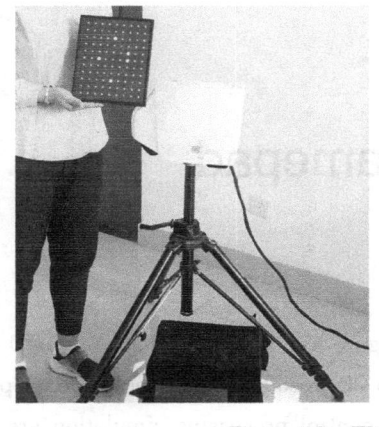
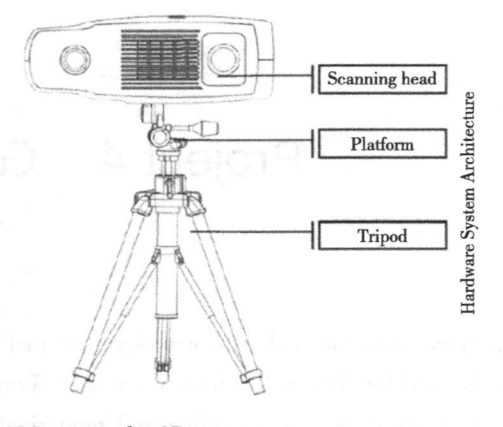

Figure 4-3　Win3DD monocular 3D scanner

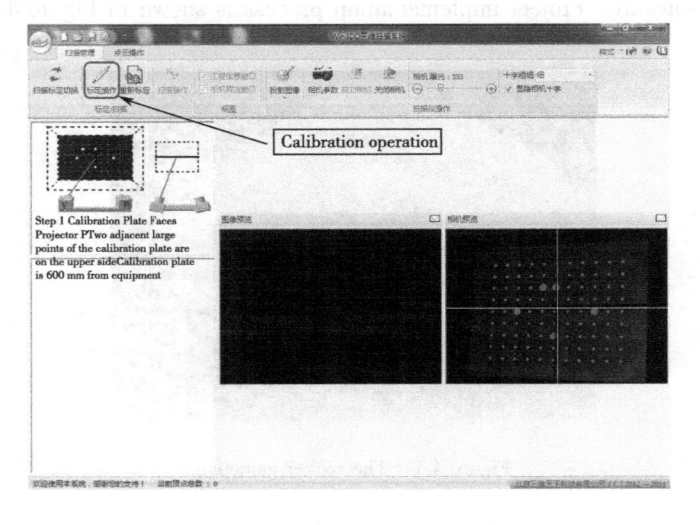

Figure 4-4　The scanning software interface

4.1.1　Scanner Calibration

The scanner should be calibrated before scanning. The calibration accuracy will directly affect the scanning accuracy of the system. If the scanner has been calibrated in the course of scanning, the next scan can be carried out without any change in the system. Calibration is required generally under the following situations.

1) The scanner is used for the first time or reused after long time in idle.

2) The scanner is involved in collision in use, resulting in camera offset.

3) The scanner is involved in serious vibration during transport.

4) In the process of scanning, there are frequent splicing errors and splicing failures, etc.

5) In the process of scanning, the scanning data is incomplete and the quality of the data is seriously degraded.

The calibration steps of Win3DD monocular 3D scanner are described below.

1) Start the Win3DD scanning system and preheat the scanning system for 5 to 10 minutes to

▶ 126

ensure that the calibration state is as close as possible to the scanning state. Open the supporting scanning software, and click the 【Scanning Calibration Switching】 button to enter the scanning calibration interface. Then conduct calibration operation according to the calibration instructions on the upper left corner of the interface.

2) Place the calibration plate in the scanning area, as shown in Figure 4-5. There are 99 points on the calibration plate, of which 5 are large points. Large points play a role in the calibration process, as shown in Figure 4-6.

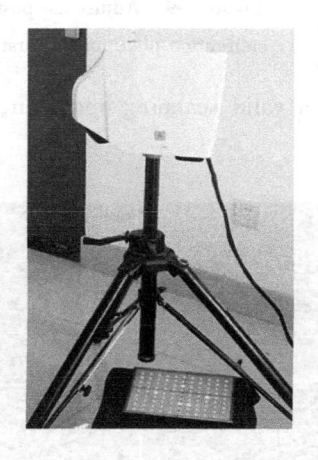

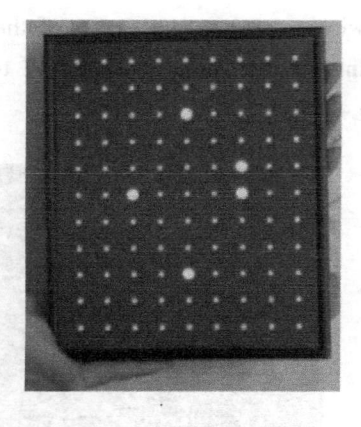

Figure 4-5　Place the calibration plate in the scanning area

Figure 4-6　Calibration plate

3) Adjust the height and pitch angle of the platform, as shown in Figure 4-7, making the black vertical lines projected on the calibration plate aligned with the white vertical lines in the software interface, as shown in Figure 4-8. Lock the platform after the adjustment is completed (Note: After every subsequent adjustment, the platform needs to be locked).

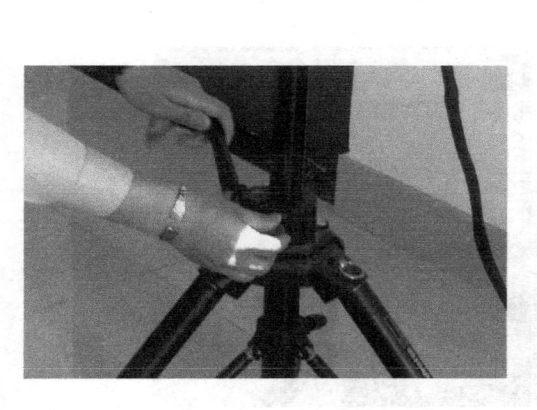

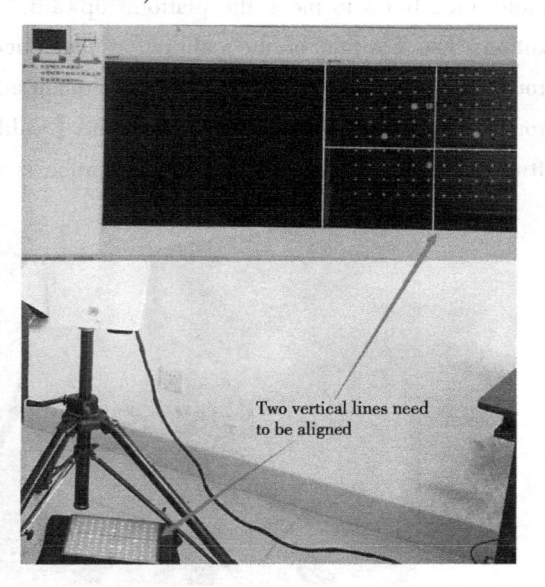

Figure 4-7　Adjust the position of the platform

Figure 4-8　Two vertical lines need to be aligned

3D Digital Design and Manufacturing

4) The first calibration. Adjust the position of the calibration plate again so that the black vertical lines run through four large points, as shown in Figure 4-9. Then click the 【Calibration Operation】 button to scan. In the process of scanning, it is necessary to pay attention to the fact that no objects block the scanning light source. After scanning, the results will be displayed on the screen. The left image shows the result of the scanning. The green cross wires represent the center of the circle. More than

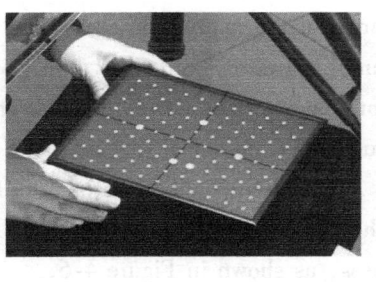

Figure 4-9　Adjust the position of the calibration plate for the first calibration

88 points per scanning are required to be recognized as a valid scanning operation, as shown in Figure 4-10.

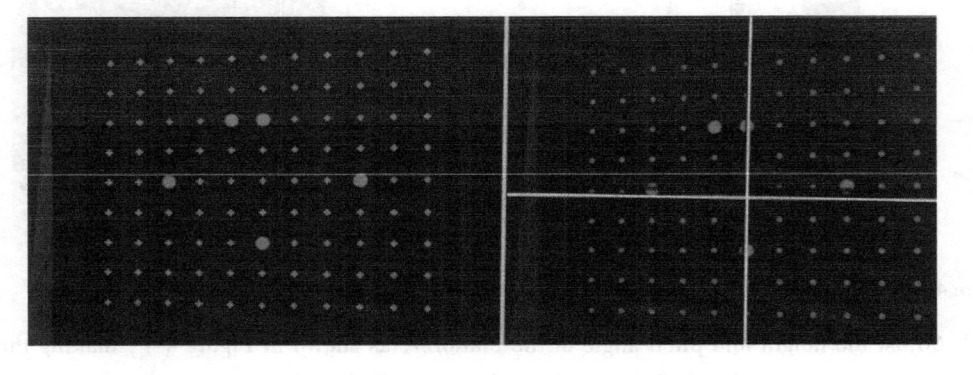

Figure 4-10　Result of the first calibration

5) The second calibration. Release the lock and remember the position of the handle. Turn the handle three times to move the platform upward. After adjustment, the cross lines may move in a position. The position of the calibration plate needs to be adjusted to make the cross lines run through four large points (After each subsequent adjustment, it is needed to make the cross lines run through the large points). Then click the 【Calibration Operation】 button. The left side of the software interface will show that the calibration is successful, as shown in Figure 4-11.

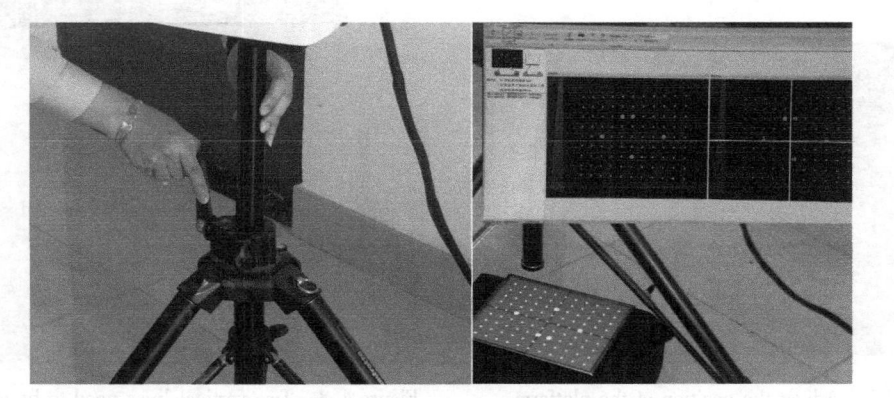

Figure 4-11　The second calibration after lifting the platform

Project 4 Gamepad

6) The third calibration. Loosen the lock and rotate the handle six times to move the platform downward. Adjust the position of the calibration plate again so that the cross lines run through four large points. Click the 【Calibration Operation】button for the third calibration, as shown in Figure 4-12.

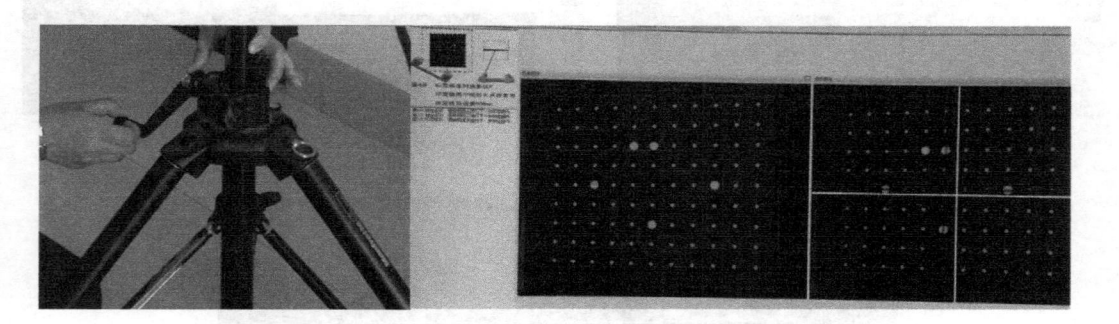

Figure 4-12 The third calibration after lowering the platform

7) The fourth calibration. Rotate the handle three times to move the platform upward and adjust it to the original position. The calibration plate is rotated 90°, and the upper left corner is padded with a sponge pad to make the cross line run through four large points. Click the 【Calibration Operation】button for the fourth calibration, as shown in Figure 4-13.

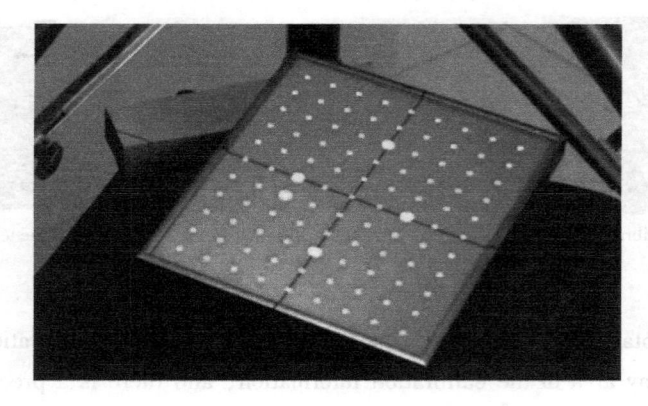

Figure 4-13 Pad up the upper left corner for the fourth calibration

8) The fifth calibration. Rotate the calibration plate 90° counterclockwise and click the 【Calibration Operation】button for the fifth calibration, as shown in Figure 4-14.

9) The sixth calibration. Again, rotate the calibration plate 90° counterclockwise and click the 【Calibration Operation】button for the sixth calibration, as shown in Figure 4-15.

10) The seventh calibration. Again, rotate the calibration plate 90° counterclockwise, and the calibration plate returns to the initial position. Place the sponge pad on the right side of the calibration plate, then click the 【Calibration Operation】button, as shown in Figure 4-16.

11) The eighth to tenth calibrations. Rotate the calibration plate 90° counterclockwise each time, and then click the【Calibration Operation】button once to calibrate. Repeat three times in total, as shown in Figure 4-17.

3D Digital Design and Manufacturing

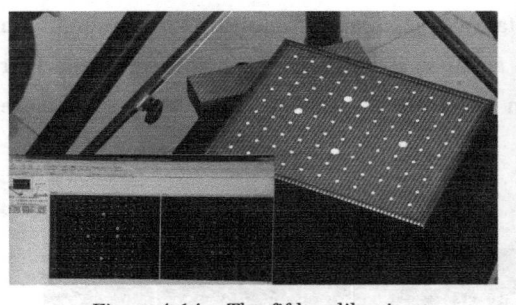

Figure 4-14　The fifth calibration

Figure 4-15　The sixth calibration

Figure 4-16　The seventh calibration

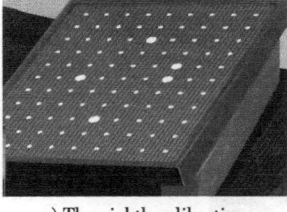

a) The eighth calibration

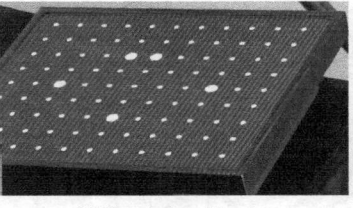

b) The ninth calibration

c) The tenth calibration

Figure 4-17　The eighth to tenth calibrations

12) So far, a total of 10 calibrations have been carried out. 【Calibration is completed】 is prompted in the display area of the calibration information, and there is a precision display below, as shown in Figure 4-18. If the calibration is unsuccessful, 【There is a large error in calibration. Please recalibrate.】 will be prompted.

> The first calibration is successful. Please follow the instructions for the next calibration operation!
> The second calibration is successful. Please follow the instructions for the next calibration operation!
> The third calibration is successful. Please follow the instructions for the next calibration operation!
> The fourth calibration is successful. Please follow the instructions for the next calibration operation!
> The fifth calibration is successful. Please follow the instructions for the next calibration operation!
> The sixth calibration is successful. Please follow the instructions for the next calibration operation!
> The seventh calibration is successful. Please follow the instructions for the next calibration operation!
> The eighth calibration is successful. Please follow the instructions for the next calibration operation!
> The ninth calibration is successful. Please follow the instructions for the next calibration operation!
> The tenth calibration is successful. Please follow the instructions for the next calibration operation!
> Calibration is completed
> Average error of calibration results: 0.027
> Click the calibration operation button to carry out a new round of calibration process!

Figure 4-18　Information prompted after calibration

Project 4　Gamepad

4.1.2　Paste the Mark Points

In order to scan three dimensional objects, it is usually necessary to affix mark points on the surface of the scanned objects, and the mark points should be pasted firmly and smoothly. Before pasting, a developer should be sprayed to enhance the scanning effect. According to the size of the gamepad and the corresponding scanning range, select the mark points with the inner diameter of 3mm, as shown in Figure 4-19.

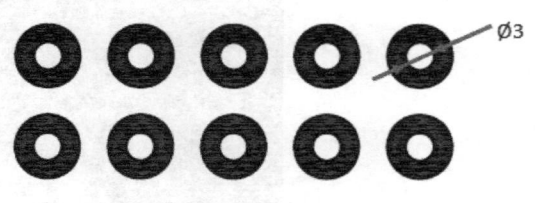

Attention should be paid to the following points when pasting mark points.

Figure 4-19　Mark points with the inner diameter of 3mm

1) The points are pasted as freely as possible. Avoid artificial grouping, as shown in Figure 4-20.

2) Avoid pasting in lines or in equilateral or isosceles triangles, as shown in Figure 4-21.

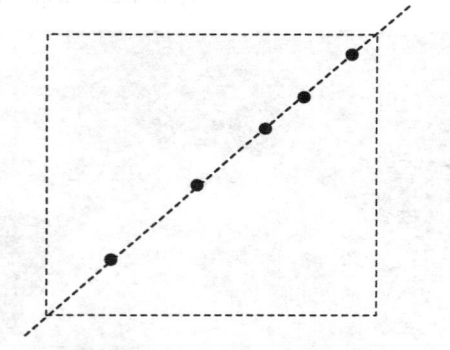

Figure 4-20　(Error) Artificial grouping mark points　　　Figure 4-21　(Error) Mark points are pasted in lines

3) Do not paste to the junction between the faces. Paste to the flat place as far as possible, as shown in Figure 4-22.

4) In order to ensure the scanning quality, corresponding mark points should be pasted on the scanning turntable, as shown in Figure 4-23.

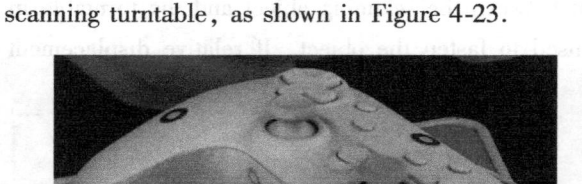

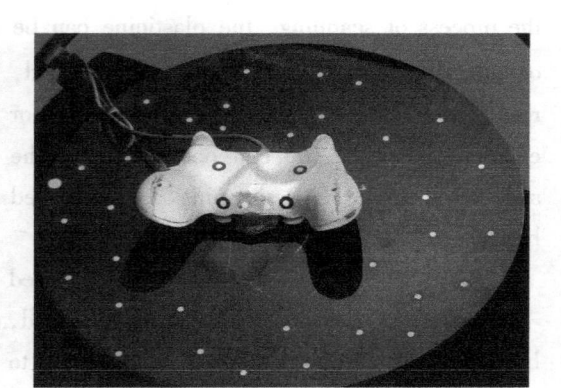

Figure 4-22　Mark points need to be avoided at the junction　　　Figure 4-23　Paste points on the turntable

131 ◀

3D Digital Design and Manufacturing

To facilitate pasting points, wipe off part of the developer on the surface of the gamepad with a cotton swab; set aside the space for pasting; and then paste the mark points, as shown in Figure 4-24. Both sides of the gamepad need to be scanned, so it is needed to paste mark points on both sides, as shown in Figure 4-25 and Figure 4-26.

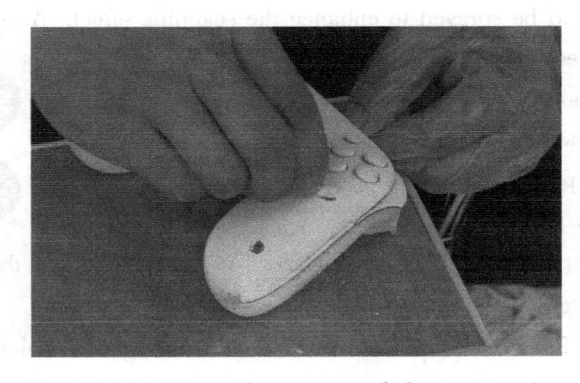

Figure 4-24　Wipe with a cotton swab for pasting points

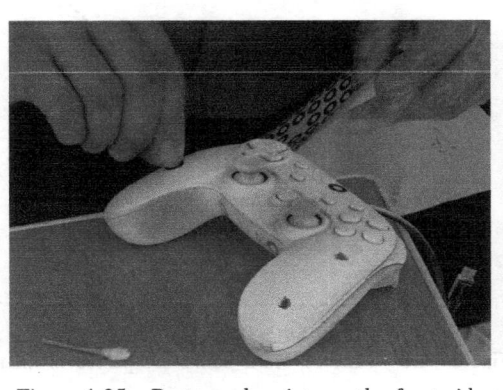

Figure 4-25　Paste mark points on the front side

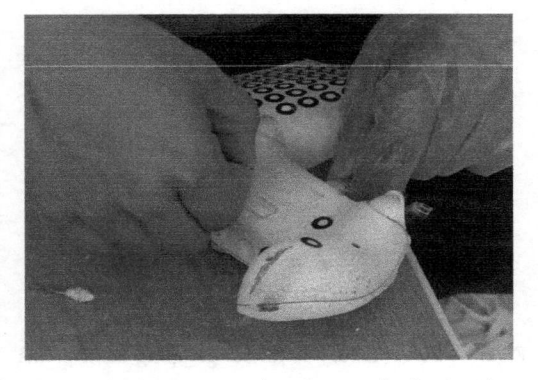

Figure 4-26　Paste mark points on the back side

4.1.3　Data Scanning

In order to prevent the relative displacement between the scanning object and the turntable in the process of scanning, the plasticine can be used to fasten the object. If relative displacement occurs, the scanning data will be stratified, resulting in a decline in scanning accuracy, or even the scanning file can not be used. The specific steps of data scanning are described below.

1) The new project is created, named "Gamepad 1" (The project can be freely named, but it is recommended that the name be related to the scanned object in order to distinguish the data.), as shown in Figure 4-27.

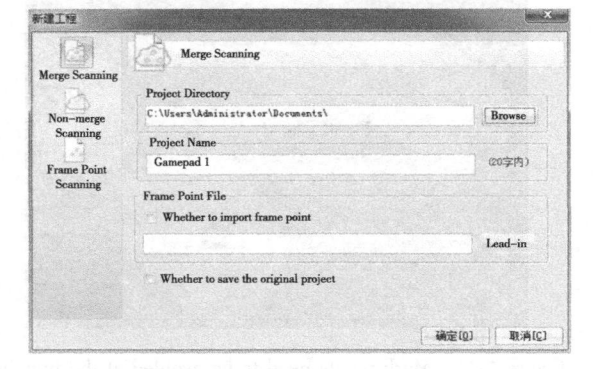

Figure 4-27　Create a new project

▶ 132

Project 4　Gamepad

2）Adjust the turntable so that the scanning object is aligned with the scanner. Click the
【Scanning Operation】button ⁂ to perform the first scanning, as shown in Figure 4-28.

Figure 4-28　The first scanning result

3）Turn the turntable gently to prevent the relative position of the scanning object and the
turntable from changing during rotation. Click the 【Scanning Operation】button again to scan.
Rotate the turntable for multiple scans until the front side data is fully scanned and the peripheral
data is complete. When the data is manually merged with the back side data, peripheral data is
needed, as shown in Figure 4-29.

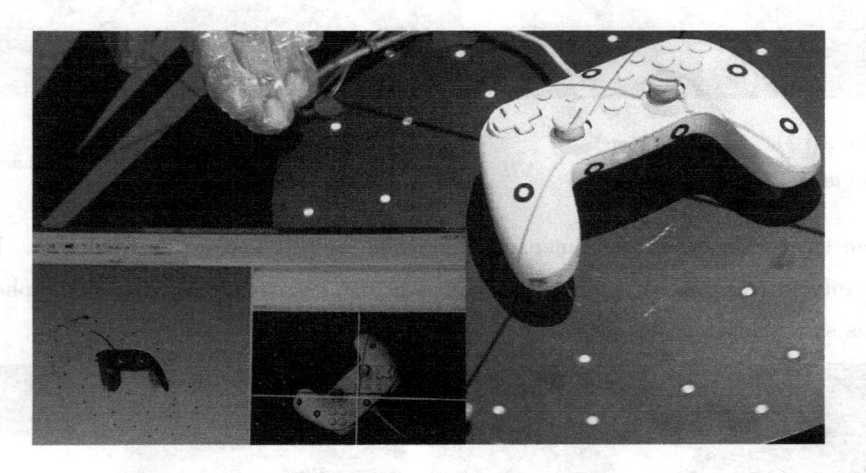

Figure 4-29　Observation the results of repeated scanning

4）After ensuring that the data is free of defects, save it as a "point cloud file (＊. asc)", as
shown in Figure 4-30.

5）Flip the gamepad, remove the plasticine pasted before, and then select the appropriate fixed
position to fasten it. Avoid fastening the gamepad in the active position, such as the rocker of the
gamepad, in order to avoid the impact on the scanned data, as shown in Figure 4-31.

133 ◀

3D Digital Design and Manufacturing

Figure 4-30   Save front side data of the gamepad

6) Similarly, a new project is created, named "Gamepad 2". Rotate the turntable and select the surface with more marking points for the first scan, as shown in Figure 4-32.

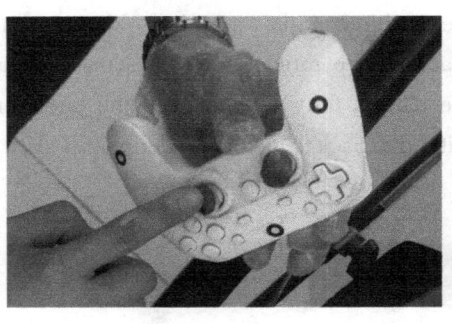

Figure 4-31   （Error position）
Avoid fastening in shaking position

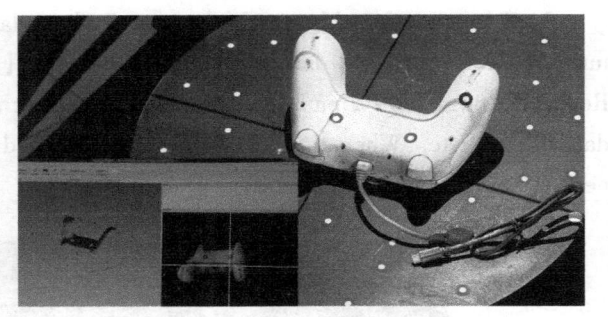

Figure 4-32   The first scanning on the back side

7) Scan the back side of the gamepad in the same way as that of the front side. The rotating turntable is rotated several times until the back side data is complete and the peripheral data is complete, as shown in Figure 4-33.

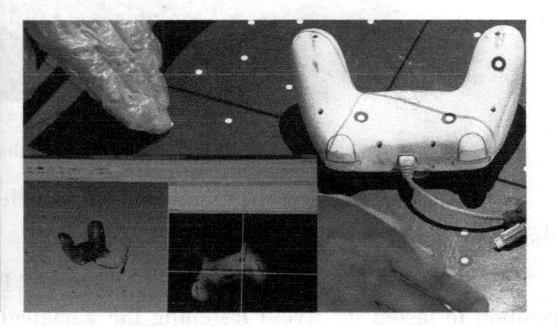

Figure 4-33   Multiple scans of back side

Project 4　Gamepad

8) Save the obtained data as "Gamepad 2. asc", and then complete the scanning of the gamepad. A total of 2 data files are shown in Figure 4-34.

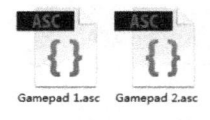

Gamepad 1.asc　Gamepad 2.asc

Figure 4-34　Data files

Task 2　Data Processing

The powerful toolbox provided by Geomagic Wrap 2017 software can directly convert 3D scanning data into 3D polygon and surface models in an easy-to-use, low-cost, fast and accurate way for manufacturing, art and industrial design, etc. Here, Geomagic Wrap 2017 software is used for manual stitching and data wrapping of the front side and back side scanning data of the gamepad.

4.2.1　Manual Stitching

1) Import the scanned data files "Gamepad 1. asc" and "Gamepad 2. asc", as shown in Figure 4-35.

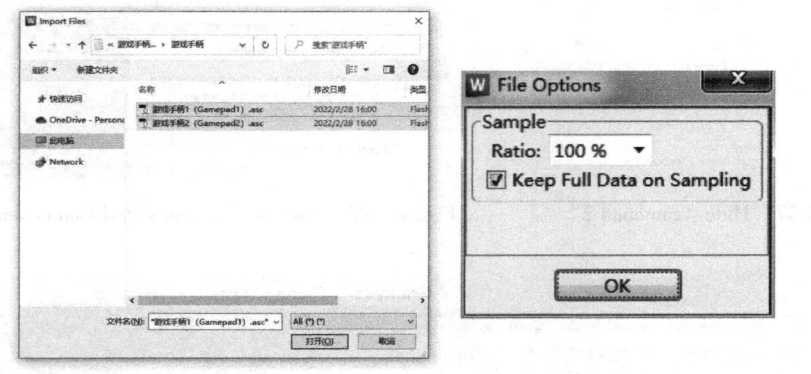

Figure 4-35　Import the scanning data

2) For ease of observation, the shading process is performed by using the 【Shade Points】 command, as shown in Figure 4-36.

3) Hide the "gamepad 2" and process the data of the "Gamepad 1", as shown in Figure 4-37.

4) Use the 【Disconnected Components】 command, as shown in Figure 4-38. After clicking 【OK】, the system automatically selects the external points and displays them in red. Press the < Delete > key or click the 【Delete】 button in the software to delete these red points, as shown in Figure 4-39.

135 ◀

3D Digital Design and Manufacturing

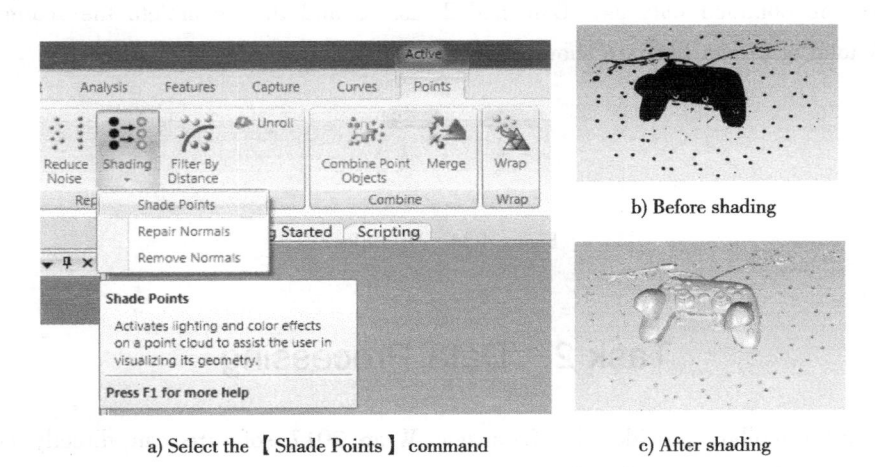

a) Select the 【Shade Points】command

b) Before shading

c) After shading

Figure 4-36　Shading process

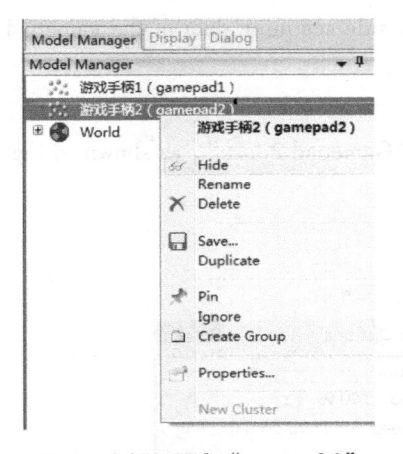

Figure 4-37　Hide "gamepad 2"

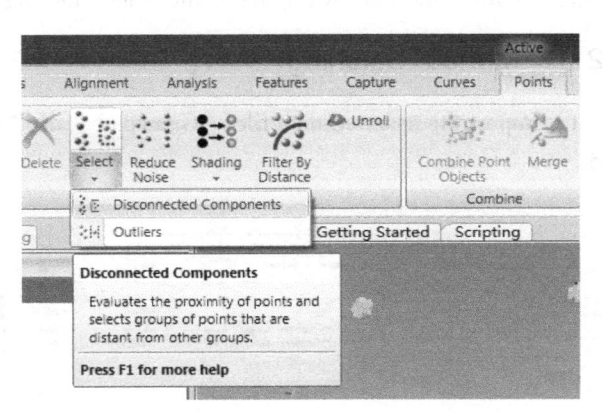

Figure 4-38　Use the【Disconnected Components】command

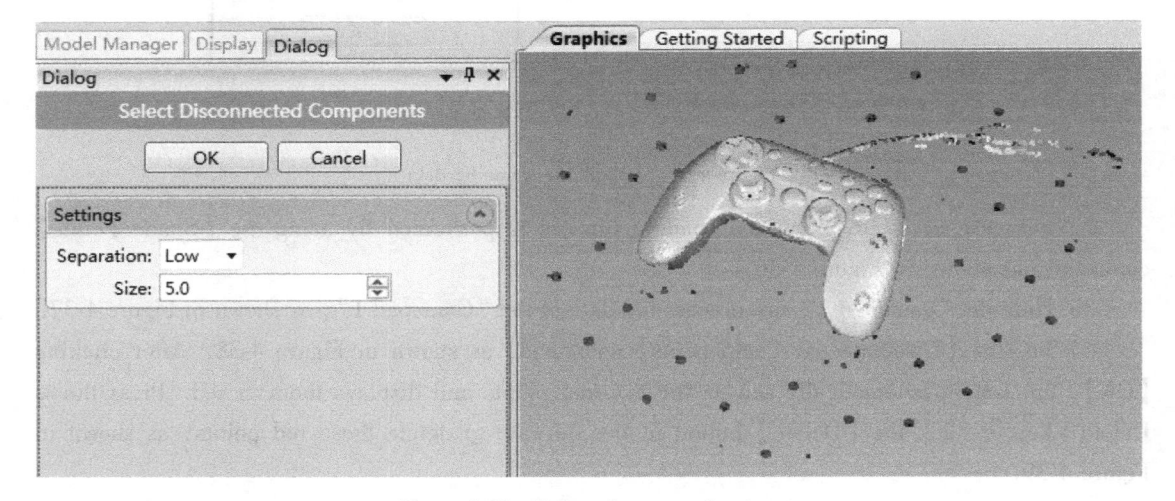

Figure 4-39　Delete the external points

136

Project 4　Gamepad

5）Next, by holding down the left mouse button to manually select the points that need to be deleted, and press the < Delete > key to remove them, as shown in Figure 4-40.

6）Process the scanning data on the back side of the gamepad in the same way, as shown in Figure 4-41.

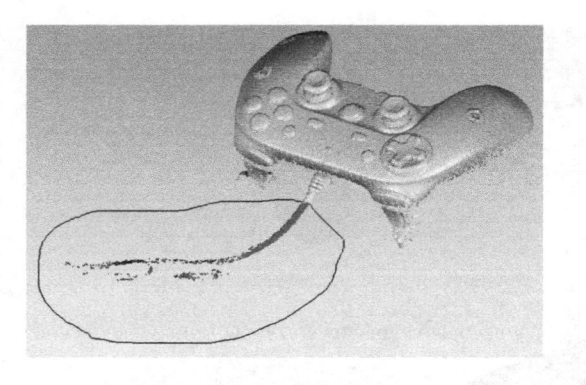

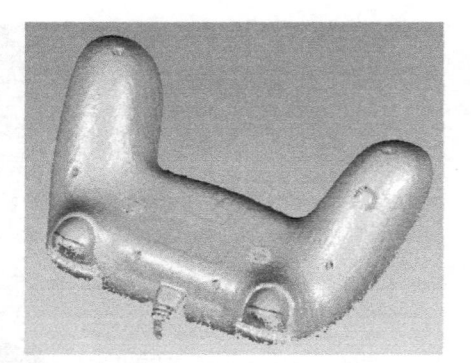

Figure 4-40　Points needs to be removed by manual selection

Figure 4-41　Scanning data on the back side after data processing

7）Select the two point cloud data in the model manager and use the【Alignment】/【Manual Registration】command to merge them, as shown in Figure 4-42.

Select 【 n-Point Registration 】 in 【Mode】; select " Gamepad 1 " in the 【 Fixed 】 of the 【 Define Sets 】; select

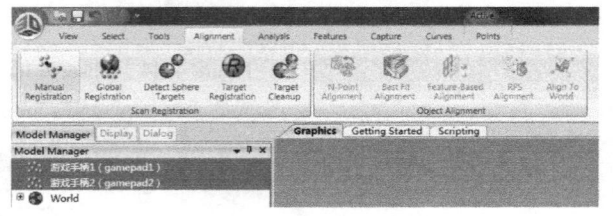

Figure 4-42　【Manual Registration】command

"Gamepad 2" in【Floating】. Three points nearby on two point cloud data are selected respectively, and the three points should be located far apart, as shown in Figure 4-43.

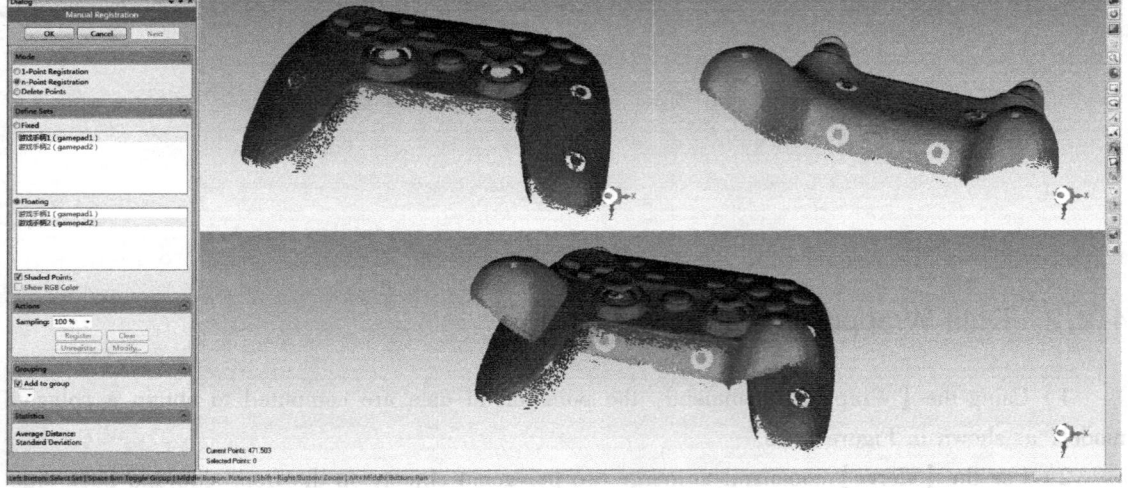

Figure 4-43　Select three points to merge

137 ◀

3D Digital Design and Manufacturing

Observe whether the effect of data merging meets expectations, and if it does, click 【OK】, as shown in Figure 4-44.

8) Use the 【Global Registration】 command to do data processing again, and the stitching is completed, as shown in Figure 4-45.

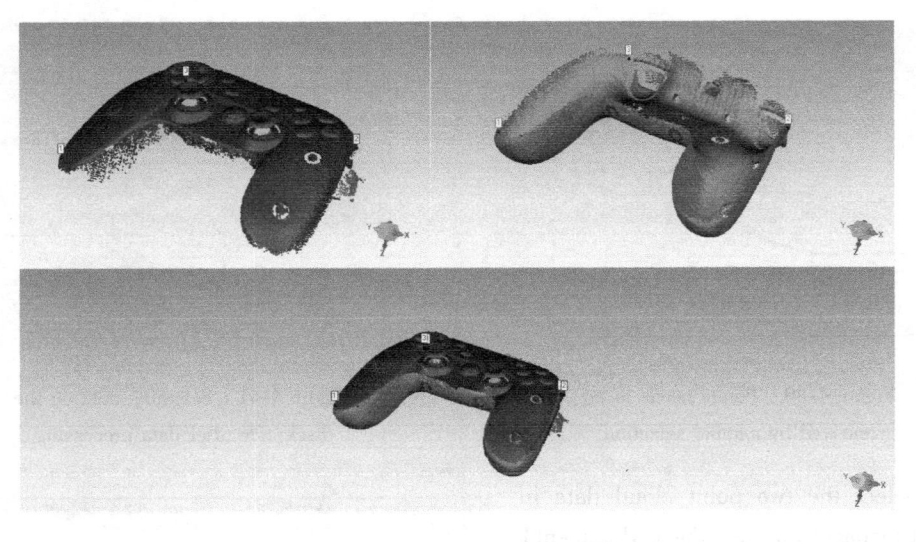

Figure 4-44　Observe the effect of merging

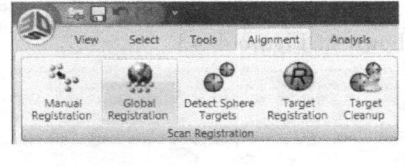

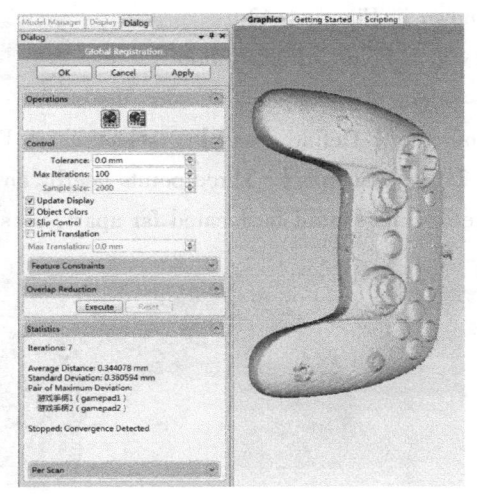

Figure 4-45　Global registration

4.2.2　Data Wrapping

1) Using the 【Wrap】 command, the point cloud data are computed to obtain a polygon model, as shown in Figure 4-46.

2) Use the 【Merge】 command to merge two polygonal objects on the front side and back side into a single object, as shown in Figure 4-47.

▶ 138

Project 4　Gamepad

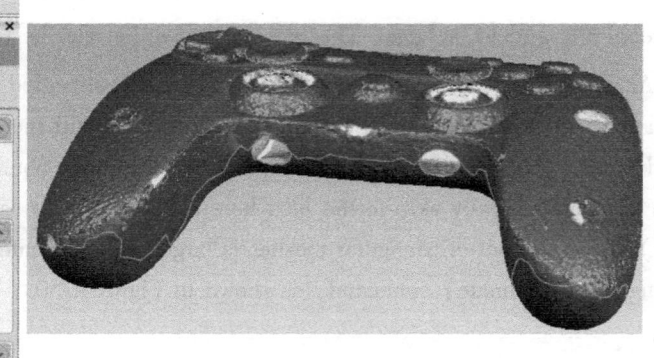

Figure 4-46　Wrap the data

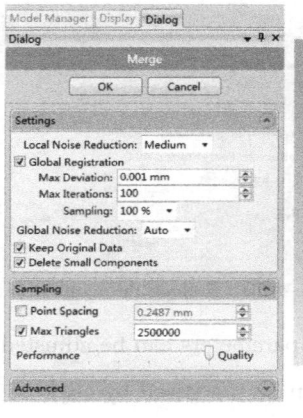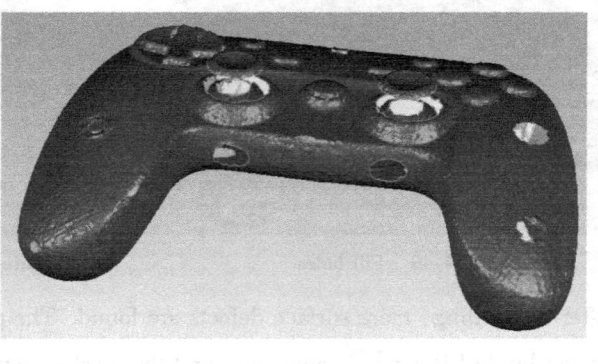

Figure 4-47　Merged sheet

3）Automatically repair the defects in polygonal meshes using the【Mesh Doctor】command, as shown in Figure 4-48.

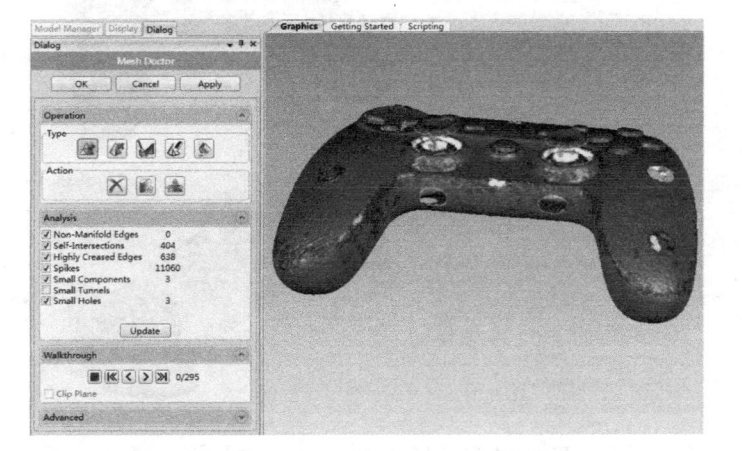

Figure 4-48　Repair by【Mesh Doctor】

139 ◀

3D Digital Design and Manufacturing

4) There are three ways to fill holes in the 【Fill Single】 command, namely, 【Curvature】, 【Tangent】 and 【Flat】. The types of holes that can be filled are【Complete】, 【Partial】. Also, two disconnected polygonal areas can be connected by 【Bridge】. Here we use the 【Curvature】 method to fill the 【Complete】holes. When fill the single hole, the mouse can be used to click on the green boundary of the hole. When the green boundary is not found, click the"Next" button and automatically skip to the next hole to be repaired, as shown in Figure 4-49.

5) If the number of triangular meshes is large, you can enter the percentage you want to reduce by using the 【Decimate】 command, as shown in Figure 4-50.

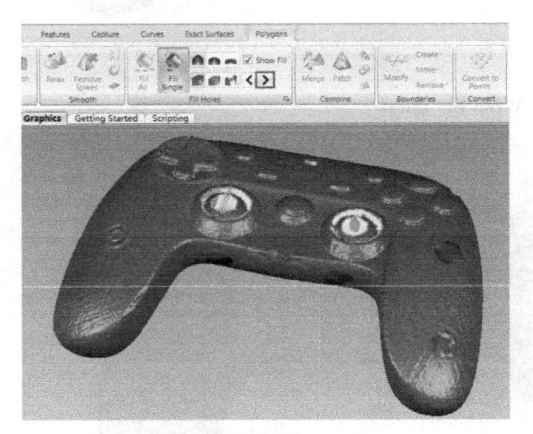

Figure 4-49　Fill holes

Figure 4-50　Reduce the number of meshes

6) After hole filling, more surface defects are found. The parameters can be adjusted according to the situation by the【Relax】 command, and the meshes can be processed by clicking 【Apply】. The angle between individual polygons can be minimized to make the polygonal meshes smoother, as shown in Figure 4-51.

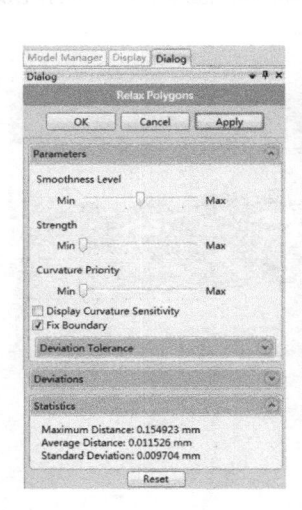

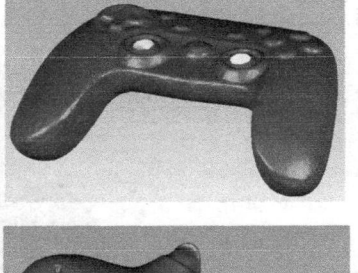

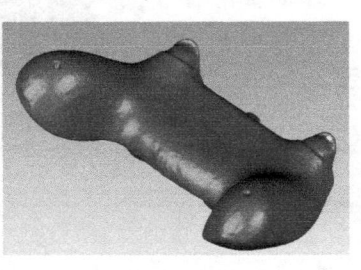

Figure 4-51　After relaxation optimization

▶ 140

Project 4　Gamepad

7) The generated triangular meshes are saved as a STL file, which is the basis of reverse modeling in NX. It is suggested that English file names and path names be used, as shown in Figure 4-52.

Figure 4-52　Save as a STL file

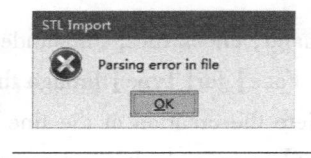

If the error prompt "Parsing error in file" occurs in the subsequent file import in NX, the possible reason is that there are Chinese characters in the export or import path, or Chinese characters in the file name.

Task 3　Reverse Modeling

In subsequent measurements of the distance between the plane and the imported planar body, the following method can be used: When enabling point capturing, uncheck【Points on Face】to select the face created in reverse modeling, and then check【Points on Face】to select the point on the imported facet body, as shown in Figure 4-53.

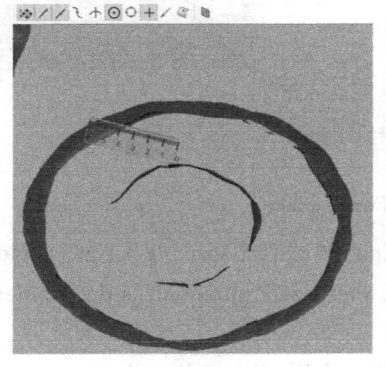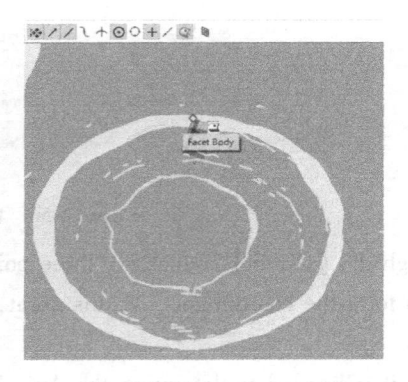

Figure 4-53　A method of measuring distance

141

3D Digital Design and Manufacturing

4.3.1 Determine Coordinates

1) Import the STL file of the gamepad, as shown in Figure 4-54.

2) The main structure of the gamepad is symmetrical. Firstly, observe the gamepad in order to determine the coordinates, and mirror surfaces should be determined, as shown in Figure 4-55.

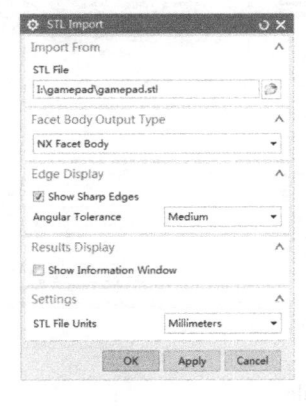

Figure 4-54　Import STL file

Figure 4-55　Observe the gamepad

3) Through the "line" function in the 【Basic Curves】command, check the 【Unbounded】(line is as long as the screen boundaries allow), and select 【Point on Face】for【Type】through the 【Point Constructor】, then select two points at the right place to complete the creation of the line. Draw a vertical line in the same way, and form a face under the 【Swept】command. It can be analyzed by measuring distances, as shown in Figure 4-56.

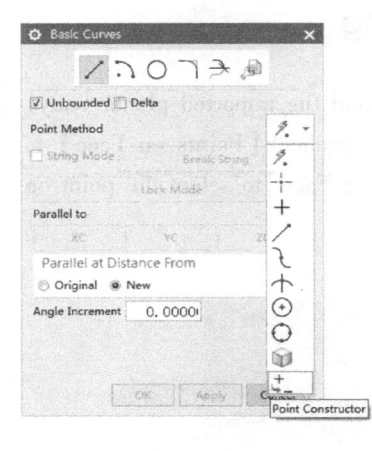

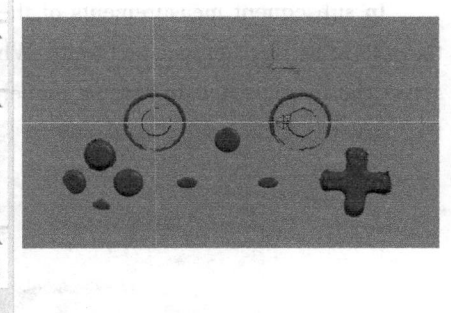

Figure 4-56　Create a face

4) Through the 【Orient】command, three points which are forming a right angle approximately are selected to form the coordinates. At this point, the Axis Z direction is determined, as shown in Figure 4-57.

5) Next, it is needed to determine the Axis X and Y. Generally, the connection between the left and right circle centers is selected for determination, as shown in Figure 4-58.

142

Project 4　Gamepad

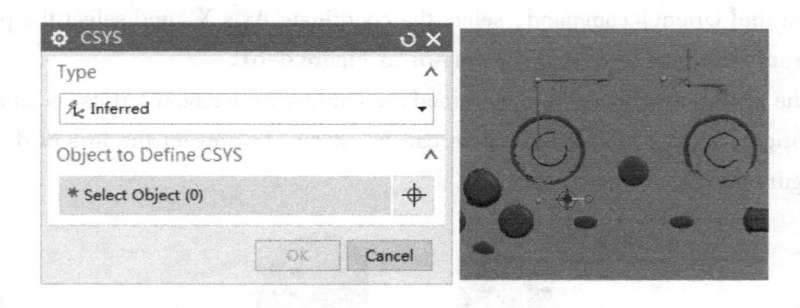

Figure 4-57　Determine Axis Z

6) By using the "Arc" function in the 【Basic Curves】, check the 【Full Circle】, and then select the type of 【Point on Face】 through the 【Point Constructor】. By imagining the shape of the gamepad, roughly select three points at the root of the fillet to draw two circles on both sides, as shown in Figure 4-59.

Figure 4-58　Determine the Axis X and Y

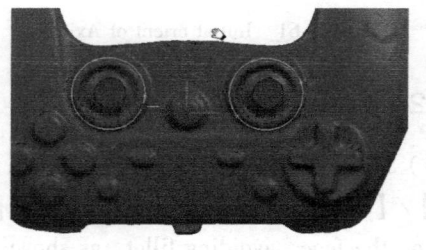

Figure 4-59　Draw two circles

7) Through the "line" function in the 【Basic Curves】, two centers of circles are selected to draw a line. The line is projected onto the plane with Axis E as a normal vector. It is suggested that the input curve in the settings is hidden so that new learners can identify which curve is the newly generated one, as shown in Figure 4-60.

Figure 4-60　Draw a line and project it

143 ◀

3D Digital Design and Manufacturing

8) Through the【Orient】command, select the coordinate Axis X, and select the projected line, to complete the initial orient of Axis X, as shown in Figure 4-61.

9) After the initial coordinates is determined, according to personal habits, a small box can be created at the origin, or the coordinate system can be saved, to prevent the loss of the coordinates, as shown in Figure 4-62.

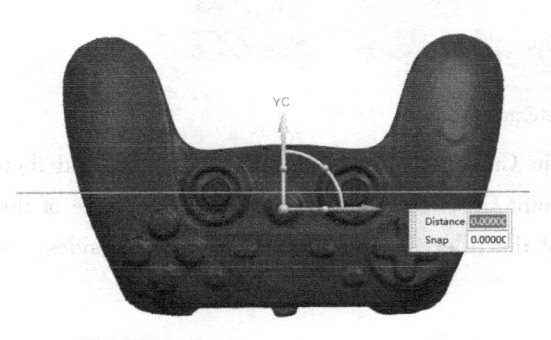

Figure 4-61   Initial orient of Axis X

Figure 4-62   Create a small box

4.3.2 Pavement

1) Select the 【Point on Face】 through the【Insert】/【Datum/ Point】/【Point】command, and select the points at the appropriate place on the face, avoiding fillet, as shown in Figure 4-63.

2) Select the created points; right-click on any point; and select 【New Group】, as shown in Figure 4-64. (Here, NX 10.0 is different from the previous versions, so special attention should be paid.)

Figure 4-63   Create points

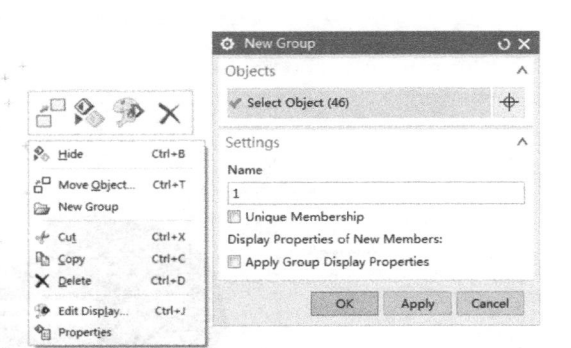

Figure 4-64   Create a group

3) Under the command of 【Fit Surface】, select any point in the group to create the surface. If the effect is not satisfactory, it can be roughly adjusted by modifying the content of 【Parameterization】. It should be noted that the greater the order is, the worse the smoothness is. It

Project 4　Gamepad

can be observed by the【Face Analysis】command, and the surface can be enlarged appropriately by the【Enlarge】 command, as shown in Figure 4-65.

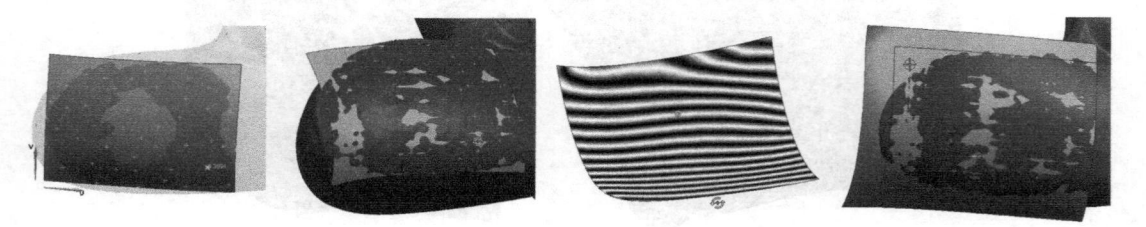

Figure 4-65　Fit the surface

4) Through the "Arc"function in the【Basic Curves】, two arcs are drawn first, then the two ends of the two arcs are used as the starting points and the end points respectively to draw the other two arcs with the other point of the arc roughly in a line. Use the 【Through Curve Mesh】 command to create a surface, as shown in Figure 4-66.

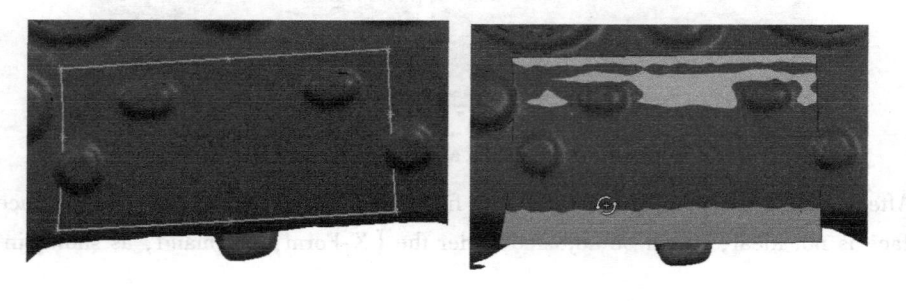

Figure 4-66　Create a surface and enlarge it

5) Two arcs are created by using the "Arc"function in the【Basic Curves】. Then create a surface by the【Swept】command, as shown in Figure 4-67.

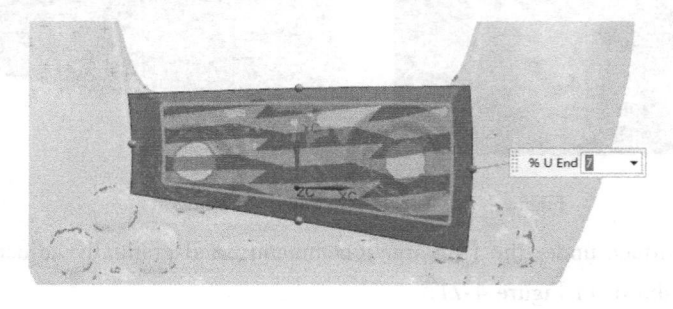

Figure 4-67　Create a surface and enlarge it

6) Fitting is done by the【Fit Surface】command, and the surface is enlarged appropriately, as shown in Figure 4-68.

7) By using the "Arc"function in the 【Basic Curves】, four arcs are drawn. A surface is created under the command of 【Through Curve Mesh】 and is enlarged appropriately, as shown in Figure 4-69.

145

3D Digital Design and Manufacturing

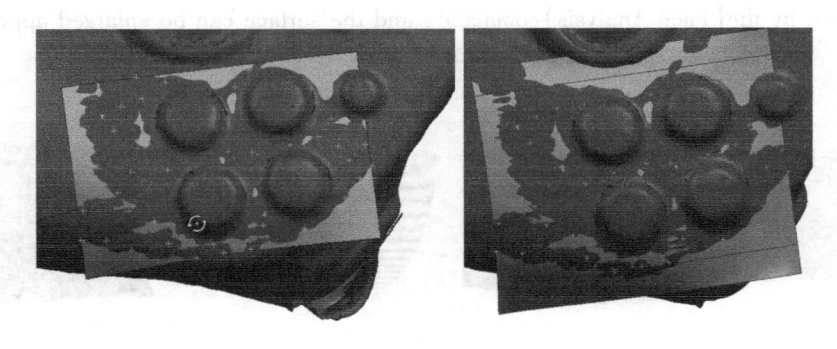

Figure 4-68　Fit a surface and enlarge it

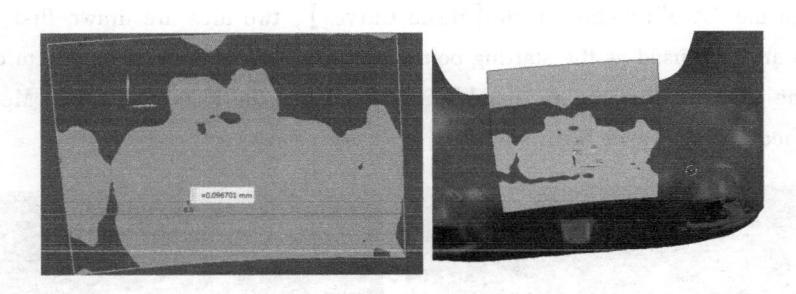

Figure 4-69　Create a surface and enlarge it

8) After creating the points, the surface is fitted under the command of 【Fit Surface】. If the fitted surface is not ideal, it can be adjusted under the 【X-Form】 command, as shown in Figure 4-70.

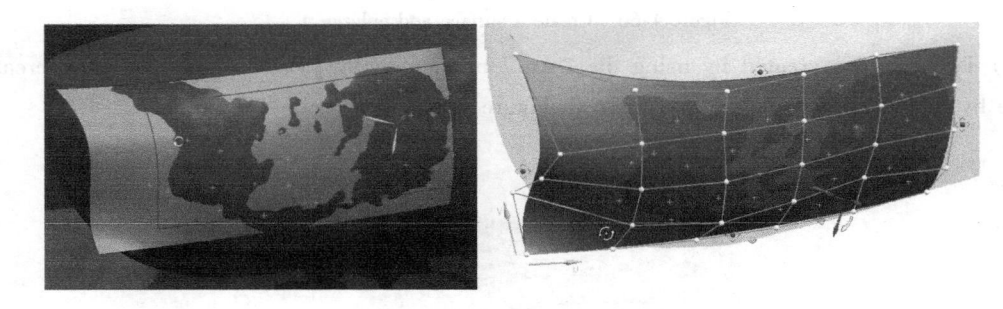

Figure 4-70　Create a surface and adjust it

9) Enlarge the surface under the 【Enlarge】 command, and gradually adjust it under the 【X-Form】 command, as shown in Figure 4-71.

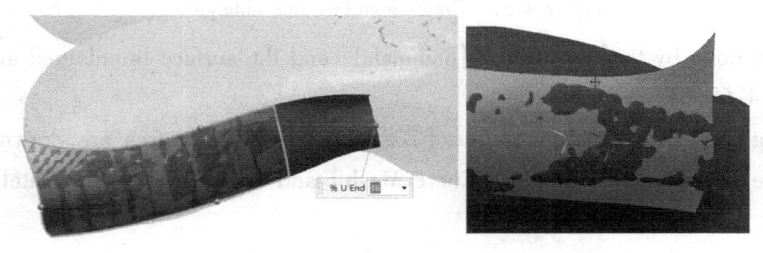

Figure 4-71　Enlarge the surface and adjust it

▶ 146

Project 4　Gamepad

10) By using the "Arc" function in the 【Basic Curves】, four arcs are drawn. A surface is created under the【Through Curve Mesh】command and is enlarged appropriately, as shown in Figure 4-72.

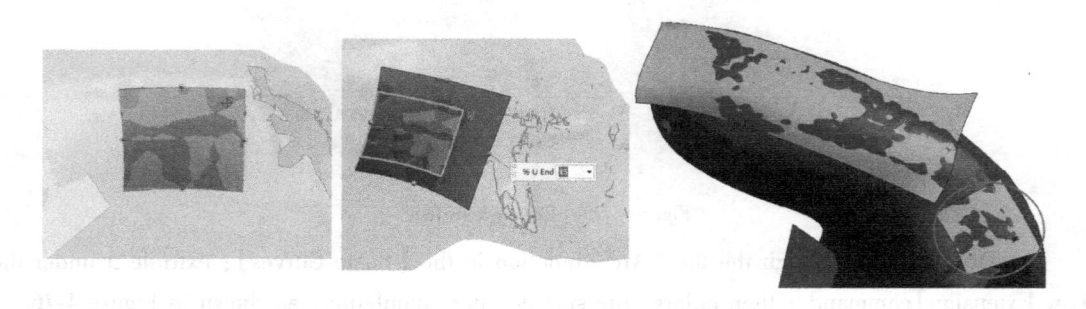

Figure 4-72　Create a surface

11) By using the "Arc" function in the 【Basic Curves】, four arcs are drawn. A surface is created under the【Through Curve Mesh】command and is enlarged appropriately. Then adjust the surface with the command of 【X-Form】, as shown in Figure 4-73.

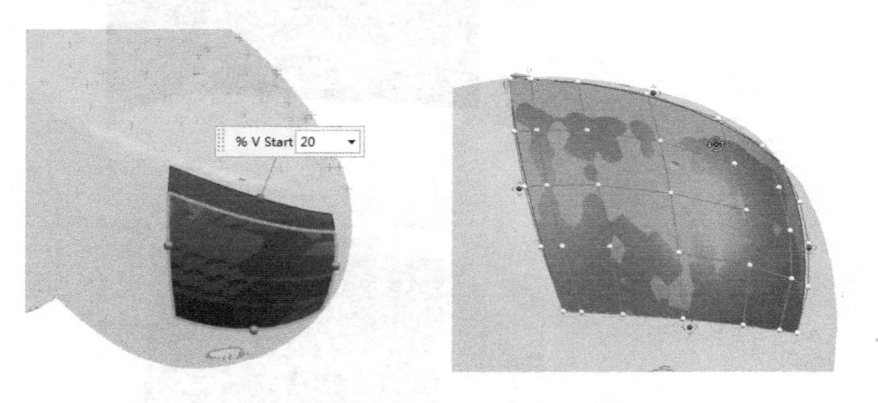

Figure 4-73　Create a surface

12) After creating the points, the surface is fitted under the 【Fit Surface】command. If the fitted surface is not ideal, it can be adjusted under the 【X-Form】 command, as shown in Figure 4-74.

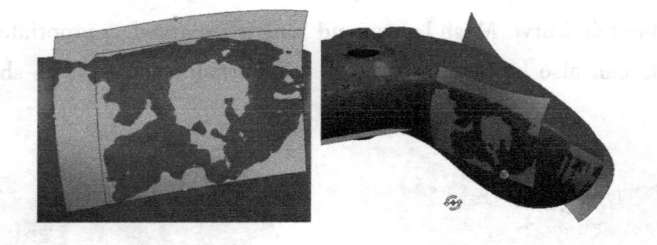

Figure 4-74　Create a surface

13) By using the "Arc" function in the 【Basic Curves】, four arcs are drawn. A surface is created under the【Through Curve Mesh】command and is enlarged appropriately. If the surface is not ideal, adjustments can also be made under the 【X-Form】 command, as shown in Figure 4-75.

147

3D Digital Design and Manufacturing

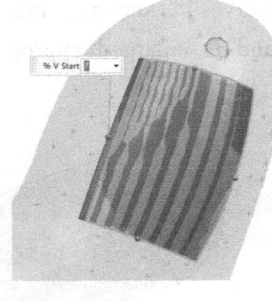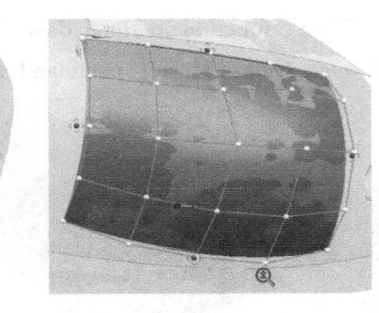

Figure 4-75　Create a surface

14）Draw the arc through the the "Arc" function in the【Basic Curves】; extrude it under the 【Law Extension】command ; then enlarge the surface after completion, as shown in Figure 4-76.

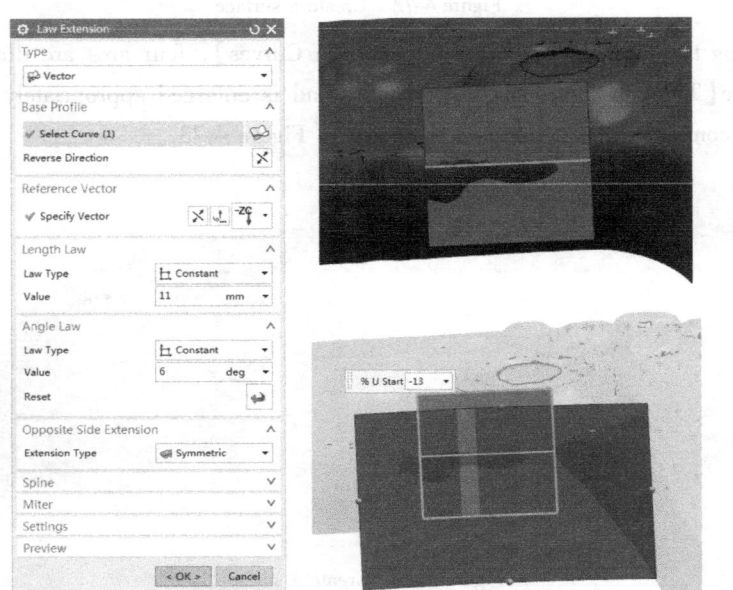

Figure 4-76　Create a surface

15）By using the "Arc" function in the【Basic Curves】, four arcs are drawn. A surface is created under the【Through Curve Mesh】command and is enlarged appropriately. If the surface is not ideal, adjustments can also be made under the【X-Form】command, as shown in Figure 4-77.

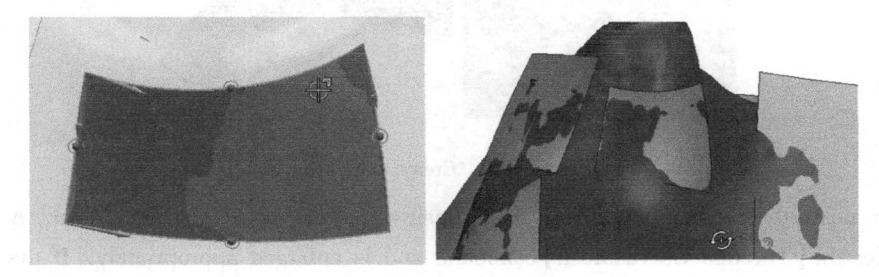

Figure 4-77　Create a surface

▶ 148

Project 4   Gamepad

16) Enlarge the surface under the 【Enlarge】 command, and gradually adjust it under the 【X-Form】 command, as shown in Figure 4-78.

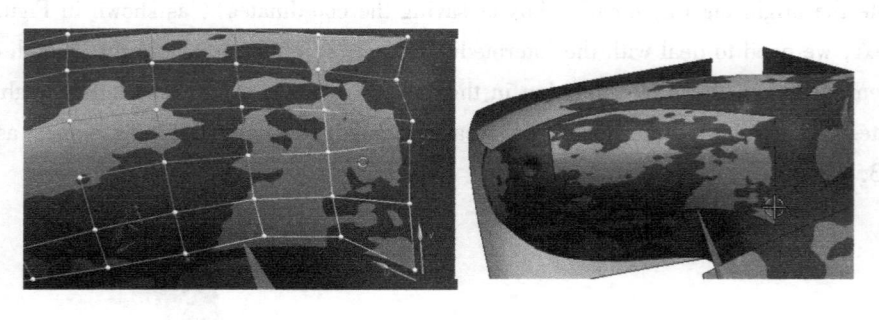

Figure 4-78   Enlarge the surface

17) Draw the arc through the "Arc" function in the 【Basic Curves】; extrude it under the 【Law Extension】command ; then enlarge the surface after completion, as shown in Figure 4-79.

18) So far, the paving stage is almost over. Trim the sheets under the 【Trim Body】 command, as shown in Figure 4-80.

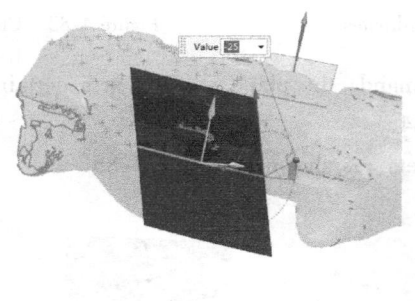

Figure 4-79   Law extension

Figure 4-80   Trimmed facets

4.3.3   Draw the Rough Shape

1) Through the 【Transformations】 command, the sheets are mirror copied through the "XC-ZC plane". Through the observation, it is found that there are some deviations on the other side, and the coordinate system needs to be adjusted. Through analysis, it is judged that the coordinate system should be rotated appropriately around the Axis Y, as shown in Figure 4-81.

2) The coordinate system is rotated and adjusted under the 【Orient】 command until the measured data meet the requirements. Because the product is an injection

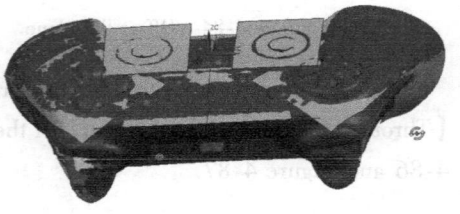

Figure 4-81   Mirror-coppied facet

149

3D Digital Design and Manufacturing

molding part, it may be deformed in the actual production, so only the large facet is more suitable, and part of the facet deviation is within the allowable range. After adjusting the coordinates, remirror and re-create the origin tag (by creating box or saving the coordinates), as shown in Figure 4-82.

3) Next, we need to deal with the intermediate mirror stitching to ensure the smooth connection between them. Through the "Line" function in the [Basic Curves], a line passing through the origin of coordinates and parallel to the Axis Y is created, and it is extruded into a plane, as shown in Figure 4-83.

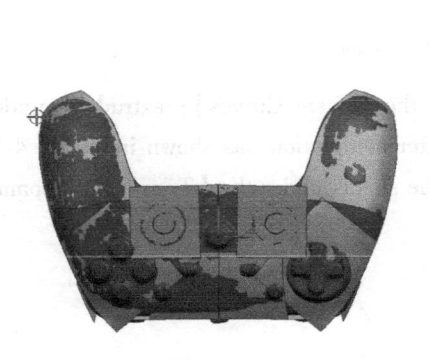

Figure 4-82　Remirror after adjusting coordinates

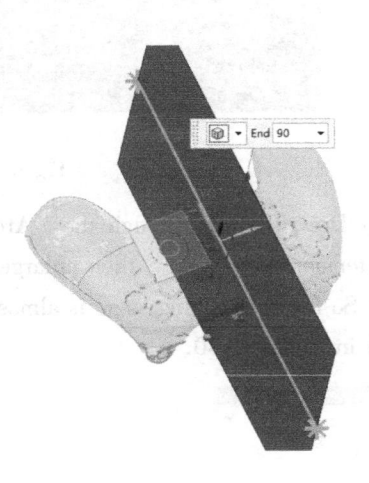

Figure 4-83　Create a plane

4) Under the [Offset Surface] command, the plane is offset at a certain distance from the left and right sides, as shown in Figure 4-84.

5) Offset plane is used to trim facets, as shown in Figure 4-85.

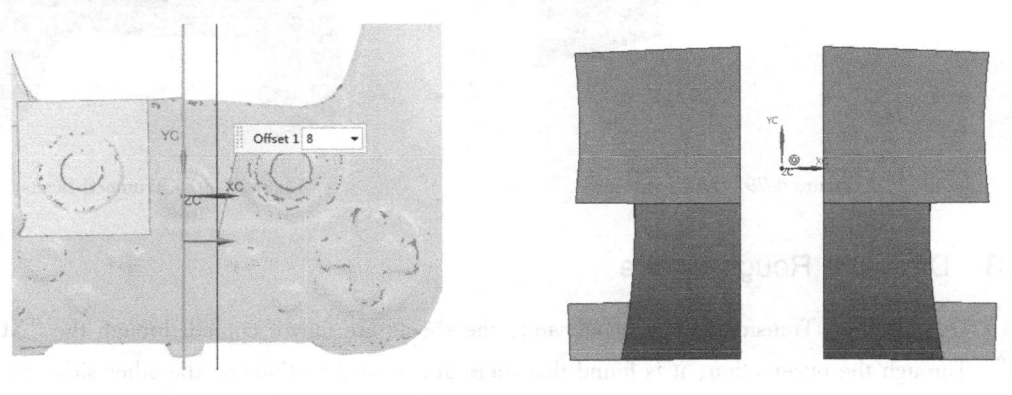

Figure 4-84　Offset the plane

Figure 4-85　Trimmed facets

6) Bridging is done under the [Bridge Curve] command, and a surface is created under the [Through Curve Mesh] command. In the same way, other facets can be created, as shown in Figure 4-86 and Figure 4-87.

▶ 150

Project 4　Gamepad

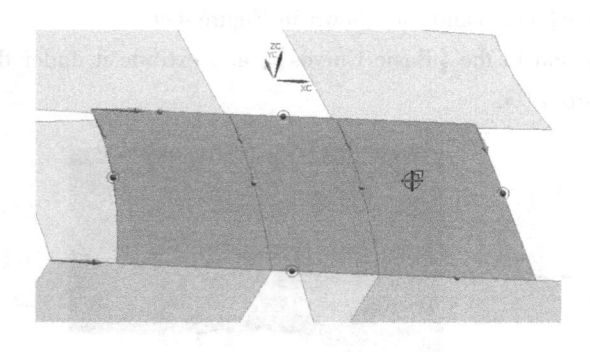

Figure 4-86　Create a surface

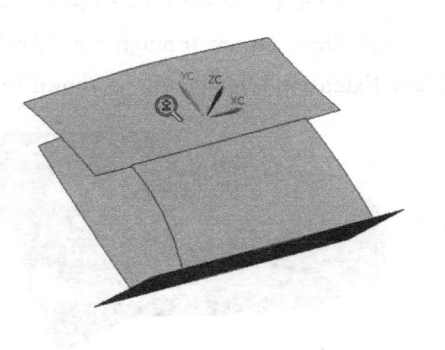

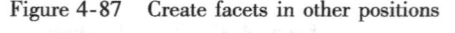

Figure 4-87　Create facets in other positions

7) So far, the coordinates have been completely determined. Cut one side of the facets, just keep one side, and mirror it to the other side after completion, as shown in Figure 4-88.

8) Draw an arc through the "Arc" function in the 【Basic Curves】, and extrude it under the 【Law Extension】command. The draft angle is adjusted according to the actual situation, as shown in Figure 4-89.

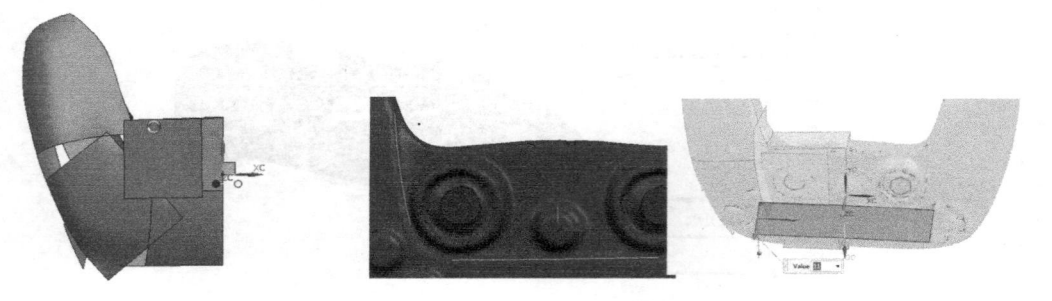

Figure 4-88　Trimmed facet

Figure 4-89　Law extension

9) In order to ensure complete smoothness in the middle after mirroring, the facets are mirrored to the other side after removal, and the two ends are captured by the "Arc" function. Then the arc is drawn and extruded again, as shown in Figure 4-90.

10) The fillet is processed under the 【Face Blend】 command, and it is trimmed under the 【Trim Sheet】 command, as shown in Figure 4-91.

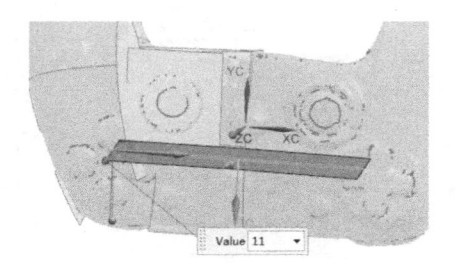

Figure 4-90　Law extension

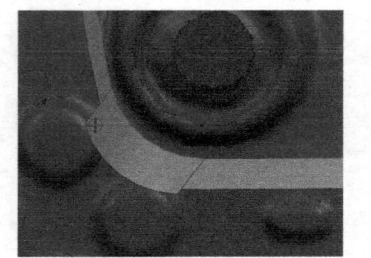

Figure 4-91　Face blend and trim

151

3D Digital Design and Manufacturing

11) Trim the sheet under the 【Trim Sheet】 command, as shown in Figure 4-92.

12) Draw an arc through the "Arc" function in the 【Basic Curves】, and extrude it under the 【Law Extension】command, as shown in Figure 4-93.

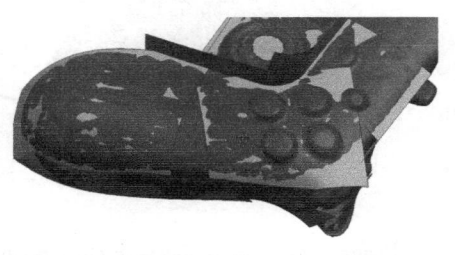

Figure 4-92　Trim the sheet

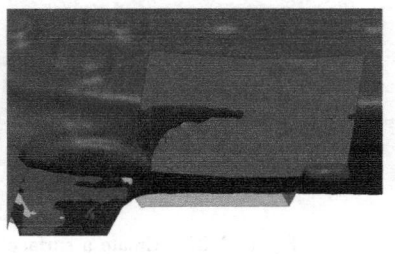

Figure 4-93　Law extension

13) Bridging is done under the【Bridge Curve】command. Adjust the parameters to make the curve close to the contourline, as shown in Figure 4-94.

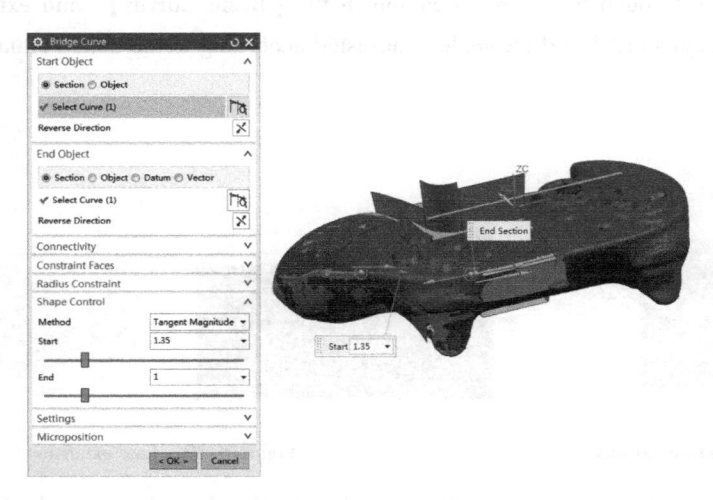

Figure 4-94　Bridge curve

14) Offset the curve under the【Offset Curve in Face】command and bridge the curve. Use the 【Through Curve Mesh】command to create a surface, as shown in Figure 4-95.

15) After proper trimming and bridging of the facets, the surface is created under the【Through Curve Mesh】command, as shown in Figure 4-96.

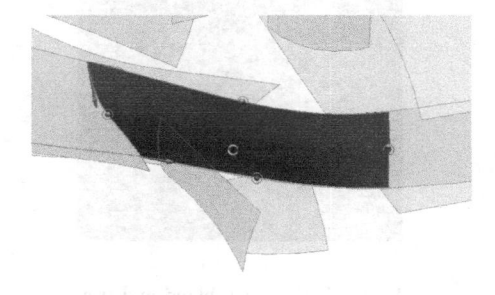

Figure 4-95　Create a surface

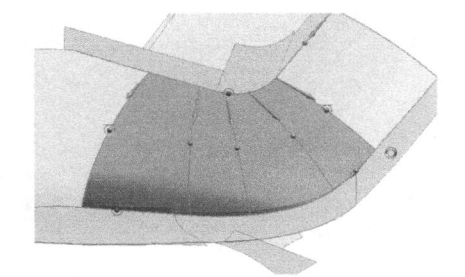

Figure 4-96　Creat a surface

▶ 152

Project 4　Gamepad

16) Trim the redundant sheets under the【Trim Sheet】command, as shown in Figure 4-97.

17) Draw an arc through the "Arc" function in the【Basic Curves】, and extrude it under the【Law Extension】command, as shown in Figure 4-98.

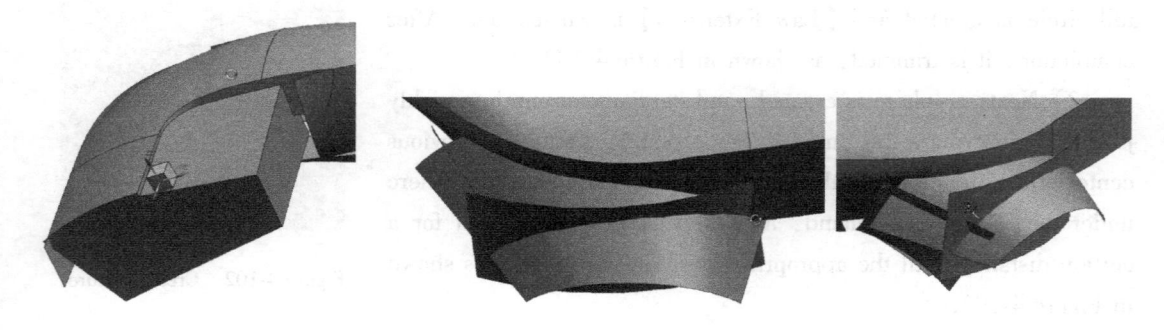

Figure 4-97　Trim sheets　　　　　　　　　　Figure 4-98　Create a sheet

18) Bridging is carried out through the【Bridge Curve】command. After proper trimming, a surface is created under the【Through Curve Mesh】command, as shown in Figure 4-99.

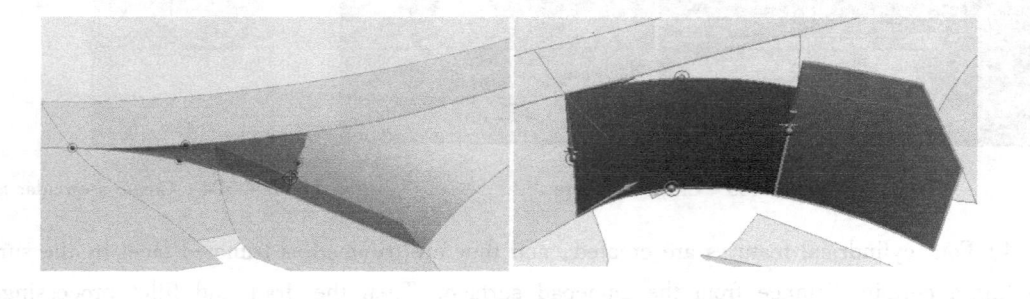

Figure 4-99　Create a surface

19) The method for other parts is similar to the previous steps. The creation of the sheets of half of the gamepad is finally completed by the【Bridge Curve】,【Through Curve Mesh】and other commands, as shown in Figure 4-100.

20) Sew the sheets into a solid body under the【Sew】command, as shown in Figure 4-101.

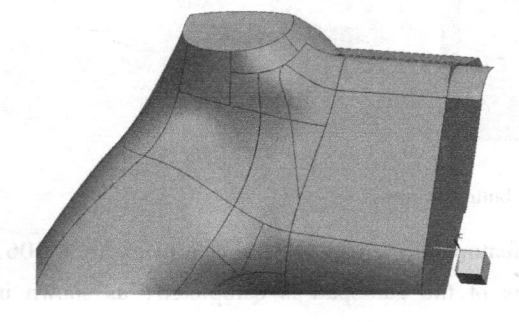 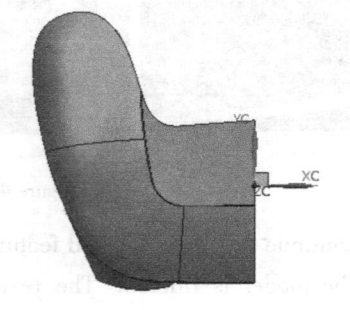

Figure 4-100　Create surfaces　　　　　　　　Figure 4-101　Sew into a solid body

153 ◄

3D Digital Design and Manufacturing

4.3.4 Draw Detail Features

1) Through the function of "Arc" in the 【Basic Curves】, a full circle is created and 【Law Extension】 is carried out. After completion, it is trimmed, as shown in Figure 4-102.

2) Next, a sphere is created, and its diameter can be roughly judged by drawing a full circle by three points. Using the previous center of the the circle as the center of sphere to create the sphere under the 【Sphere】 command, and then move it downward for a certain distance until the appropriate position is reached, as shown in Figure 4-103.

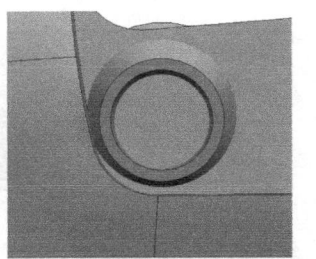

Figure 4-102　Create features

3) Create intermediate circular features, as shown in Figure 4-104.

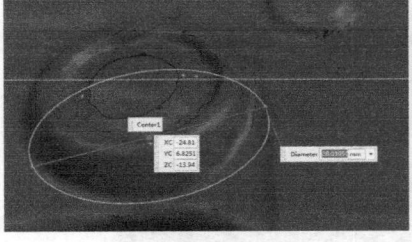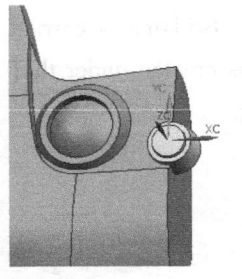

Figure 4-103　Create a sphere　　　　　　Figure 4-104　Create a circular table

4) Four cylindrical features are created, and they are trimmed as trimmed facet by the surface offsetting a certain distance from the gamepad surface. Then the draft and fillet processing are carried out, as shown in Figure 4-105.

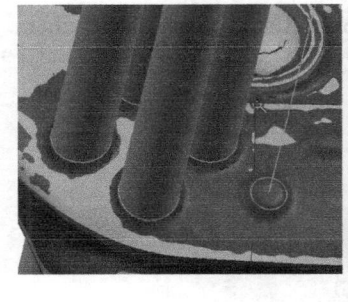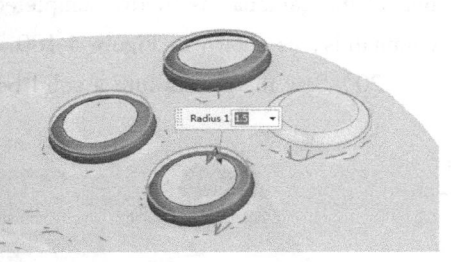

Figure 4-105　Create button features

5) Continue to create detailed features until all features are created, as shown in Figure 4-106.

6) The model is filleted. The reverse modeling of the gamepad is completed, as shown in Figure 4-107.

154

Project 4　Gamepad

Figure 4-106　Gamepad　　　　Figure 4-107　Complete the reverse modeling of the gamepad

Task 4　Machining Programming

4.4.1　Process Analysis

Considering the large stock of the back side of the gamepad, the back side is machined first in order to ensure the rigidity of the workpiece, as shown in Figure 4-108a. Then lime is poured into the box, and the back side is clamped to machine the front side. After both sides are machined, the joint is polished at last.

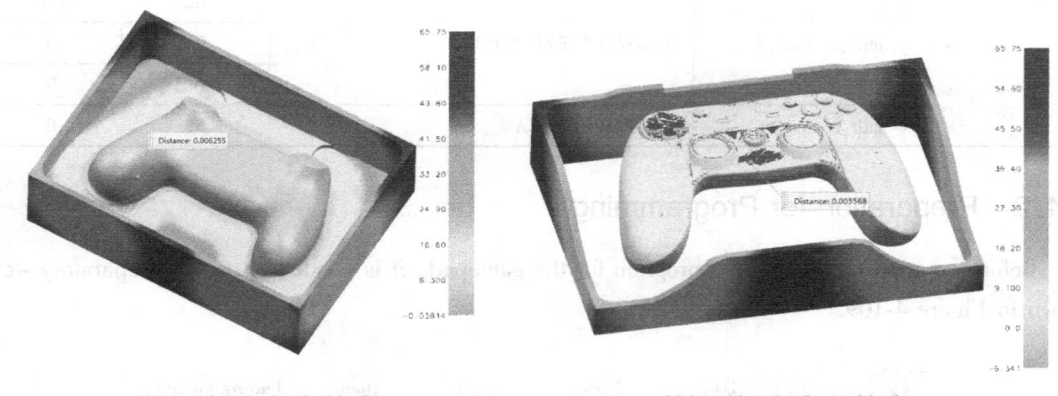

a) Machine the back side first　　　　　　　b) Machine the front side then

Figure 4-108　Machining scheme

In order to save machining time, first use three-axis roughing, then use five-axis to machine local features. Machining process is shown in Table 4-1.

155 ◀

3D Digital Design and Manufacturing

Table 4-1 Machining process of the gamepad

Process	Type	Subtype of process	Tool	Component stock/mm
Back side roughing	mill_contour	CAVITY_MILL	D20R0.8	0.3
	mill_contour	CORNER_ROUGH	D8	0.3
Back side semi-finishing	mill_contour	CONTOUR_AREA	R3	0.2
	mill_contour	STREAMLINE	R3	0.2
Back side finishing	mill_contour	CONTOUR_AREA	R3	0
	mill_contour	STREAMLINE	R3	0
Front side roughing	mill_contour	CAVITY_MILL	D20R0.8	0.3
			D6R0.5	0.3
			D3R0.5	0.3
			D1	0.3
Front side semi-finishing	mill_contour	CONTOUR_AREA	R3	0.2
Front side finishing	mill_contour	CONTOUR_AREA	D6R0.5	0
	mill_contour	ZLEVEL_PROFILE	D6R0.5	0
	mill_contour	CONTOUR_AREA	D3R0.5	0
	mill_multi-axis	VARIABLE_CONTOUR	R3	0
	mill_contour	CONTOUR_AREA	D6R0.5	0
	mill_multi-axis	VARIABLE_STREAMLINE	D2	0
	mill_multi-axis	CONTOUR_PROFILE	D8	0
			D2	0
	mill_contour	FLOWCUT_REF_TOOL	R2	0.07
			R1	0
			R0.5	0
	mill_contour	CONTOUR_AREA	R1	0

4.4.2 Preparation for Programming

Before creating the machining program for the gamepad, it is needed to do the preparatory work shown in Figure 4-109.

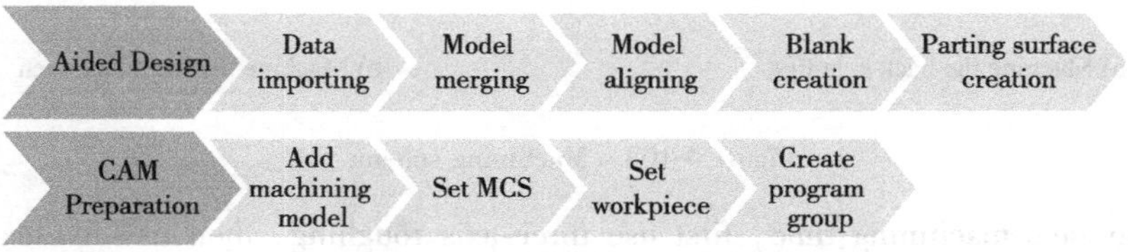

Figure 4-109 Preparation for programming

▶ 156

Project 4   Gamepad

1. Aided Design

1) Data importing. Import the gamepad model data using the 【File】/【Import】/【Parasolid】 command.

2) Model merging. Move the three solid bodies shown in Figure 4-110a to other layer, and then use the 【Unite】 command to merge the remaining solid bodies, as shown in Figure 4-110b. The color of the model is changed to grey to facilitate the observation of the tool path in the later stage.

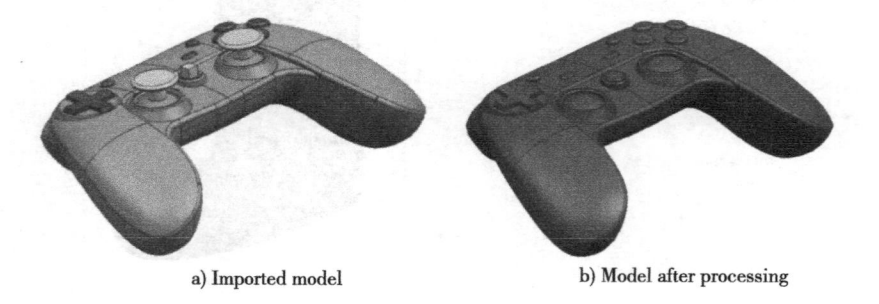

a) Imported model   b) Model after processing

Figure 4-110   Process model data

3) Model aligning. Because the position of the gamepad model is not aligned with the absolute coordinate system, it needs to be re-aligned. Create the plane, line and arc shown in Figure 4-111.

Click the 【Insert】/【Datum/Point】/【Datum CSYS】 command to pop up the 【Datum CSYS】 dialog box. Select 【Plane, X-Axis. Point】 in 【Type】, then select the plane, line and the center of arc created before in turn to create the datum coordinate shown in Figure 4-112. Save the reference coordinate system by using the 【Format】/【WCS】/【Save】 command.

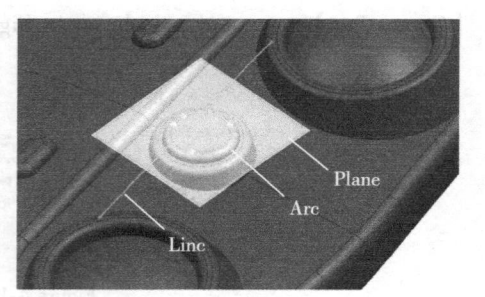

Figure 4-111   Create plane, line and arc

Figure 4-112   Create reference CSYS

Click the 【Edit】/【Move Object】 command, and then select all model data. Select 【CSYS to CSYS】 in 【Motion】; Set 【Specify From CSYS】 as the newly created reference coordinate system

157

3D Digital Design and Manufacturing

and【Specify To CSYS】as "Absolute CSYS". The result is shown in Figure 4-113. The position of the model has been re-alinged.

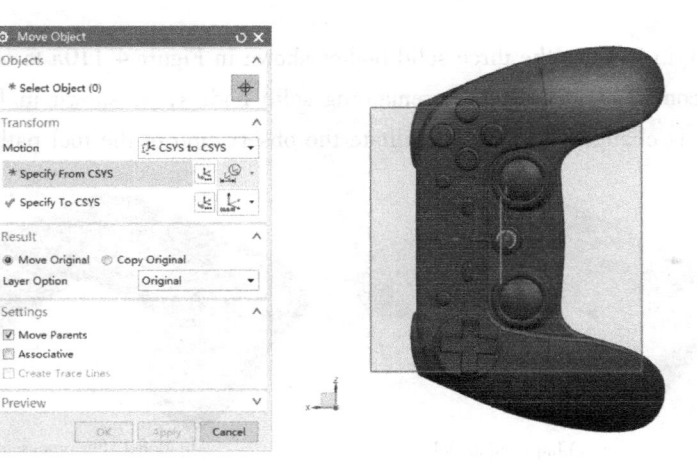

Figure 4-113   Model re-aligned

4）Blank creation. Select the【Electrode Design】in the【Application】, and then select the 【Create Box】command, as shown in Figure 4-114.

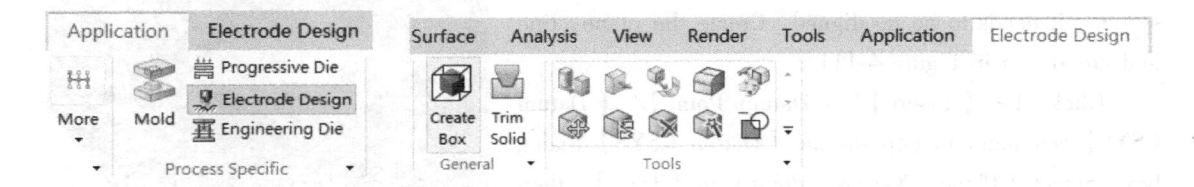

Figure 4-114   【Create Box】command

Considering that the diameter of roughing tool is 20 mm and the stock of about 5 mm is left around the blank, the input【Clearance】value is "25mm", as shown in Figure 4-115.

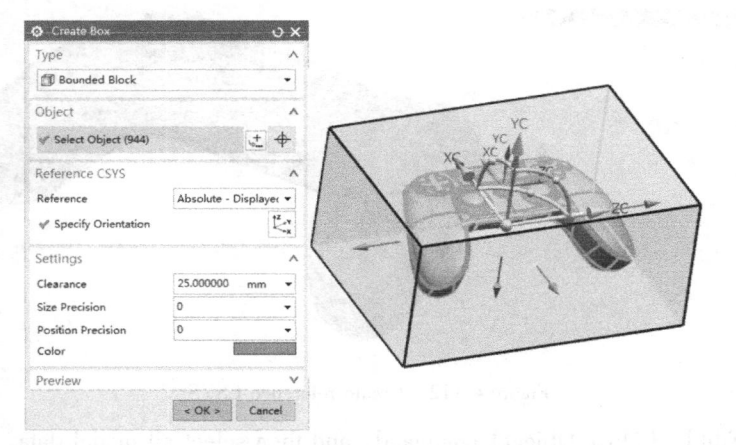

Figure 4-115   【Create Box】dialog box

▶ 158

Project 4　Gamepad

The upper and lower sides do not need to have too much stock left, and it can be retained about 5mm. So the 【Offset Face】 command is used to offset both sides inward by "20mm", as shown in Figure 4-116.

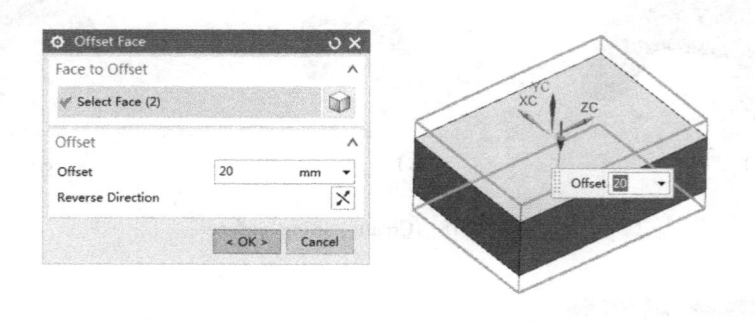

Figure 4-116　Offset the faces

5) Parting surface creation. Use the command of 【Analysis】/【Shape】/【Slope】 to judge the approximate position of the parting surface after observation, as shown in Figure 4-117.

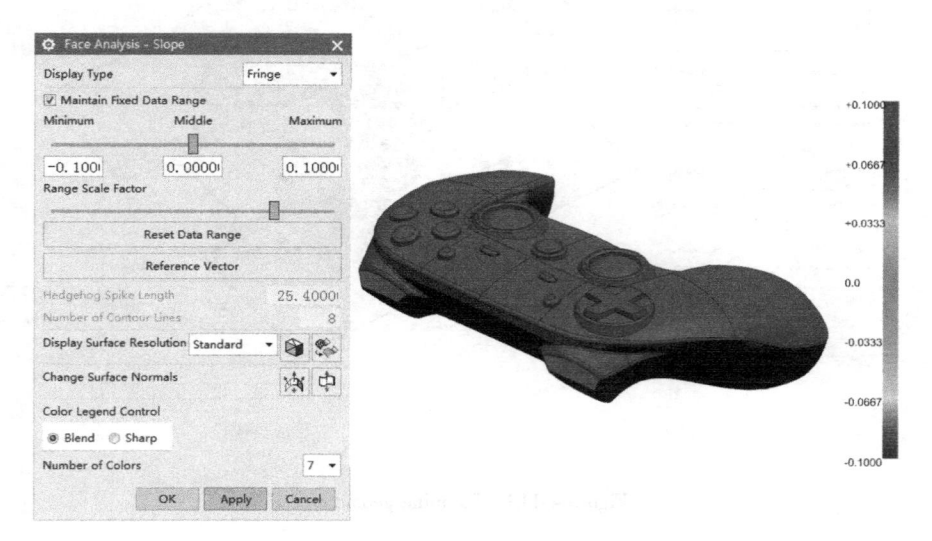

Figure 4-117　Slope analysis

With the command of 【Extrude】, the edge located at the parting surface is selected for extruding, and the extruding length can be increased appropriately to ensure that it exceeds the blank. Since the product is a symmetrical part, after creating half of the parting surface, copy the mirror image to the other side using the 【Transformations】 command, and sew it with the 【Sew】 command, as shown in Figure 4-118.

Using the command of 【Analysis】/【Examine Geometry】, it is confirmed that the boundary of parting surface is composed of inner and outer circles, and the outer circle boundary is beyond the blank geometry, as shown in Figure 4-119. The 【Move to Layer】 command is used to move the parting surface to Layer 3.

159

3D Digital Design and Manufacturing

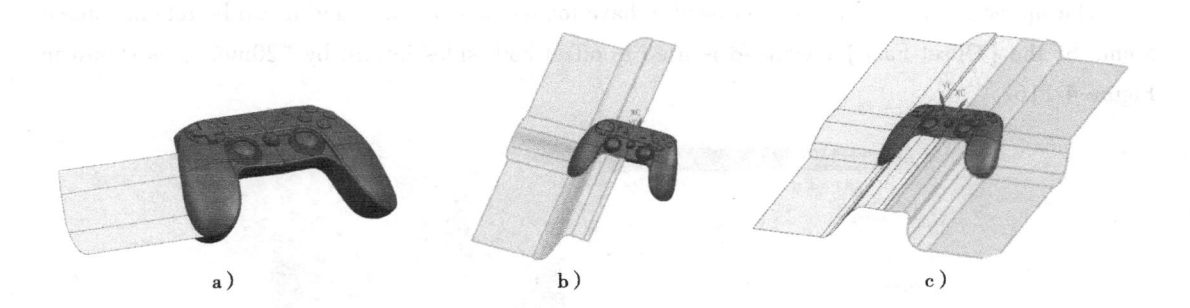

a)   b)   c)

Figure 4-118   Create parting surface

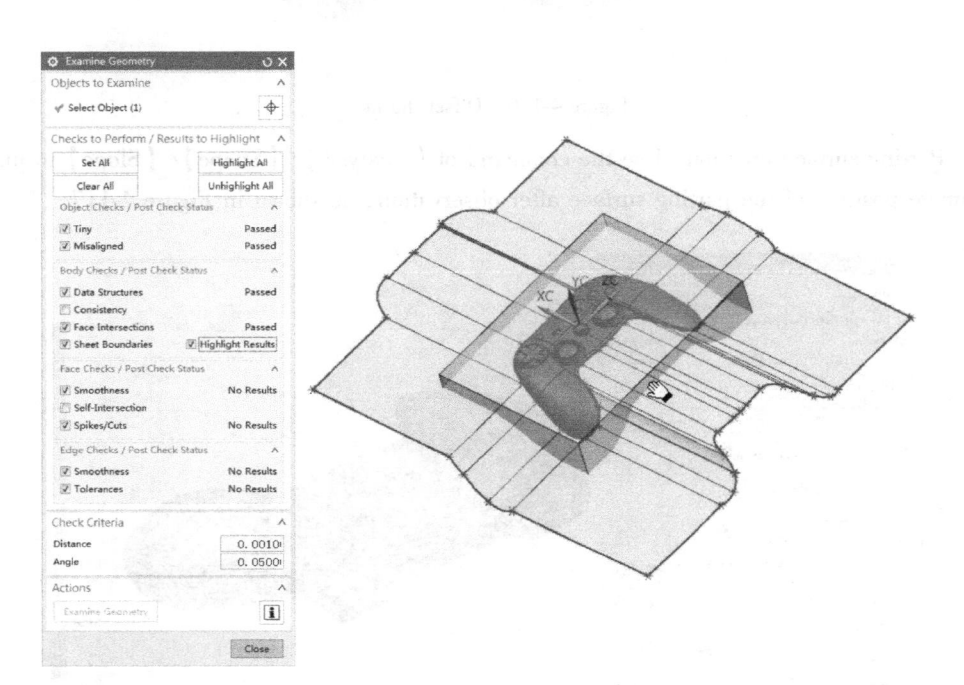

Figure 4-119   Examine geometry

2. CAM Preparation

1) Add machining model. Enter the 【Manufacturing】 module, and pop up the 【Machining Environment】 dialog box, then select 【cam_general】 and 【mill_contour】, as shown in Figure 4-120.

2) Set MCS. Switch to the geometric view of the operation navigator, then rename 【MCS_MILL】 to 【MCS_MILL-F】, and rename 【WORKPIECE】 to 【WORKPIECE-F】 (F denotes the back side), as shown in Figure 4-121.

Double-click 【MCS_MILL-F】 to pop up the 【MCS Mill】 dialog box, then click the "CSYS dialog box" button ⛌ to pup up the【CSYS】dialog box. Select the center of the top surface of the blank as the coordinate origin, and adjust the direction of coordinate axis to make the Axis ZM upward and align the Axis XM with the long edge of blank, as shown in Figure 4-122.

▶ 160

Project 4　Gamepad

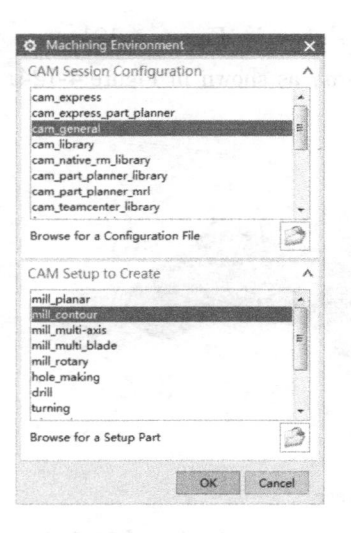

Figure 4-120　【Machine Environment】dialog box

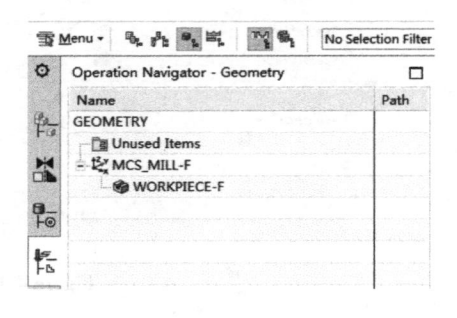

Figure 4-121　Geometric view

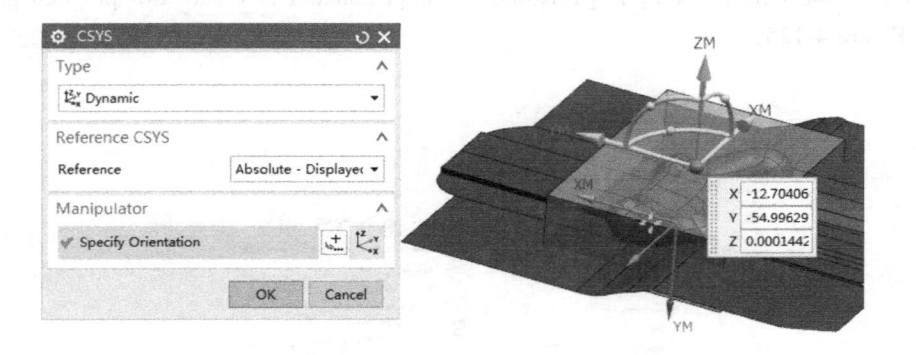

Figure 4-122　Adjust MCS

In the【MCS Mill】dialog box, select【Plane】in【Clearance Option】, then select the top surface of the blank and offset it 50 mm upward to specify it as the safety plane, as shown in Figure 4-123.

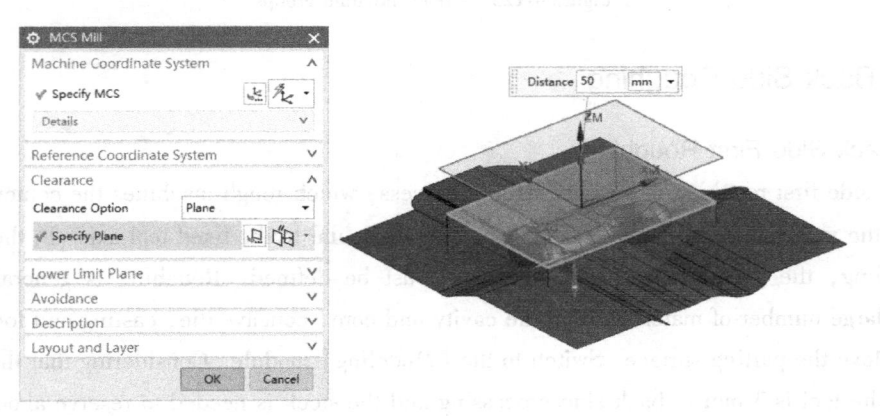

Figure 4-123　Safety plane settings

161

3D Digital Design and Manufacturing

3) Set workpiece. Double-click 【WORKPIECE-F】 shown in Figure 4-121 to pop up the 【Workpiece】 dialog box, then specify part and blank in turn, as shown in Figure 4-124.

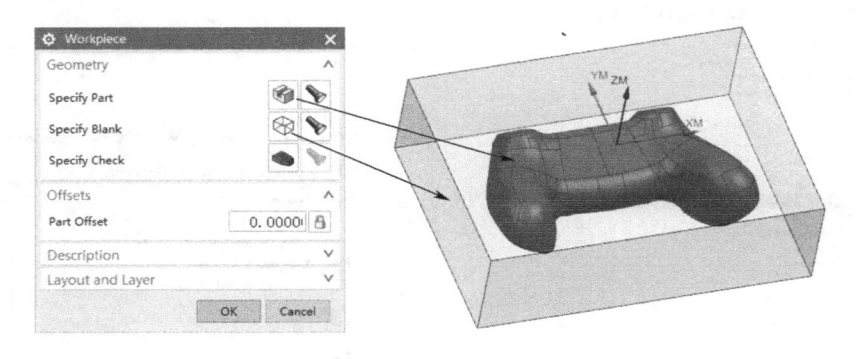

Figure 4-124　Set workpiece

4) Create program group. Switch to the program order view, and select 【PROGRAM】. Then right-click and select the【Insert】/【Program Group】command to create two program groups, as shown in Figure 4-125.

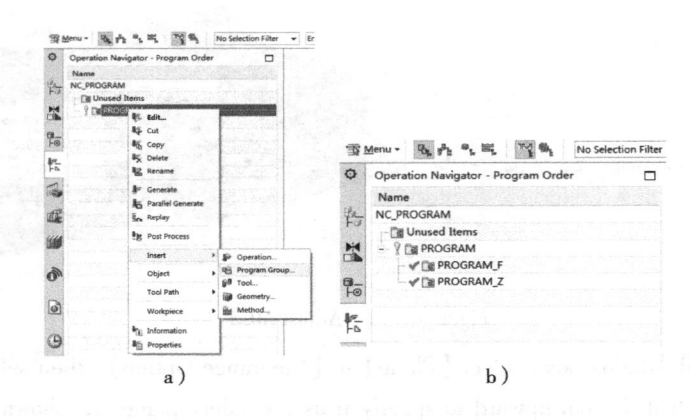

a)　　　　　　　　　　　b)

Figure 4-125　Create program groups

4.4.3　Back Side Roughing

1. Back Side First Roughing

Back side first roughing is a cavity milling process, which rough machines the contour shape by removing the material in the plane cutting layer perpendicular to the fixed tool axis. In the process of cavity milling, the geometry of part and blank must be defined. Roughing is generally used to remove a large number of materials from die cavity and core, concave die, casting and forging parts.

1) Move the parting surface. Switch to the 【Modeling】 module. Considering that the maximum radius of the tool is 3 mm in back side processing and the stock is needed to reserve about 0.3 mm, so use the 【Move Object】 command, and select 【Copy Original】, to move the parting surface downward (to the front direction of the gamepad) by 3.3 mm, as shown in Figure 4-126. The

▶ 162

Project 4   Gamepad

【Move to Layer】 command is used to place the moved parting surface on Layer 4.

2) Create operation. Click the 【Create Operation】 command; select 【mill_contour】 in 【Type】; select "Cavity Mill" in 【Operation Subtype】; select 【PROGRAM_F】 in 【Program】; select 【NONE】 in 【Tool】; select 【WORKPIECE-F】 in 【Geometry】, as shown in Figure 4-127. Click 【OK】 in the end.

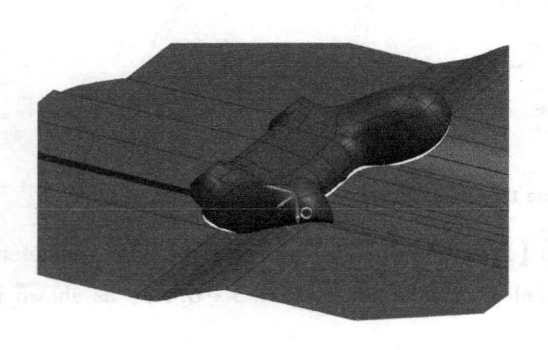

Figure 4-126   Move the parting surface

Figure 4-127   Create the cavity milling process

3) Set examine geometry. After the 【Cavity Mill】 dialog box shown in Figure 4-128 is popped up, click the "Specify Check" button, and select the moved parting surface in Step 1) as the examine geometry.

4) Create a tool. Click the "create new" button in the 【Tool】 group in the 【Cavity Mill】 dialog box, as shown in Figure 4-129, to pop up the 【New Tool】 dialog box. Enter the name of the tool, then click 【OK】 to pop up the 【Milling Tool-5 Parameters】 dialog box. Enter parameters shown in Figure 4-130, then click 【OK】.

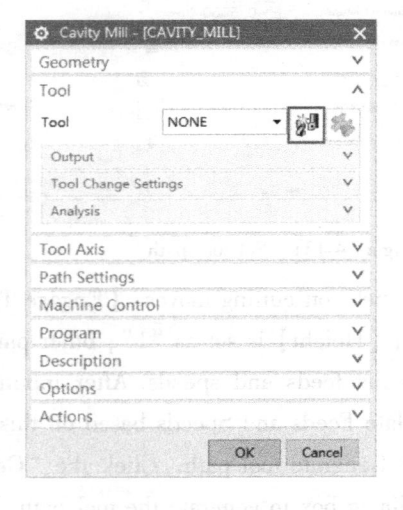

Figure 4-128   【Cavity Mill】 dialog box

Figure 4-129   Create a new tool

163

3D Digital Design and Manufacturing

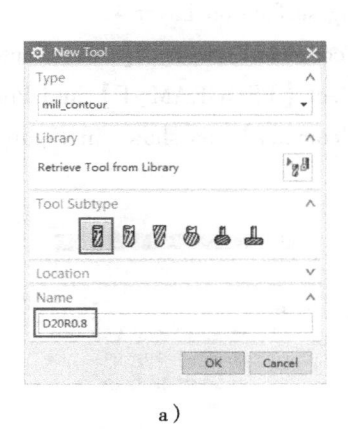
a)

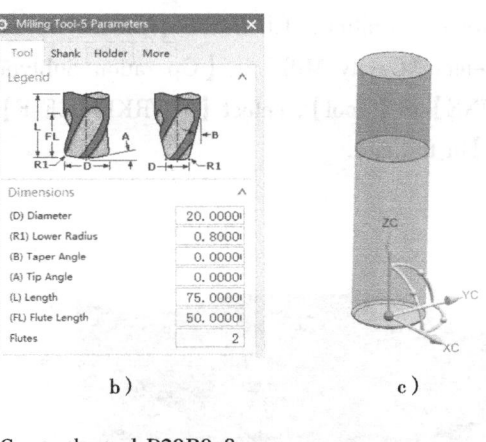
b)

c)

Figure 4-130　Create the tool D20R0.8

5) Set the tool path. 【Cut Pattern】 is set to 【Follow Periphery】; 【Percent of Flat Diameter】 is "70"; 【Maximum Distance】 is "0.5" (for steel parts it is generally 0.3 ~ 0.5), as shown in Figure 4-131.

6) Set cutting parameters. 【Cut Order】 is set to 【Depth First】, 【Island Cleanup】 is checked, and the part side stock is set to "0.3", as shown in Figure 4-132.

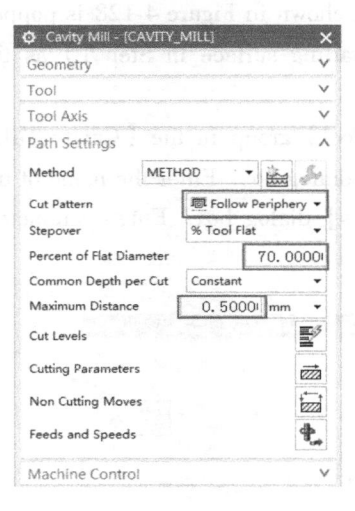

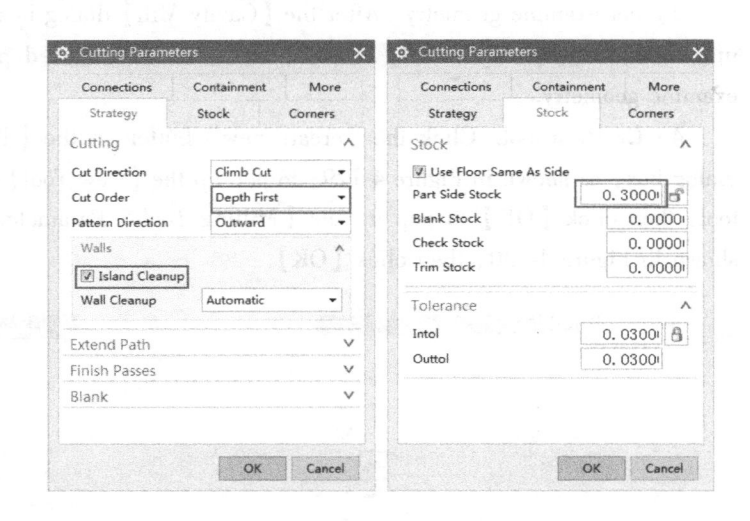

Figure 4-131　Set tool path

Figure 4-132　Set cutting parameters

7) Set non cutting moves. 【Engage Type】 is set as 【Ramp on Shape】; 【Ramp Angle】 is set as "3"; 【Height】 is set as "1"; other parameters are set as shown in Figure 4-133.

8) Set feeds and speeds. After inputting the spindle speed and cutting feed rate, click the "Calculate Feeds and Speeds based on this value" button, as shown in Figure 4-134.

9) Generate tool path. Click the "Generate" button in the 【Actions】 group of the 【Cavity_Mill】 dialog box to generate the tool path, as shown in Figure 4-135. The tool path is beyond the blank range, and the rim that should be left is also milled away, so it needs to be modified.

▶ 164

Project 4　Gamepad

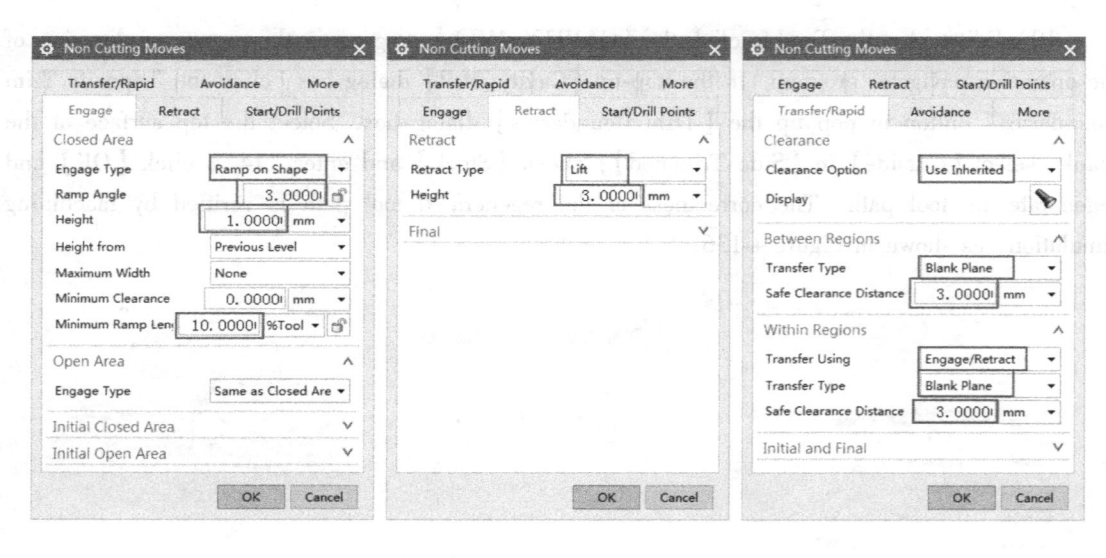

Figure 4-133　Set non cutting moves

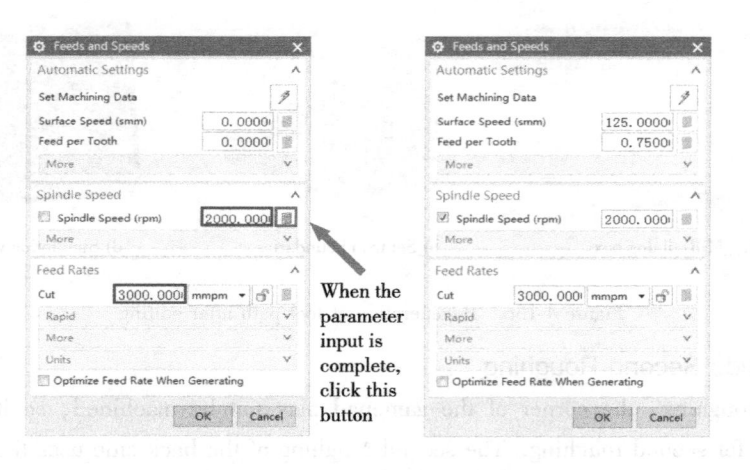

When the parameter input is complete, click this button

Figure 4-134　Set feeds and speeds

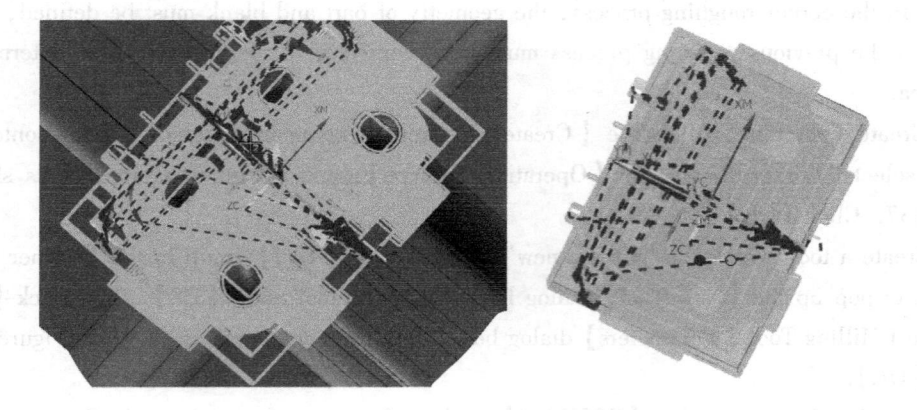

Figure 4-135　Initially generated tool path

165 ◄

3D Digital Design and Manufacturing

10) Edit tool path. Double-click the 【CAVITY_MILL】 program in the program order view of the operation navigator program. In the pop-up 【Cavity_Mill】 dialog box, click the "Specify Trim Boundaries" button to pop up the 【Trim Boundaries】 dialog box. Select the top surface of the blank; select 【Outside】 in 【Side Trimmed】; check 【Stock】 and enter "14"; click 【OK】 and regenerate the tool path. The correctness of the regenerated tool path is verified by machining simulation, as shown in Figure 4-136.

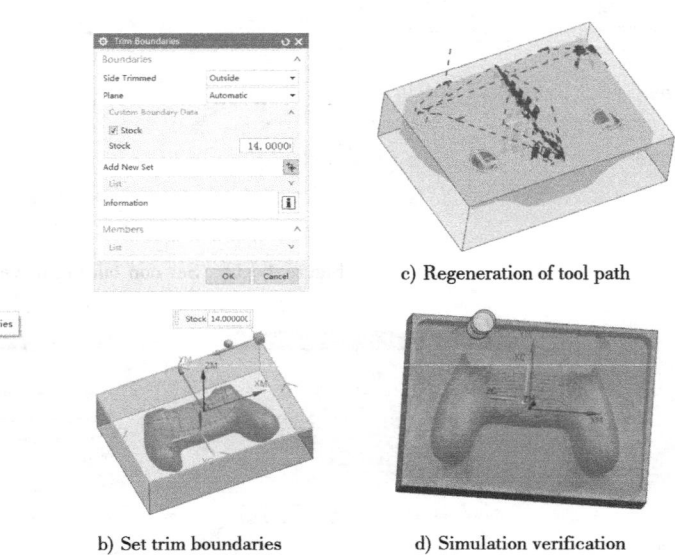

c) Regeneration of tool path

a) (Cavity_Mill) dialog box   b) Set trim boundaries   d) Simulation verification

Figure 4-136　Regeneration of tool path after editing

2. Back Side Second Roughing

After first roughing, the corner of the gamepad may not be machined, so it is necessary to create a process for second roughing. The second roughing of the back side uses the corner roughing process to cut the remaining material that the tool cannot handle in the corner during the first roughing. In the corner roughing process, the geometry of part and blank must be defined, and the tool used in the previous roughing process must be specified as "reference tool" to determine the cutting area.

1) Create operation. Click the 【Create Operation】 command; select 【mill_contour】 in 【Type】; select "Corner Rough" in 【Operation Subtype】; and set other parameters as shown in Figure 4-137. Click 【OK】.

2) Create a tool. Click the "Create new" button in the 【Tool】 group in the 【Corner Rough】 dialog box to pop up the 【New Tool】 dialog box. Enter the tool name 【D8】, then click 【OK】to pop up the 【Milling Tool-5 Parameters】 dialog box. Input the parameters as shown in Figure 4-138, and click【OK】.

3) Set reference tool. Select 【D20R0.8】 as the reference tool, as shown in Figure 4-139.

▶ 166

Project 4　Gamepad

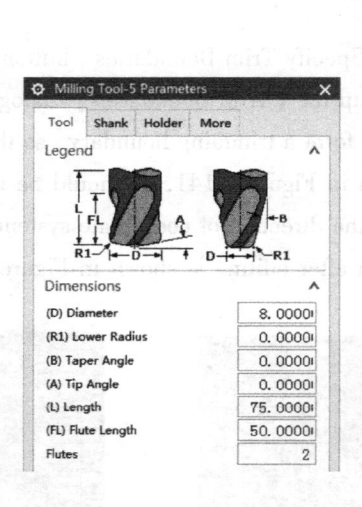

Figure 4-137　Create corner roughing process

Figure 4-138　Create a tool

Figure 4-139　Set reference tool

4) Tool path settings. The angle is set to"45"; the stock is set to"0. 3"; the spindle speed is set to"3000"; the cutting feed rate is set to "2500"; the other parameters are set as shown in Figure 4-140.

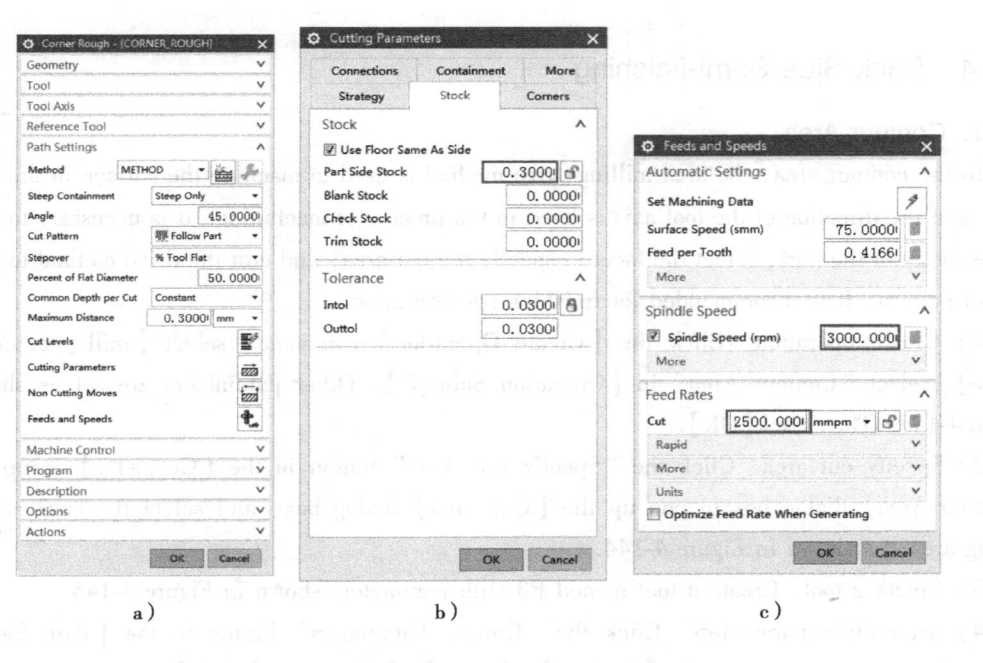

a)　　　　　　　　　　b)　　　　　　　　　　c)

Figure 4-140　Tool path settings

167

3D Digital Design and Manufacturing

5) Generate tool path. Click the "Generate" button in the [Actions] group to generate the tool path shown in Figure 4-141. Some tool paths are not needed after observation, so they need to be trimmed.

6) Edit tool path. Click the "Specify Trim Boundaries" button in the [Geometry] group of the [Corner Rough] dialog box to pop up the [Trim Boundaries] dialog box. Select[Internal] in [Side Trimmed] and click a few points to form a trimming boundary, so that the tool path that needs to be trimmed can be included, as shown in Figure 4-141. It should be noted that the trimmed boundary is created on the XC-YC plane, so the direction of coordinate system should be adjusted according to the situation. Regenerated tool path after editing is shown in Figure 4-142.

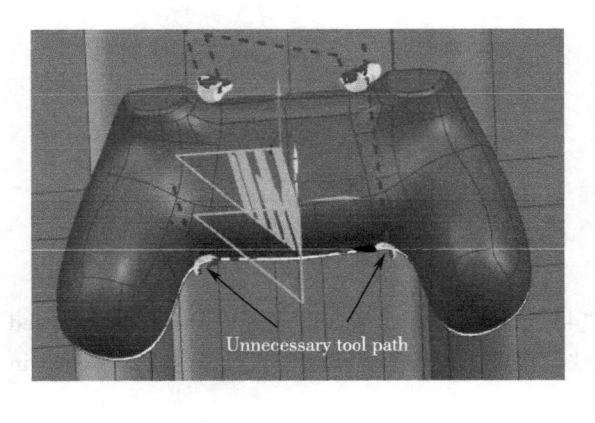 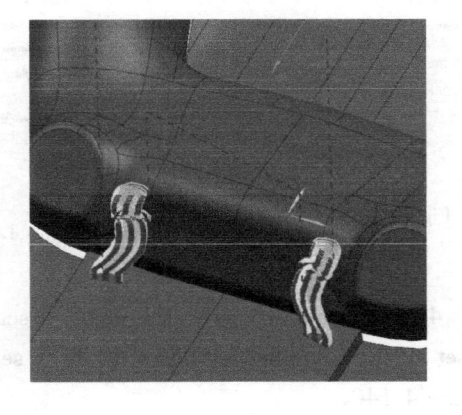

Figure 4-141   The first generation of tool path   Figure 4-142   Regeneration of tool path after editing

4.4.4   Back Side Semi-finishing

1. Contour Area

In the contour area, the area milling drive method is used to machine the surface in the cutting area, and the direction of the tool axis is fixed in the process of machining. It is necessary to specify the geometry of the part, select the facet to specify the cut area, and edit the drive method to specify the cut pattern. It is recommended for finishing specific areas.

1) Create operation. Click the [Create Operation] command; select [mill_contour] in [Type]; select "Contour Area" in [Operation Subtype]. Other parameters are set as shown in Figure 4-143. Then click[OK].

2) Specify cut area. Click the "Specify Cut Area" button in the [Geometry] group of the [Contour Area] dialog box to pop up the [Cut Area] dialog box, and select the surface as the cutting area, as shown in Figure 4-144.

3) Create a tool. Create a tool named R3 with parameters shown in Figure 4-145.

4) Set cutting parameters. Click the "Cutting Parameters" button in the [Path Settings] group, then set parameters in the [Strategy], [Stock], [Clearances] and [More] tabs, as shown in Figure 4-146.

▶ 168

Project 4　Gamepad

Figure 4-143　Create contour area process

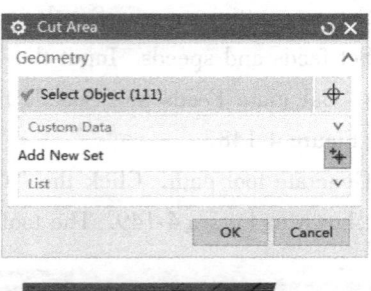

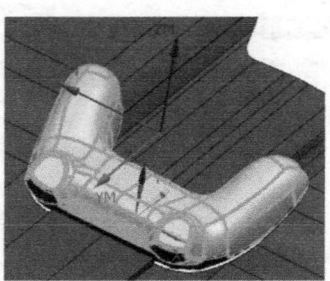

Figure 4-144　Specify cut area

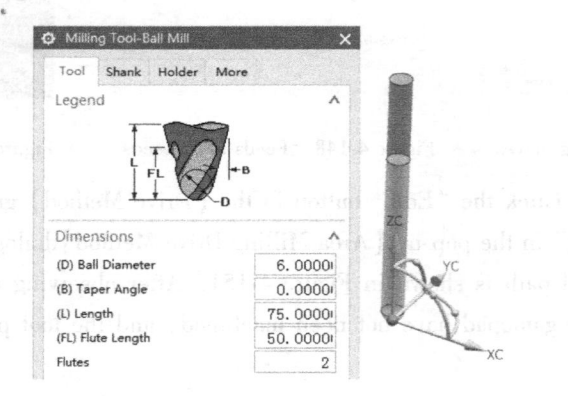

Figure 4-145　Create a tool with a ball diameter of 6mm

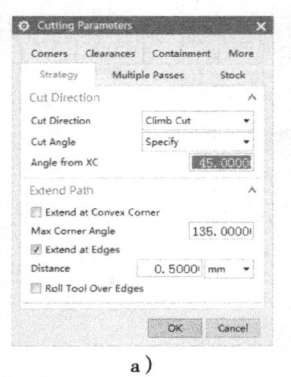

a)

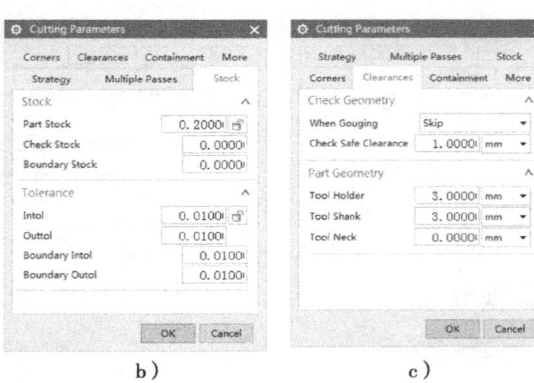

b)

c)

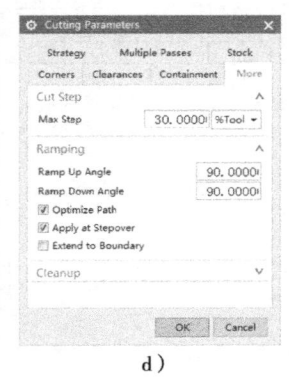

d)

Figure 4-146　Set cutting parameters

169

3D Digital Design and Manufacturing

5) Set non cutting moves. Set the engage radius to "20", as shown in Figure 4-147.

6) Set feeds and speeds. Input the spindle speed as "6500" and cutting feed rate as "1500". Click the "Calculate Feeds and Speeds based on this value" button behind the spindle speed, as shown in Figure 4-148.

7) Generate tool path. Click the "Generate" button in the 【Actions】 group to generate tool path, as shown in Figure 4-149. The tool path will not penetrate the parting surface after checking.

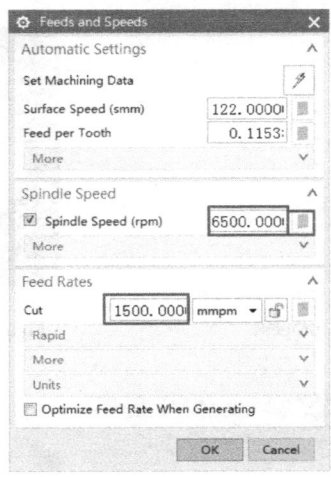 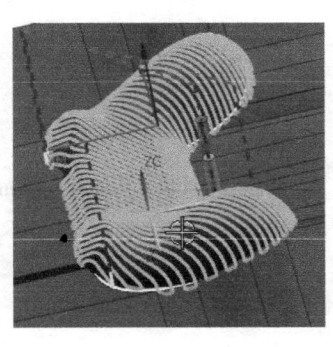

Figure 4-147　Non cutting moves　　　Figure 4-148　Feeds and speeds　　　Figure 4-149　Observe tool path

8) Edit tool path. Click the "Edit" button in the 【Drive Method】 group, and the 【Maximum Distance】 is set to "0.3" in the pop-up 【Area Milling Drive Method】 dialog box, as shown in Figure 4-150. Regenerated tool path is shown in Figure 4-151. After observing the tool path, it is found that two positions of the gamepad have not been machined, and the tool path needs to be added to machine it.

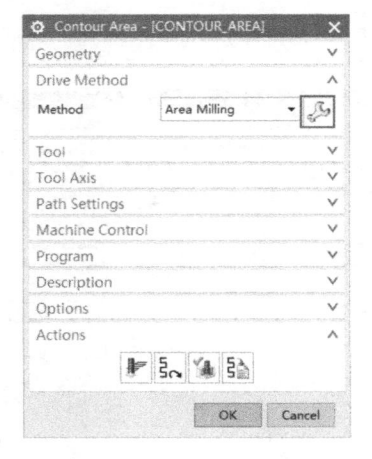 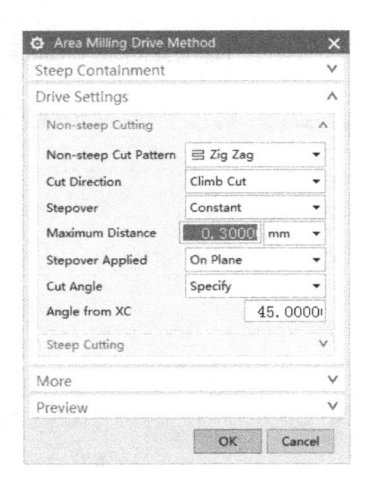

Figure 4-150　Area milling drive method

▶ 170

Project 4　Gamepad

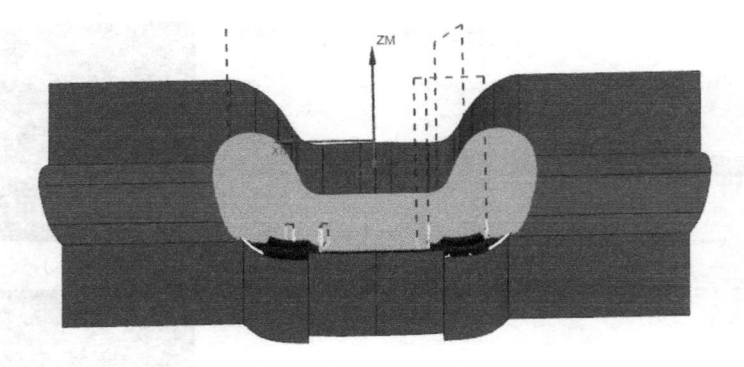

Figure 4-151　Observe the regenerated tool path

2. Streamline

Streamline refers to the contour milling process of fixed axis surface using flow curve and cross curve to guide cut pattern and follow the shape of drive geometry. When using streamline, the flow and direction of smoothing cut pattern should be controlled.

1) Create operation. Click the 【Create Operation】 command; select 【mill＿contour】 in 【Type】; select "Streamline"in 【Operation Subtype】; and set other parameters as shown in Figure 4-152. Click 【OK】in the end.

2) Specify cut area. Click the "Specify Cut Area" button in the 【Geometry】 group of the 【Streamline】dialog box, and select Area 1, as shown in Figure 4-153.

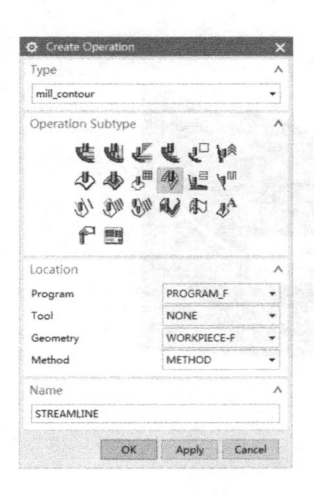

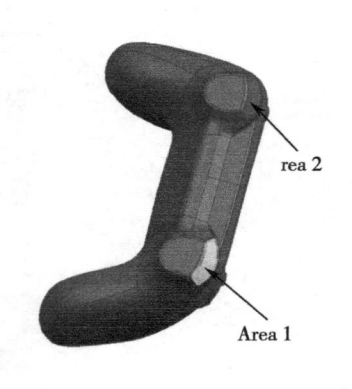

Figure 4-152　Create streamline processes　　　　　　Figure 4-153 Set cut area

3) Set streamline drive method. Click the "Edit" button in the 【Drive Method】 group, and set 【Automatic】 in 【Selection Method】. Click "Specify Cut Direction" button, and select one of the arrows in the circle shown in Figure 4-154c, indicating that the cut pattern is from top to bottom. 【Tool Position】 is set as 【Tanto】; 【Cut Pattern】 is modified to 【Zig Zag】; and the number of stepovers is set to "30". Modify 【Cut Step】 to 【Tolerances】, as shown in Figure 4-154.

171

3D Digital Design and Manufacturing

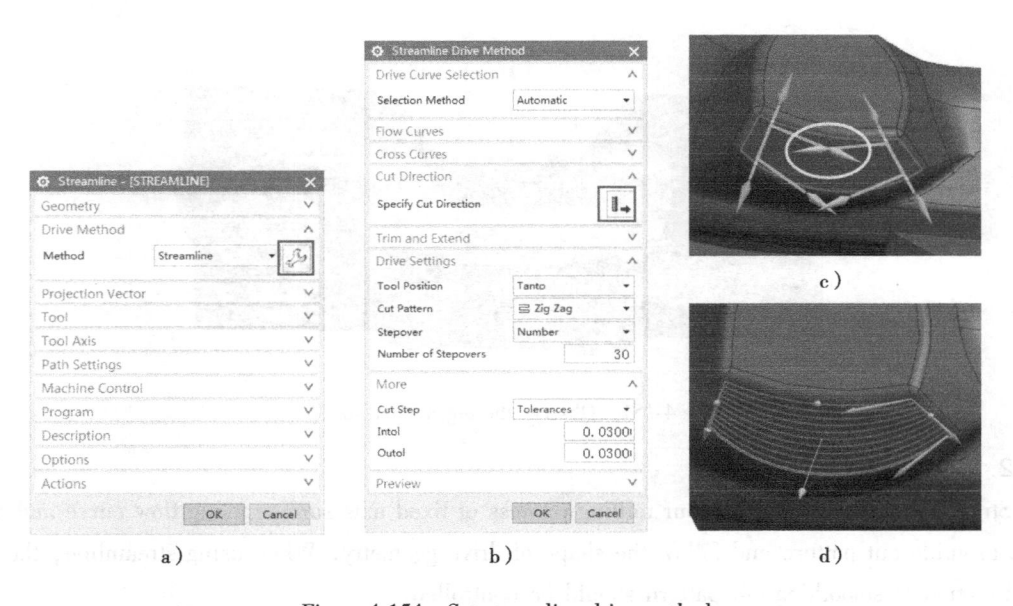

Figure 4-154　Set streamline drive method

4）Set projection vector. Select【Normal to Driver】in【Vector】, as shown in Figure 4-155.

5）Set tool. Select ball-end milling cutter【R3】in【Tool】, as shown in Figure 4-155.

6）Set tool axis. Select【Dynamic】in【Axis】, and tilt the coordinate axis by 5°, as shown in Figure 4-155.

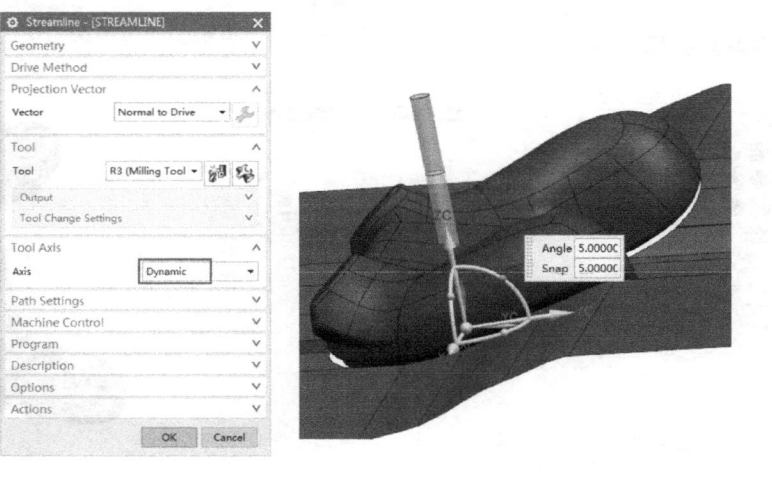

Figure 4-155　Set projection vector and tool axis

7）Set cutting parameters. Click the "Cutting Parameters" button in the【Path Settings】group; input【Part Stock】to "0.2"; input【Intol】and【Outtol】to "0.01"; set【When Gouging】to【Skip】; check【Optimize Path】, as shown in Figure 4-156.

8）Set feeds and speeds. Input the spindle speed as "6500" and cutting feed rate as "1500". Then click the "Calculate Feeds and Speeds based on this value" button behind the spindle speed,

▶ 172

Project 4   Gamepad

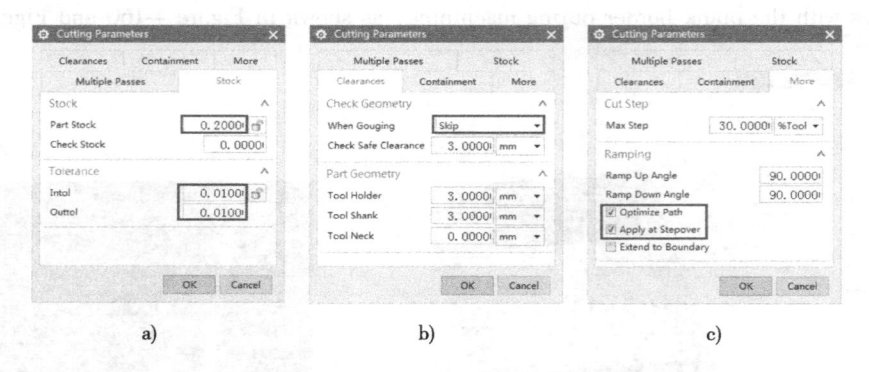

a)   b)   c)

Figure 4-156   Set cutting parameters

as shown in Figure 4-157.

9) Generate tool path. Click the "Generate" button in the 【Actions】 group to generate the tool path, as shown in Figure 4-158.

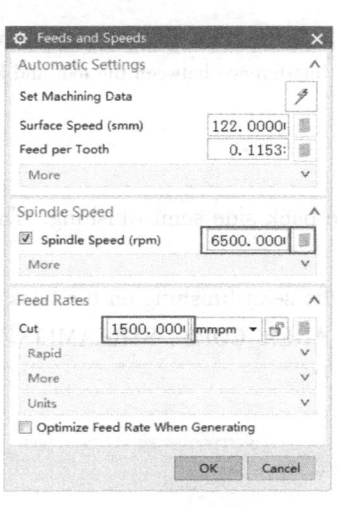

Figure 4-157   Set feeds and speeds

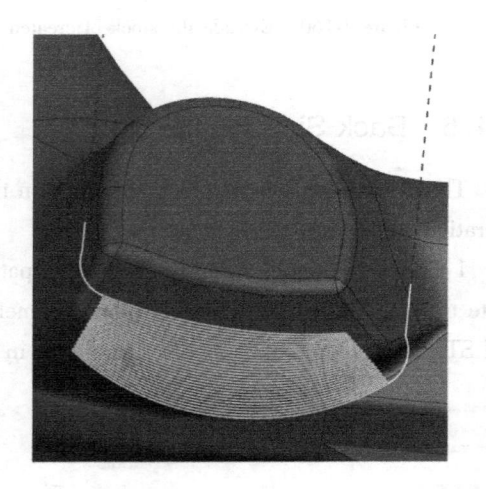

Figure 4-158   Generate tool path

10) Copy tool path. Copy the newly created streamline tool path, and place the cutting area from Area 1 shown in Figure 4-153 to Area 2 to check whether other parameters are set correctly, especially the cutting direction driven by streamline. Click the "Generate" button in the 【Actions】 group to regenerate the tool path. The tool path for two streamline programs is created as shown in Figure 4-159.

11) Interference check. Switch to the 【Modeling】 module and use the 【Extrude】 command to create a border for simulating the blank stock, in order to check whether the

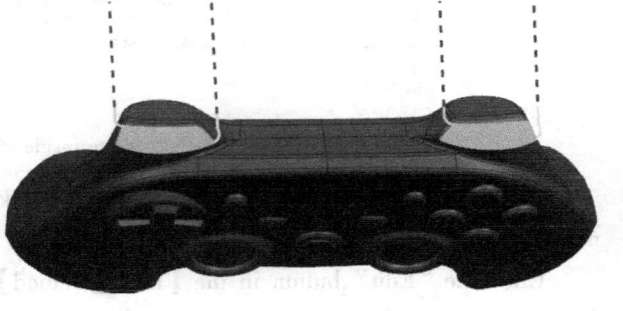

Figure 4-159   Tool path of two streamline programs

173

3D Digital Design and Manufacturing

tool interferes with the blank border during machining, as shown in Figure 4-160 and Figure 4-161.

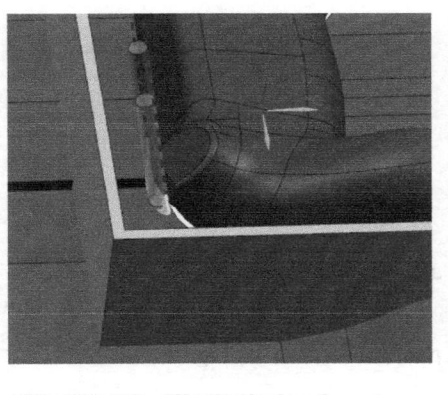

Figure 4-160   Extrude the stock of created blank

Figure 4-161   Check whether there is an interference between the tool and blank stock

4.4.5 Back Side Finishing

The back side finishing is carried out on the basis of the back side semi-finishing. The specific operation steps are described below.

1) Copy tool path. Copy the three tool paths created in the semi-finishing on the back side, and paste them inside to get three tool paths, namely CONTOUR_AREA_COPY, STREAMLINE_COPY_1 and STREAMLINE_COPY_COPY, as shown in Figure 4-162.

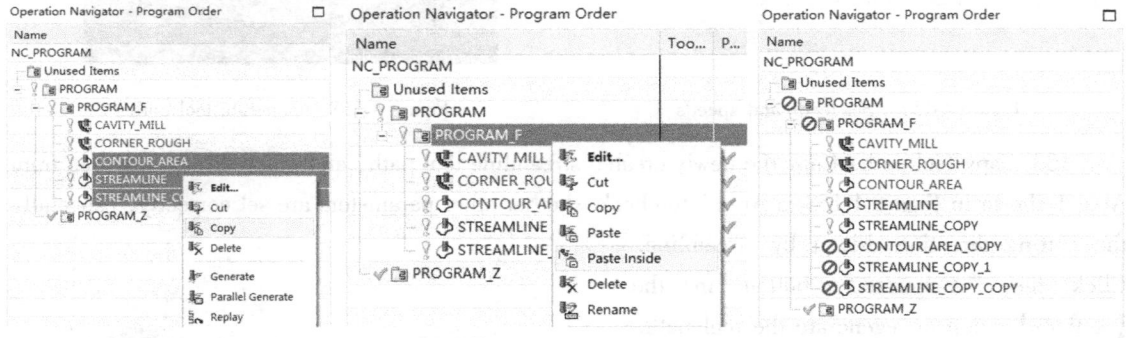

Figure 4-162   Paste inside after copying tool path

2) Modify contour area tool path. Double-click CONTOUR _ AREA _ COPY in the operation navigator.

Click the "Edit" button in the 【Drive Method】group to modify the 【Maximum Distance】, 【Stepover Applied】and 【Angle from XC】, as shown in Figure 4-163.

In the 【Path Settings】group, set the 【Part Stock】to "0"; set the spindle speed to "6500" and

▶ 174

Project 4　Gamepad

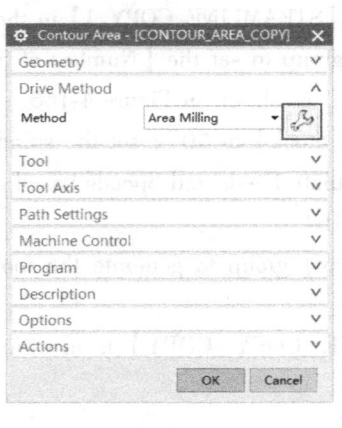

Figure 4-163　Area milling drive method

cutting feed rate to "1000"; click the "Calculate Feeds and Speeds based on this value" button behind the spindle speed, as shown in Figure 4-164.

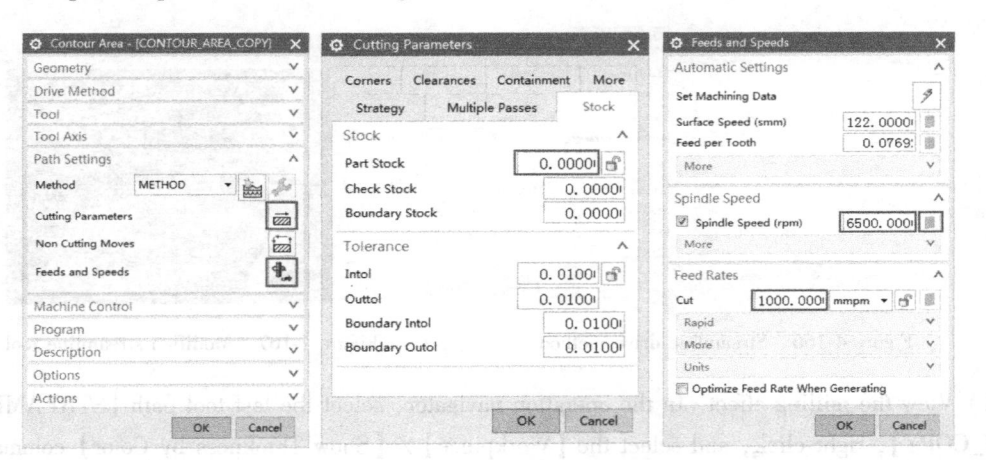

Figure 4-164　Stock and cutting feed rate settings

Click the "Generate" button in the 【Actions】 group to generate the tool path, as shown in Figure 4-165.

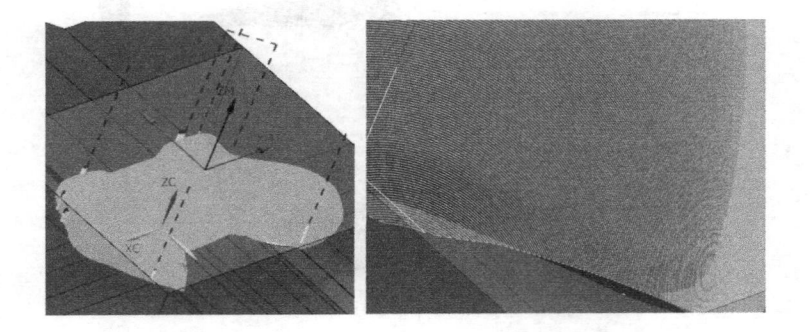

Figure 4-165　Generate tool path

175

3D Digital Design and Manufacturing

3) Modify streamline tool path. Double-click 【STEAMLINE_COPY_1】 in the operation navigator. Click the "Edit" button in the 【Drive Method】 group to set the 【Number of Stepovers】 to "50", and set both 【Intol】 and the 【Outol】 to "0.01", as shown in Figure 4-166.

In the 【Path Settings】 group, set the 【Part Stock】 to "0"; set the spindle speed to "6500" and cutting feed rate to "1000"; click the "Calculate Feeds and Speeds based on this value" button behind the spindle speed, as shown in Figure 4-164.

Click the "Generate" button in the 【Actions】 group to generate the tool path, as shown in Figure 4-167.

Another streamline tool path 【STREAMLINE_COPY_COPY】 is modified in the same way, which will not be repeated here.

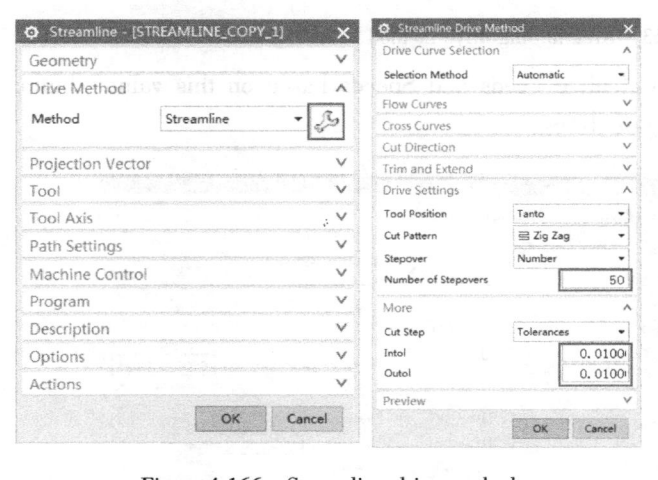

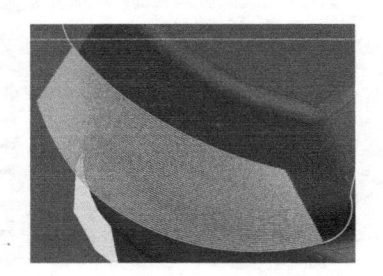

Figure 4-166　Streamline drive method　　　　Figure 4-167　Modified streamline tool path

4) View the milling effect. In the operation navigator, select the last tool path 【STREAMLINE_COPY_COPY】, right-click, and select the 【Workpiece】 / 【Show Thickness by Color】 command to view the milling effect, as shown in Figure 4-168.

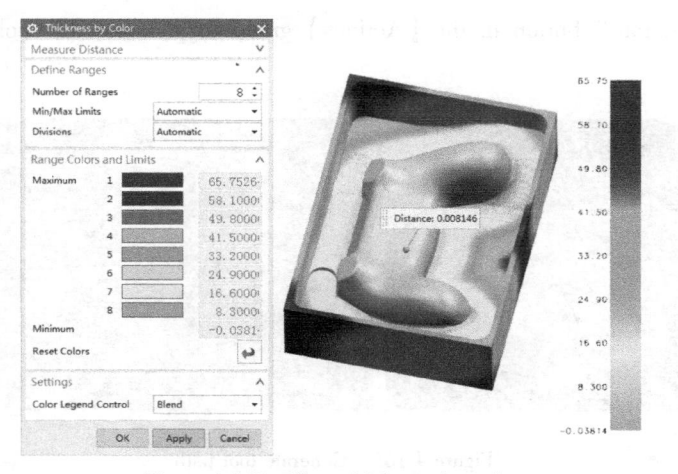

Figure 4-168　Show thickness by color

▶ 176

Project 4　Gamepad

4.4.6　Front Side Geometry Setting

1）Set MCS. Switch to the geometric view of the operation navigator. Click the【Create Geometry】command; select "MCS" for【Geometry Subtype】; and enter the name【MCS-Z】, as shown in Figure 4-169. Click【OK】in the end.

In the pop-up【MCS】dialog box, select【Plane】in【Clearance Option】; select the top surface of the blank and offset it 50 mm upward to specify it as the safety plane, as shown in Figure 4-170.

Click the "CSYS Dialog" button 📐 in the【Machine Coordinate System】group; select the center of the top surface of the blank as the coordinate origin; adjust the direction of the coordinate axis to make ZM axis upward; and align the Axis XM with the long edge of the blank.

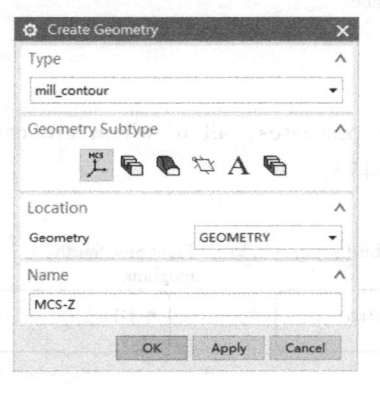

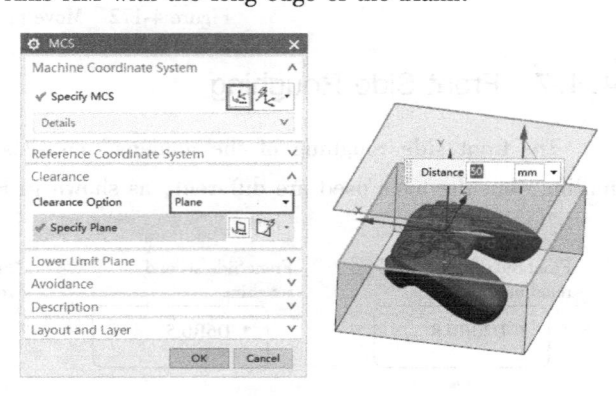

Figure 4-169　Create geometry

Figure 4-170　Set safety plane

2）Set workpiece. Click the【Create Geometry】command; select "WORKPIECE" in【Geometry Subtype】; and enter the name【WORKPIECE-Z】, as shown in Figure 4-171. Click【OK】to pop up the【Workpiece】dialog box. Select the gamepad model as the【Specify Part】; select the square as the【Specify Blank】, and click【OK】.

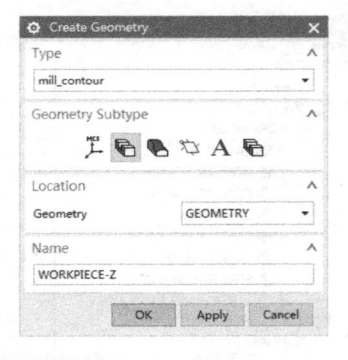

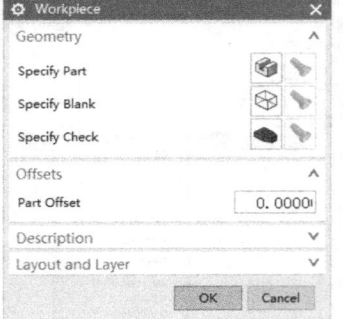

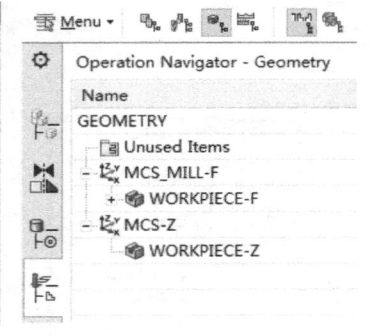

Figure 4-171　Set workpiece

3）Move parting surface. Open the initial parting surface in Layer 3; use the【Move Object】command; select【Copy Original】to move the parting surface downward (in the opposite direction of

177

3D Digital Design and Manufacturing

the gamepad) by 3.3mm, as shown in Figure 4-172. Then the 【Move to Layer】 command is used to place the moved process parting surface on Layer 5.

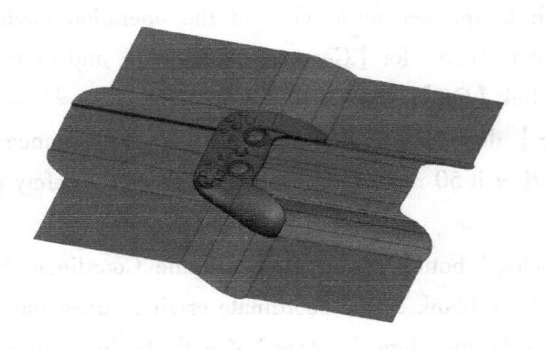

Figure 4-172   Move parting surface

4.4.7   Front Side Roughing

The front side roughing of the gamepad consists of four processes, all of which are cavity milling, but the tools used are different, as shown in Figure 4-173.

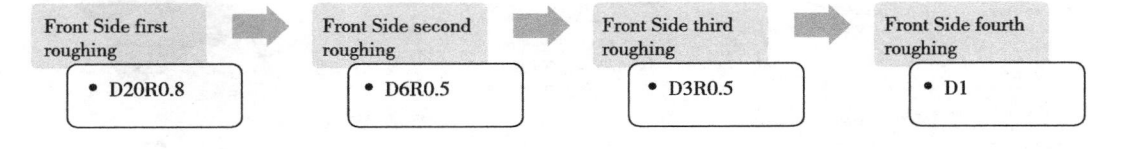

Figure 4-173   Front side roughing process

1. Front Side First Roughing

1) Copy tool path. Copy the back side roughing tool path【CAVITY_MILL】, then paste it into 【WORKPIECE-Z】 by 【Paste Inside】 command, and get the tool path 【CAVITY_MILL_COPY】, as shown in Figure 4-174.

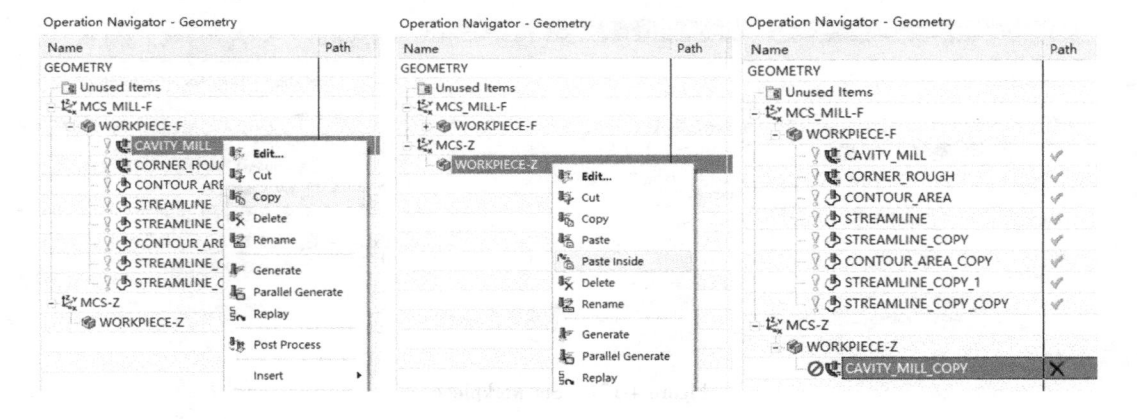

Figure 4-174   Copy the roughing tool path and then paste it inside

▶ 178

2) Edit tool path. Double-click【CAVITY_MILL_COPY】to pop up the【Cavity Mill】dialog box.

Click the "Specify Check" button to select the moved parting surface as the examine geometry. Click the "Specify Trim Boundaries" button to cancel the selection of【Stock】in the pop-up【Trim Boundaries】dialog box, as shown in Figure 4-175.

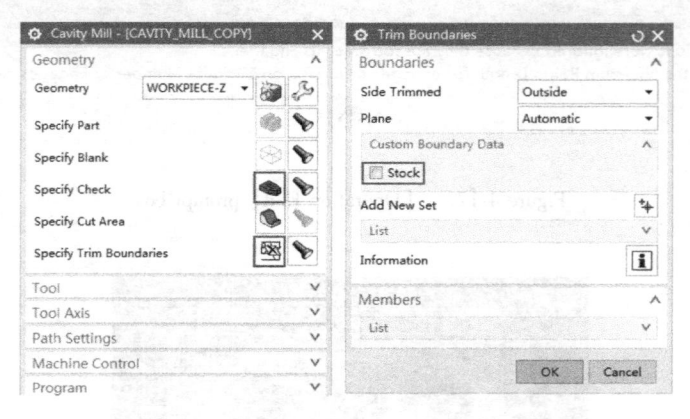

Figure 4-175　Set geometry

Click the "Cutting Parameters" button, then set the【Pattern Direction】to【Inward】, as shown in Figure 4-176.

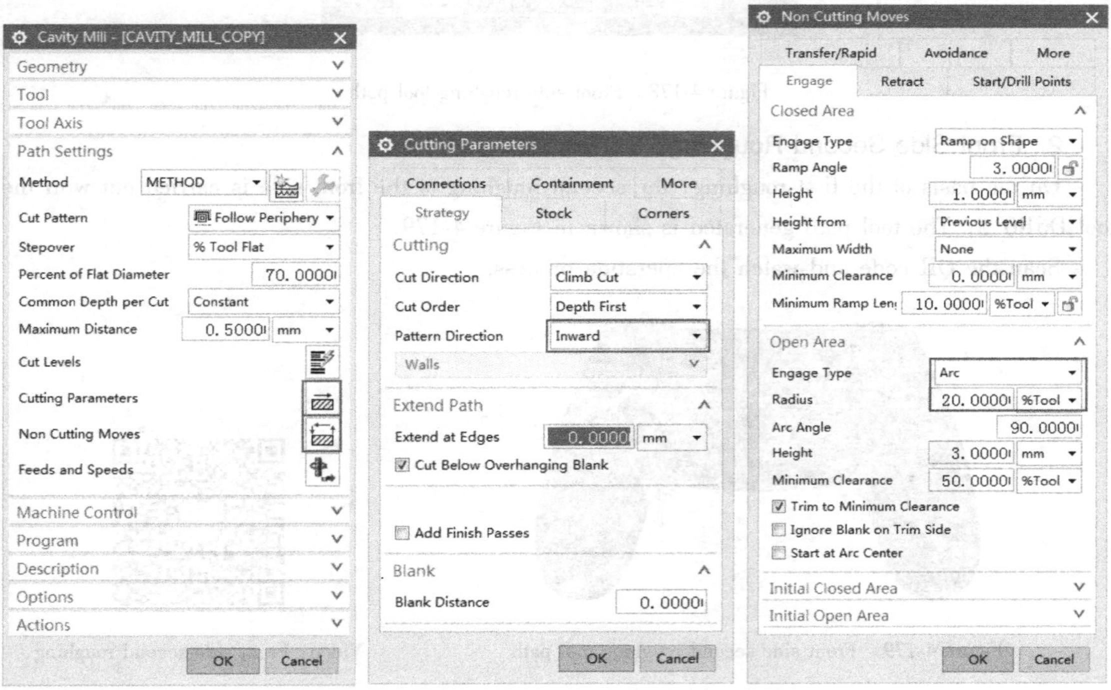

Figure 4-176　Cutting parameters

3D Digital Design and Manufacturing

Click the "Non Cutting Moves" button; Set the 【Engage Type】in open area to【Arc】and enter the 【Radius】 to 20% of the tool diameter, as shown in Figure 4-176.

3) Generate tool path. Click the"Generate" button in the 【Actions】group to pop up the prompt box, as shown in Figure 4-177. Click 【OK】, and generate the tool path shown in Figure 4-178.

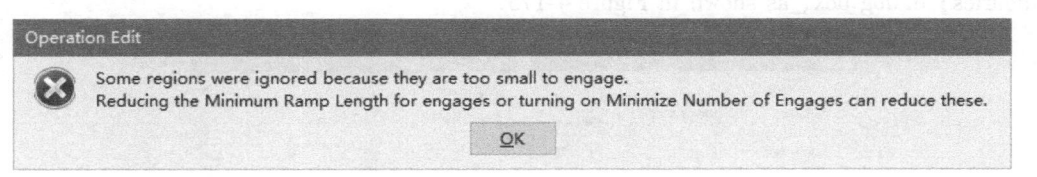

Figure 4-177　　【Operation Edit】prompt box

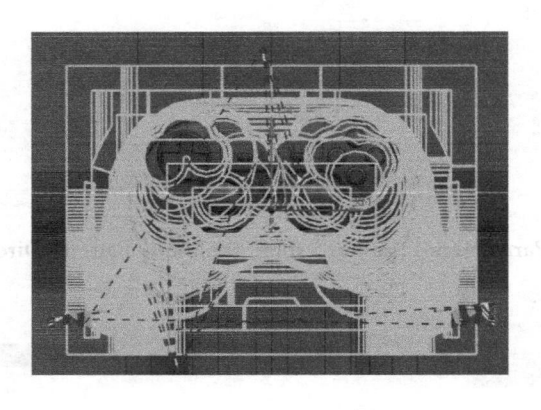

Figure 4-178　　Front side roughing tool path

2. Front Side Second Roughing

On the basis of the first roughing, the second roughing of the front side is carried out with the tool D6R0.5. The tool path generated is shown in Figure 4-179.

Scan the QR code and watch the operation process.

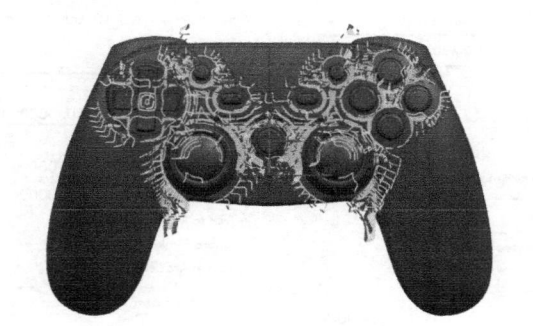

Figure 4-179　　Front side second roughing tool path　　　　　Video: Front side second roughing

3. Front Side Third Roughing

On the basis of the second roughing, the third roughing of the front side is carried out with the

▶ 180

tool D3R0.5. The tool path generated is shown in Figure 4-180.

Scan the QR code and watch the operation process.

Figure 4-180   Front side third roughing tool path   Video: Front side third roughing

4. Front Side Fourth Roughing

On the basis of the third roughing, the fourth roughing of the front side is carried out with the tool D1. The tool path generated is shown in Figure 4-181.

Scan the QR code and watch the operation process.

Figure 4-181   Front side fourth roughing tool path   Video: Front side fourth roughing

4.4.8   Front Side Semi-finishing

1) Copy tool path. Copy the back side semi-finishing tool path 【CONTOUR_AREA】, then paste it into 【WORKPIECE-Z】 by the【Paste Inside】command and get the tool path 【CONTOUR_ AREA_COPY_1】.

2) Specify check. Click the "Specify Check" button, then select the moved parting surface as the examine geometry.

3) Select the cut area. Click the "Specify Cut Area" button, then specify the cut area again, as shown in Figure 4-182.

4) Generate tool path. Click the "Generate" button in the 【Actions】group to generate the tool path, as shown in Figure 4-183.

3D Digital Design and Manufacturing

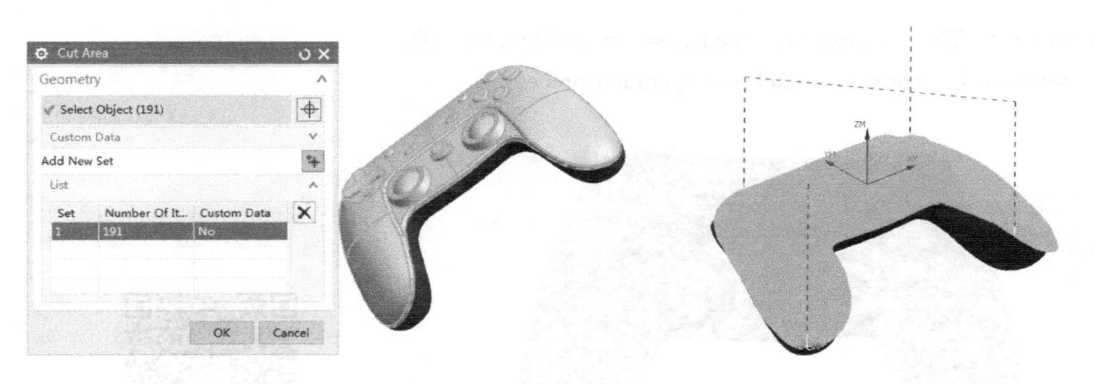

Figure 4-182   Specify cut area   Figure 4-183   Front side semi-finishing tool path

4.4.9   Front Side Finishing

The process of the front side finishing of the gamepad is shown in Figure 4-184.

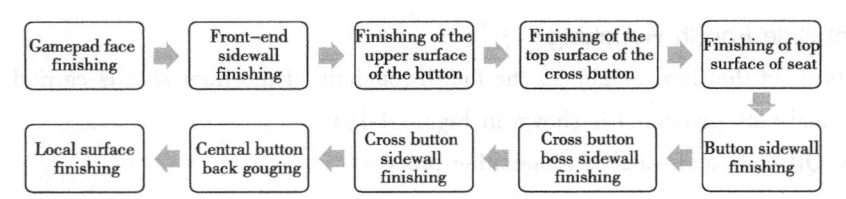

Figure 4-184   The process of the front side finishing of the gamepad

1. Finishing of the Gamepad Face

1) Create operation. Click the 【Create Operation】command; select 【Mill_Contour】in 【Type】; select "Contour Area" in 【Operation Subtype】; set other parameters as shown in Figure 4-185. Click 【OK】to pop up the【Contour Area】dialog box.

2) Specify cut area. Click the "Specify Cut Area" button to specify the cut area shown in Figure 4-186.

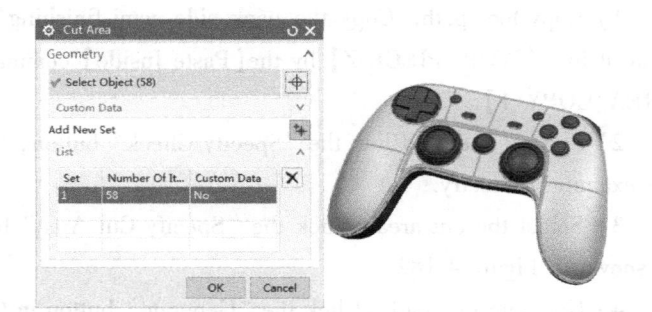

Figure 4-185   Create operation   Figure 4-186   Specify cut area

▶ 182

Project 4 Gamepad

3) Set drive method. Click the "Edit" button in the 【Drive Method】group; set【Stepover】to 【Constant】; set 【Maximum Distance】to "0.12"; set the rest of the parameters as shown in Figure 4-187.

4) Set cutting parameters. Click the "Cutting Parameters" button in the 【Path Settings】 group; set【Part Stock】to "0"; set both 【Intol】and 【Outtol】to "0.01"; check【Optimize Path】, as shown in Figure 4-188.

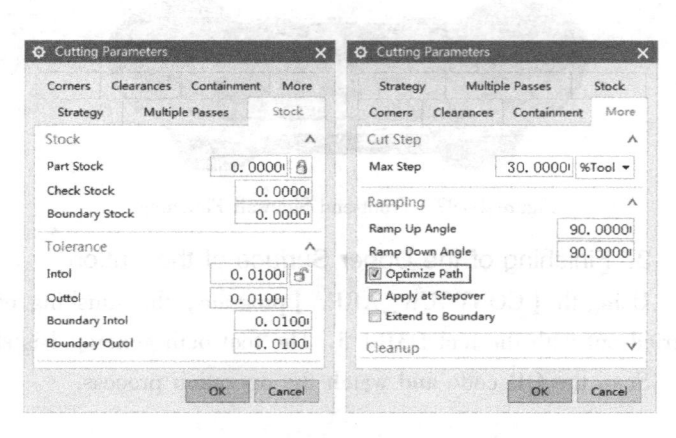

Figure 4-187　Set drive method　　　　Figure 4-188　Set cutting parameters

5) Set non cutting moves. Click the "Non Cutting Moves" button in the 【Path Settings】group to set the 【Radius】is 10% of the tool diameter, as shown in Figure 4-189.

6) Set feeds and speeds. Click the "Feeds and Speeds" button in the 【Path Settings】group. Set the spindle speed to "8000" and the cutting feed rate to "1000", and click the "Calculate Feeds and Speeds based on this value" button behind the 【Spindle Speed】input box, as shown in Figure 4-190.

7) Generate tool path. Click the "Generate" button in the 【Actions】group to generate tool path, as shown in Figure 4-191.

 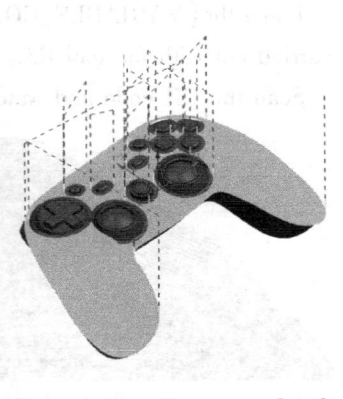

Figure 4-189　Set non cutting moves　Figure 4-190　Set feeds and speeds　Figure 4-191　Generate tool path

183

3D Digital Design and Manufacturing

2. Front-end Sidewall Finishing

Using the【ZLEVEL_PROFILE】program, the finishing of the front-end sidewall is carried out with the tool D6R0.5. The tool path generated is shown in Figure 4-192.

Scan the QR code and watch the operation process.

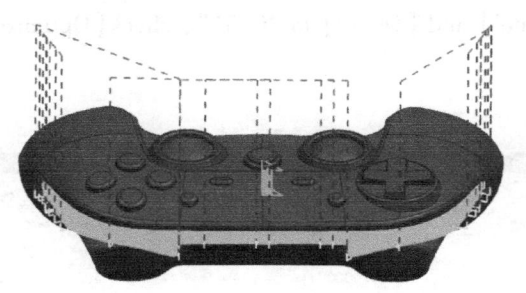

Figure 4-192　Front-end Sidewall Finishing　　　　Video: Front-end sidewall finishing

3. Finishing of the Upper Surface of the Button

Using the【CONTOUR_AREA】program, the finishing of the upper surface of the button is carried out with the tool D3R0.5. The tool path generated is shown in Figure 4-193.

Scan the QR code and watch the operation process.

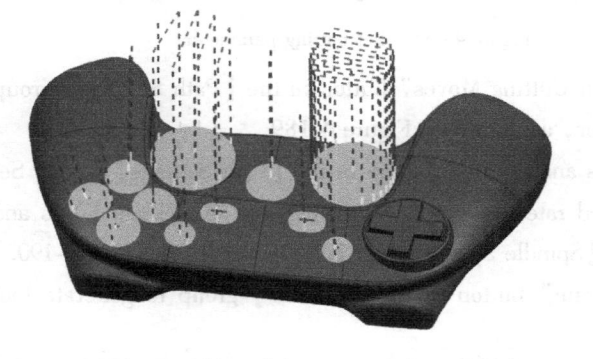

Figure 4-193　Finishing of the upper surface of the button　　Video: Finishing of the upper surface of the button

4. Finishing of the Top Surface of the Cross Button

Using the【VARIABLE_CONTOUR】program, the finishing of the top surface of the cross button is carried out with the tool R3. The tool path generated is shown in Figure 4-194.

Scan the QR code and watch the operation process.

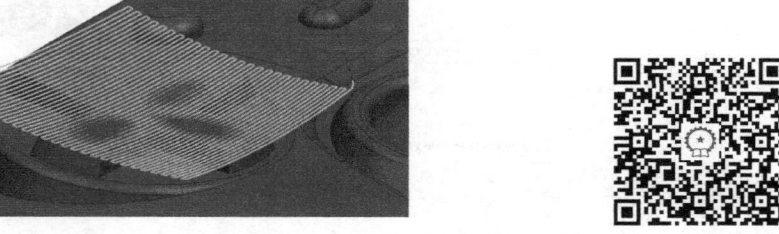

Figure 4-194　Finishing of the top surface of the cross button　　Video: Finishing of the top surface of the cross button

Project 4　Gamepad

5. Finishing of the Top Surface of the Seat

Using the【CONTOUR_AREA】program, the finishing of the top surface of the seat is carried out with the tool D6R0.5. The tool path generated is shown in Figure 4-195.

Scan the QR code and watch the operation process.

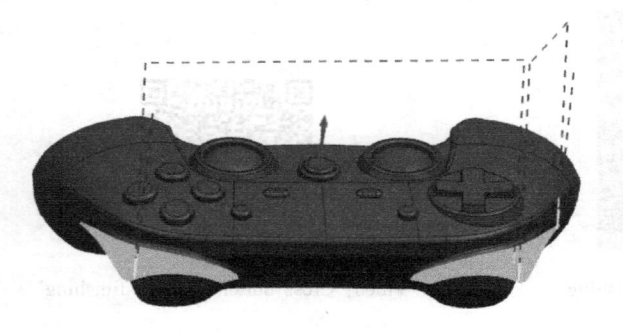

Figure 4-195　Finishing of top surface of seat　　　　Video: Finishing of the top surface of the seat

6. Button Sidewall Finishing

Using the【VARIABLE_STREAMLINE】program, the finishing of the button sidewall is carried out with the tool D2. The tool path generated is shown in Figure 4-196.

Scan the QR code·and watch the operation process.

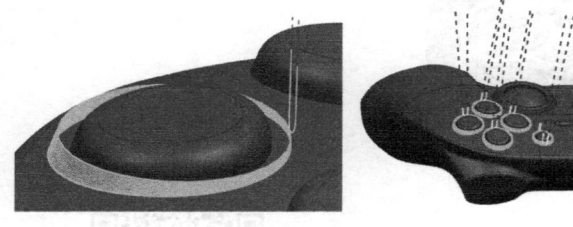

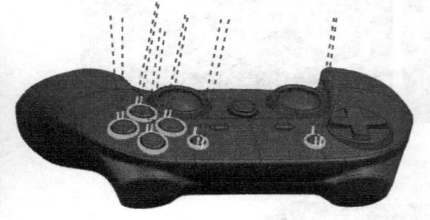

Figure 4-196　Button sidewall finishing　　　　Video: Button sidewall finishing

7. Cross Button Boss Sidewall Finishing

Using the【CONTOUR_PROFILE】program, the finishing of the cross button boss sidewall is carried out with the tool D8. The tool path generated is shown in Figure 4-197.

Scan the QR code and watch the operation process.

Figure 4-197　Cross button boss sidewall Finishing　　　　Video: Cross button boss sidewall finishing

185 ◀

3D Digital Design and Manufacturing

8. Cross Button Sidewall Finishing

Using the【CONTOUR_PROFILE】program, the finishing of the cross button sidewall is carried out with the tool D2. The tool path generated is shown in Figure 4-198.

Scan the QR code and watch the operation process.

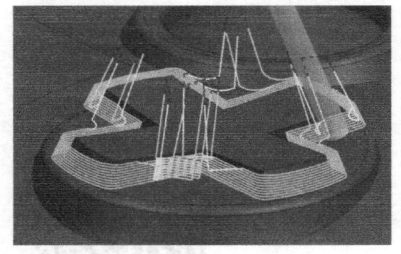

Figure 4-198　Cross button sidewall finishing　　　　Video: Cross button sidewall finishing

9. Central Button Back Gouging

Using the【FLOWCUT_REF_TOOL】program, the back gouging of the central button is carried out for three times with the tool R2, R1 and R0. 5. The tool path generated is shown in Figure 4-199.

Scan the QR code and watch the operation process.

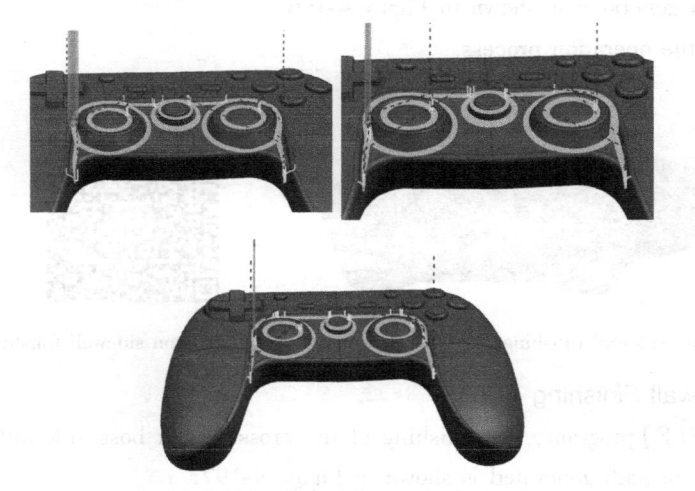

Figure 4-199　Central button back gouging　　　　Video: Central button back gouging

10. Local Surface Finishing

1) Check the milling effect. In the operation navigator, select the 【FLOWCUT_REF_TOOL_COPY_COPY】. Right-click and select the 【Workpiece】/ 【Show Thickness by Color】 command. After inspection, it is found that the wall faces of the cross button and the side of the two large circular buttons still have an stock of about 0. 2 mm, as shown in Figure 4-200.

2) Create operation. Click the 【Create Operation】; select 【mill_contour】 in 【Type】; select "Contour Area" in 【Operation Subtype】; other parameters are set as shown in Figure 4-201. Click 【OK】to pop up the【Contour Area】dialog box.

3) Specify cut area. Click the "Specify Cut Area" button in the 【Geometry】 group to specify the cut area shown in Figure 4-202.

▶ 186

Project 4 Gamepad

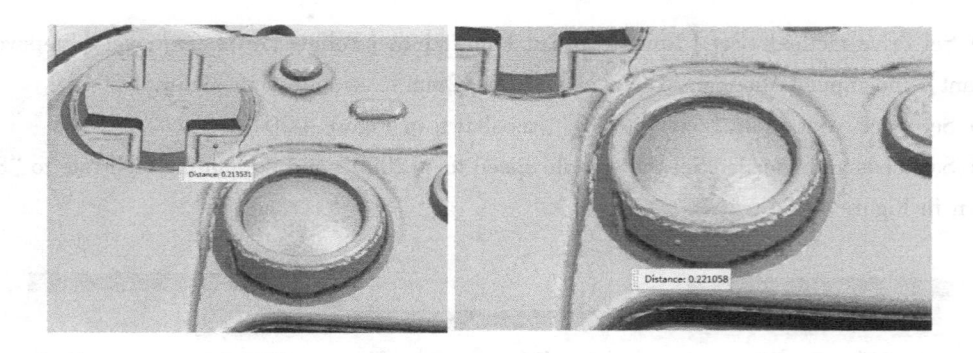

Figure 4-200 Check the milling effect

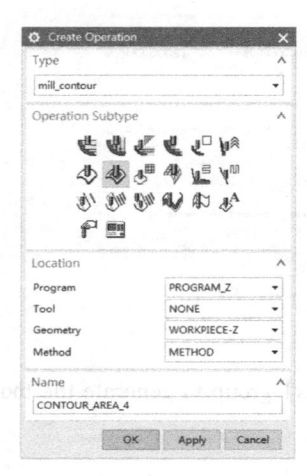

Figure 4-201 Create operation

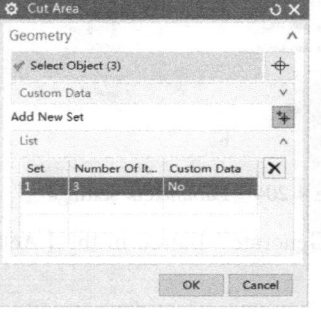 

Figure 4-202 Specify cut area

4) Specify trim boundaries. Click the "Specify Trim Boundaries" button in the 【Geometry】 group. Set【Side Trimmed】to【Outside】to represent trimming the tool path outside the trimming boundary. The trimming boundary at the cross button is specified by【Curve】, and the trimming boundary at the two big buttons is specified by【Points】, as shown in Figure 4-203.

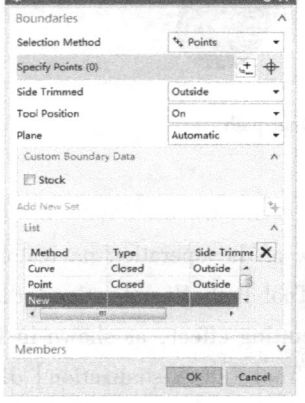 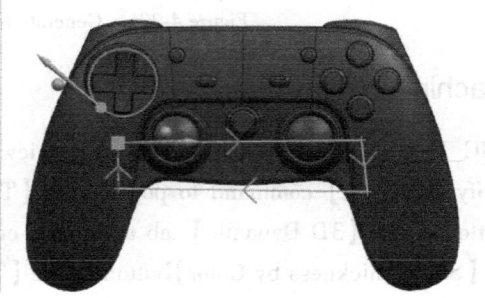

Figure 4-203 Specify trim boundaries

187

3D Digital Design and Manufacturing

5) Set drive method. Set 【Non-steep Cut Pattern】 to 【Follow Periphery】; set 【Stepover】 to 【Constant】 and input 【Maximum Distance】 to "0. 12mm", as shown in Figure 4-204a.

6) Set stock. Set 【Part Stock】 to "0", as shown in Figure 4-204b.

7) Set feeds and speeds. Set the spindle speed to "12000" and the cutting feed rate to "800", as shown in Figure 4-204c.

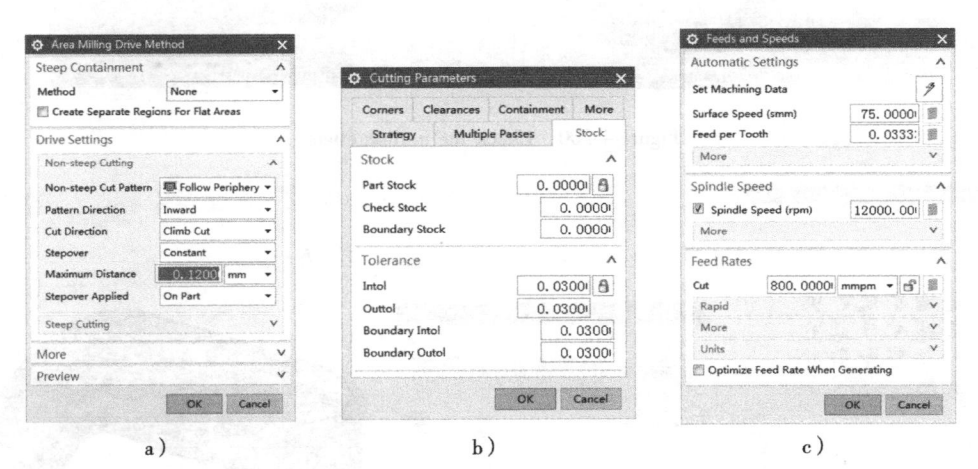

a)　　　　　　　　　　b)　　　　　　　　　　c)

Figure 4-204　Parameters settings

8) Generate tool path. Click the "Generate" button in the 【Actions】 group to generate the tool path, as shown in Figure 4-205.

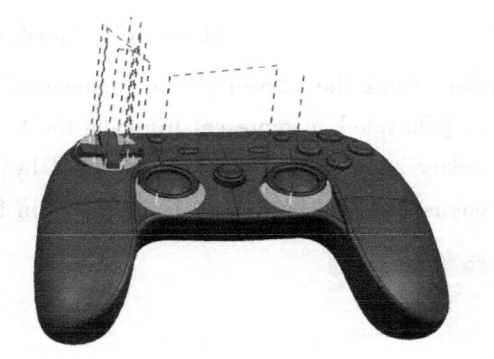

Figure 4-205　Generate tool path

4.4.10　Machining Simulation

Select 【NC_PROGRAM】 in the program order view of the operation navigator program, then click the 【Verify Tool Path】 command to pop up the 【Tool Path Visualization】 dialog box. Click the "Play" button in the 【3D Dynamic】 tab to see the cutting effect, as shown in Figure 4-206.

Click the 【Show Thickness by Color】button in the 【Tool Path Visualization】 dialog box to view the blank stock after machining, as shown in Figure 4-207.

Project 4　Gamepad

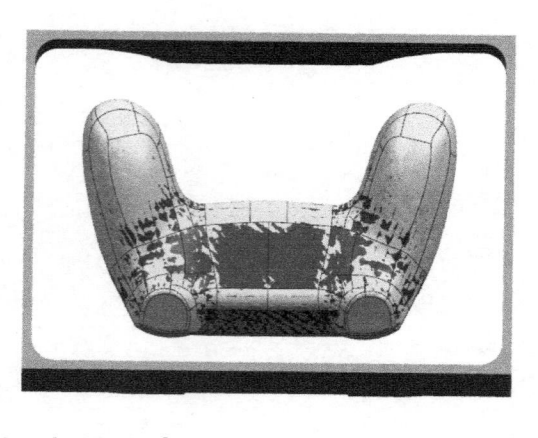

Figure 4-206　Effect of machining simulation

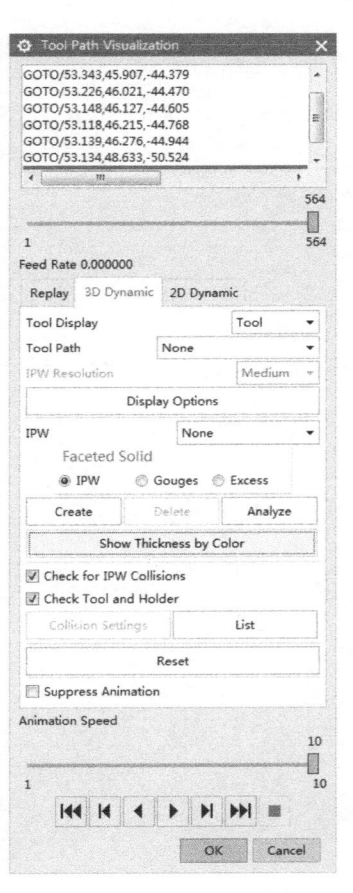

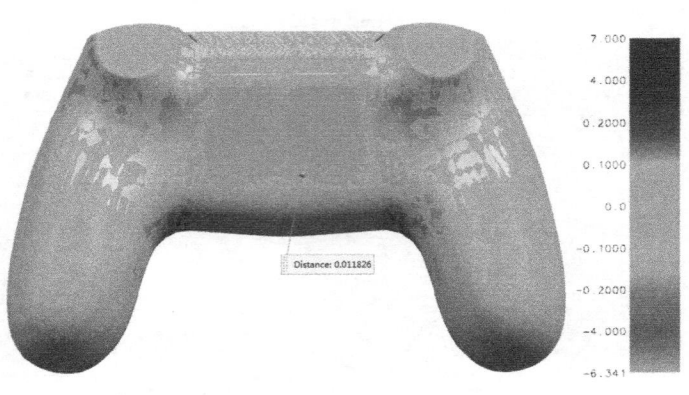

Figure 4-207　Show thickness by color

189